LONDON MATHEMATICAL SOCIETY LECTURE NOTE SERIES

Managing Editor: Professor N.J. Hitchin, Mathematical Institute,
University of Oxford, 24–29 St Giles, Oxford OX1 3LB, United Kingdom

The titles below are available from booksellers, or, in case of difficulty, from Cambridge University Press.

London Mathematical Society Lecture Note Series. 268

Spectral Asymptotics in the Semi-Classical Limit

Mouez Dimassi
Université de Paris-Nord

Johannes Sjöstrand
Ecole Polytechnique

CAMBRIDGE
UNIVERSITY PRESS

CAMBRIDGE UNIVERSITY PRESS
Cambridge, New York, Melbourne, Madrid, Cape Town, Singapore, São Paulo

Cambridge University Press
The Edinburgh Building, Cambridge CB2 2RU, UK

Published in the United States of America by Cambridge University Press, New York

www.cambridge.org
Information on this title: www.cambridge.org/9780521665445

© Cambridge University Press 1999

First published 1999

A catalogue record for this publication is available from the British Library

ISBN-13 978-0-521-66544-5 paperback
ISBN-10 0-521-66544-2 paperback

Transferred to digital printing 2007

Contents

0. Introduction

A new branch of mathematical analysis, so-called microlocal analysis, started to be more systematically developed about 30 years ago by Kohn–Nirenberg, Hörmander, Maslov and Sato, soon followed by many others. Originally the motivations came from problems in partial differential equations, but it soon became increasingly clear that many aspects of microlocal analysis are reminiscent of quantum mechanics, and for instance the Heisenberg uncertainty principle plays a fundamental role in both theories. Mathematically, a version of this principle says that if $u \in L^2(\mathbf{R}^n)$ and we define the Fourier transform by

$$\widehat{u}(\xi) = \int e^{-ix \cdot \xi} u(x) dx, \tag{0.1}$$

so that Parseval's relation

$$\|u\|^2 = \frac{1}{(2\pi)^n} \|\widehat{u}\|^2,$$

holds, where the norms are those of L^2, then if we take $n = 1$ for simplicity, and let $x_0, \xi_0 \in \mathbf{R}$:

$$\|u\|^2 \leq 2\|(\cdot - x_0)u\| \frac{1}{\sqrt{2\pi}} \|(\cdot - \xi_0)\widehat{u}\|, \ u \in \mathcal{S}(\mathbf{R}). \tag{0.2}$$

Here $\mathcal{S}(\mathbf{R})$ is the Schwartz space of smooth functions on $\mathbf{R}$ which decay rapidly at infinity together with all their derivatives. A rough interpretation of this is that if most of the L^2-norm ('energy') of u is concentrated to an interval of length a and most of the energy of $\widehat{u}$ is concentrated to an interval of length b, then:

$$ab \geq 2\pi. \tag{0.3}$$

The reason for putting this precise numerical constant comes from well-known asymptotic formulas for the counting of eigenvalues (of Weyl type) which can be interpreted by saying that each eigenfunction occupies a volume (2π) in phase space.

Another similarity between the two theories is the interplay between classical and quantum objects. In microlocal analysis, the quantum objects are given by pseudodifferential and Fourier integral operators etc. and the classical ones by those of symplectic geometry: canonical transformations, Poisson brackets etc. In quantum mechanics the same duality appears in the semi-classical limit. If we consider for instance the stationary Schrödinger operator

$$-h^2 \Delta + V(x), \tag{0.4}$$

when h becomes very small, then the quantum objects are wave functions, eigenvalues etc., while the classical ones are given by the classical trajectories

Spectral Asymptotics in the Semi-Classical Limit

of the associated classical Hamiltonian $p := \xi^2 + V(x)$, i.e. the integral curves of the corresponding Hamilton field $H_p = 2\xi \cdot (\partial/\partial x) - V'(x) \cdot (\partial/\partial \xi)$.

Thanks to microlocal analysis it has been possible to get refined results about the distributions of eigenvalues for differential operators (mostly elliptic ones) on compact manifolds and in bounded domains (Hörmander, Duistermaat–Guillemin, Ivrii and others), and while the Weyl asymptotics gives the leading terms in such results and is simply a phase space volume, the further terms or remainder estimates depend on dynamical properties of the Hamilton flow.

These notes are about the analogous developments for the semiclassical limit. The motivation among specialists (such as Chazarain, Helffer–Robert and later many others) was that microlocal analysis should provide a tool for a more rigorous understanding of many spectral problems also in this field. To some extent the early work consisted of carrying over the above mentioned spectral results to the study of, say, (0.4), but the area turned out to be much richer and new problems and results appeared, and the microlocal analysis itself has received new impulses from these efforts.

The contents of these notes are:

1. Local symplectic geometry. Here we develop some of the standard theory, following closely one of the chapters in [GrSj].

2. The WKB-method. We discuss the construction of local asymptotic solutions of $(P - E)u = 0$, where P is the operator (0.4), and get an example of the interplay between classical and quantum objects.

3. The WKB-method for a potential minimum. Here we follow some work by Helffer and one of the authors, and show how to construct asymptotic eigenvalues and eigenfunctions near a non-degenerate minimum of the potential.

4. Self-adjoint operators. This is mostly a compilation of abstract spectral theory, and at the end of the chapter, we determine the low-lying eigenvalues for potentials with a non-degenerate minimum. This also justifies the more complete asymptotics of eigenvalues obtained in Chapter 3.

5. The method of stationary phase. We followed closely the presentation of [GrSj], based on the classical work of Hörmander [Höl]. A small variation leads to some refined remainder estimates, which may be new. This method is one of the fundamental ingredients of microlocal analysis, even though in the present notes we choose not to appeal explicitly to this method when presenting the theory of pseudodifferential operators. ([GrSj] shows how to get everything from stationary phase.)

6. Tunnel effect and interaction matrix. This chapter is devoted to exponentially small corrections to eigenvalues of (0.4), due to the interaction of potential wells through the classically forbidden region. An essential tool is the use of exponentially weighted L^2-estimates, developed for second order operators by Lithner and Agmon. We have followed some work of Helffer and one of the authors.

7. h-pseudodifferential operators. In this chapter the basic theory of pseudodifferential operators is developed, without trying to reach maximal generality or refinement. These operators are of the form $P(x, hD; h)$, where $P(x, \xi; h)$ belongs to some suitable space of symbols. The most standard case is when $P(x, \xi; h)$ is uniformly bounded together with all its derivatives, uniformly with respect to h. In the case $n = 1$ (for simplicity) the symbol $P(x, h\xi; h)$ varies only a little in rectangles of the form $I_x \times J_\xi$ if I_x and J_ξ are intervals of length ϵ_0 and ϵ_0/h respectively, for some small but fixed constant $\epsilon_0 > 0$. The area of $I_x \times J_\xi$ is ϵ_0^2/h, and the uncertainty principle is satisfied with a good margin, when h is small enough. The symbolic calculus is developed and in particular it is established that h-pseudodifferential operators form an algebra, and the symbols of the composition of two operators is the product of the symbols plus an error which is roughly of the order h smaller.

8. Functional calculus for pseudodifferential operators. We base this calculus on a functional formula using almost analytic extensions, and a semi-classical version of an important lemma of Beals which permits us to characterize pseudodifferential operators. One of the main results (which is due to Helffer and Robert in the semi-classical case) says that if P is a self-adjoint h-pseudodifferential operator (from now on sometimes called pseudor for short) and $f \in C_0^\infty$, then $f(P)$ is again a h-pseudor with leading symbol $f(p(x, \xi))$, where $p(x, \xi)$ is the leading symbol of P. We follow some joint papers of Helffer and one of the authors. This approach to the calculus is also extended to the case of several commuting self-adjoint operators.

9. Trace class operators and applications of the functional calculus. Here we derive asymptotic expansions for the trace and get as a corollary the leading (Weyl-)asymptotics for the number of eigenvalues in an interval.

10. More precise spectral asymptotics for non-critical Hamiltonians. Here we study the unitary evolution group as a Fourier integral operator and the singularity of its trace near the time 0. This leads to an estimate of the remainder in the spectral asymptotic formula, which in general is optimal.

11. Improvement when the periodic trajectories form a set of measure 0. Here we estimate the trace of the evolution group also for large times. The methods and the results of this chapter as well as the preceding one are fairly standard, first due to Hörmander, Guillemin–Duistermaat in the non-semiclassical case, then extended to the semi-classical case (and improved) by Ivrii, Petkov and Robert.

12. A more general study of the trace. Here we extend the results of Chapter 10 to the case of microhyperbolic systems. The presentation is inspired by works of Ivrii, which avoid explicit constructions (which might be impossible anyway), but we have used a stationary approach, which in later work by one of the authors has been extended to situations with an implicit dependence of the spectral parameter. Such implicit spectral problems appear frequently when making so-called Grushin reductions of a spectral problem.

13. Spectral theory for perturbed periodic problems. For slowly varying perturbations of periodic Schrödinger operators, one can make a reduction to the study of an h-pseudor, a so called effective Hamiltonian, and it then becomes possible to obtain asymptotic results about the eigenvalues of the perturbed operator in a gap of the spectrum of an unperturbed one. We have followed work by Gérard–Martinez–Sjöstrand and Dimassi, related to earlier works by Buslaev, Guillot–Ralston–Trubowitz and Helffer–Sjöstrand. The reduction used is an example of a so-called Grushin reduction, a technique which has turned out to be extremely useful in many situations, in particular when combined with functional formulas of the type given in Chapters 8, 9 and 12.

14. Normal forms for some scalar pseudodifferential operators. Here we return to non-degenerate potential wells, studied in Chapters 3, 4 and 6, and establish a quantum Birkhoff normal form, which permits (under a non-resonance condition) to obtain complete asymptotic expansions of all eigenvalues in an interval $[0, h^\delta]$, where $\delta > 0$ is arbitrary. This chapter is based on a work of Sjöstrand, in a cirle of ideas developed by Lazutkin, Colin de Verdière, Graffi–Paul, Bellissard–Vittot, Iantchenko and many others.

15. Spectrum of operators with periodic bicharacteristics. When the Hamilton flow is periodic there is a phenomenon of clustering of eigenvalues, that we study, following works by Colin de Verdière, Weinstein, Helffer–Robert and others.

We hope that these notes may serve as an introduction to a still very active subject, and they correspond largely to a course given by the authors at the universities of Rennes (,where the first impulse to write these notes was received), Paris Sud, Paris Nord, as well as the Ecole Polytechnique. They cover more recent material than the now classical book by Robert [Ro1], but remain hopefully at an introductory level. A fairly large portion can be covered in a one semester course. For further and deeper study, we can recommend the recent book by Ivrii [I1]. See also the book of

Safarov–Vassiliev [SaVa] which deals with asymptotics of large eigenvalues for boundary value problems.

We would like to thank N. Lerner and G. Métivier for giving one of the authors the original impulse to write these notes. We have also profited more or less directly from a long collaboration and many stimulating discussions with B. Helffer, who we thank particularly. We also thank A. Grigis for the permission to use two chapters from [GrSj] with only minor changes, and one of the referees who indicated some important references.

1. Local symplectic geometry

We assume that the reader is familiar with some basic objects of differential geometry, such as manifolds, tangent and cotangent vectors, differential forms and vector bundles. We shall, however, review briefly some of these notions. For a smooth manifold X of dimension n we shall denote by $C^k(X)$ the space of k times continuously differentiable complex valued functions on X if $k \in \mathbf{N}$, and we set $C^\infty(X) = \cap_{k \in \mathbf{N}} C^k(X)$.

Tangent and cotangent vectors. Let X be a smooth manifold of dimension n. Let $x_0 \in X$. If $\gamma, \widetilde{\gamma} :]-1, 1[\to X$ are two C^1 curves with $\gamma(0) = \widetilde{\gamma}(0) = x_0$, we say that γ, $\widetilde{\gamma}$ are equivalent if $\|\gamma(t) - \widetilde{\gamma}(t)\| = o(t)$, $t \to 0$. (Here we choose some local coordinates $x_1, \ldots, x_n$ near x_0, so that $\|\gamma(t) - \widetilde{\gamma}(t)\|$ is well defined, and we notice that the choice of local coordinates and of the corresponding norm does not influence the definition.) The equivalence class of γ will be denoted by $\gamma'(0)$ or $\frac{d}{dt}\gamma(0)$, and will be called a tangent vector at x_0. The set of all tangent vectors at a point x_0 is denoted by $T_{x_0}X$ and is called the tangent space of X at x_0. It is easy to see (by working in a system of local coordinates) that $T_{x_0}X$ is a real vector space of dimension n.

If $f, \widetilde{f} : X \to \mathbf{R}$ are two C^1 functions, we say that f, $\widetilde{f}$ are equivalent if $(f(x) - f(x_0)) - (\widetilde{f}(x) - \widetilde{f}(x_0)) = o(\|x - x_0\|)$, $x \to x_0$. We let $df(x_0)$ (called the differential of f at x_0) denote the equivalence class of f. It is (by definition) a differential 1 form at x_0, also called a cotangent vector at x_0. The set $T^*_{x_0}X$ of cotangent vectors at a point x_0 is a real vector space of dimension n. It is called the cotangent space at x_0 of X.

There is a natural duality between $T^*_{x_0}X$ and $T_{x_0}X$, given by

$$\langle df(x_0), \gamma'(t) \rangle = \left(\frac{d}{dt}\right)_{t=0} f(\gamma(t)).$$

If $x_1, \ldots, x_n$ are local coordinates defined in a neighborhood of x_0, then $dx_1(x_0), \ldots, dx_n(x_0)$ (or $dx_1, \ldots, dx_n$ for short) form a basis of $T^*_{x_0}X$. A corresponding dual basis in $T_{x_0}X$ is given by $\frac{\partial}{\partial x_1}, \ldots, \frac{\partial}{\partial x_n}$, where $\frac{\partial}{\partial x_j}$ is the tangent vector induced by the curve $t \mapsto x_0 + te_j$. Here we work in the local coordinates above, and e_j denotes the jth unit vector in $\mathbf{R}^n$. It is easy to check that $df = \sum_1^n \frac{\partial f}{\partial x_j} dx_j$, $\gamma'(0) = \sum_1^n \frac{d\gamma_j}{dx_j} \frac{\partial}{\partial x_j}$ at the point x_0.

The sets $TX = \cup_{x_0 \in X} T_{x_0}X$ and $T^*X = \cup_{x_0 \in X} T^*_{x_0}X$ are vector bundles and in particular C^∞ manifolds. If $x_1, \ldots, x_n$ are local coordinates on X, then we get the corresponding local coordinates $(x, t) = (x_1, \ldots, x_n, t_1, \ldots, t_n)$ on TX and $(x, \xi) = (x_1, \ldots, x_n, \xi_1, \ldots, \xi_n)$ on T^*X by representing $\nu \in TX$ and $\rho \in T^*X$ by their base point x (given by the coordinates $(x_1, \ldots, x_n)$) and the corresponding tangent vector $\sum t_j \frac{\partial}{\partial x_j}$ and cotangent vector $\sum \xi_j dx_j$. If

$y_1, \ldots, y_n$ is a second system of local coordinates, then in the intersection of the two open sets in X parameterized by the two systems of local coordinates, we have the point-wise relations $t = \frac{\partial x}{\partial y} s$, $\eta = {}^t(\frac{\partial x}{\partial y})\xi$ for the corresponding local coordinates (x, t), (y, s) on TX and (x, ξ), (y, η) on T^*X. Here $\frac{\partial x}{\partial y} = (\frac{\partial x_j}{\partial y_k})_{1 \leq j,k \leq n}$ is the standard Jacobian matrix.

If $\rho \in T^*X$, we let $\pi(\rho) \in X$ be the associated base point so that $\pi : T^*X \to X$ is the natural projection map. A section in T^*X is a right inverse of π. The same definitions can be given for TX and more generally for any vector bundle. Sections in T^*X are called differential 1 forms, and sections in TX are called vector fields. Most of the time we will only consider sections of class C^∞ and we will also most of the time only consider locally defined sections (i.e. sections of T^*U and TU where U is some small open subset of X).

A vector field can be written in local coordinates as $\nu = \sum t_j(x)\frac{\partial}{\partial x_j}$ and a differential 1 form as $\omega = \sum \xi_j(x)dx_j$.

If Y is a second manifold and $f : Y \to X$ a map of class C^1, $y_0 \in Y$, $x_0 = f(y_0) \in X$, then we have a natural map $f_* = df : T_{y_0}Y \to T_{x_0}X$, which in local coordinates is given by the ordinary Jacobian matrix $(\frac{\partial x_j}{\partial y_k})$. The adjoint is $f^* : T_{x_0}^*X \to T_{y_0}^*Y$ and we note that if u is a C^1 function on X and $\gamma : I \to Y$ is a C^1 curve, with $\gamma(0) = y_0$, 0 in the interior Int I of the interval I, then $(f \circ \gamma)'(0) = f_*(\gamma'(0))$, $d(u \circ f)(y_0) = f^*(du(x_0))$. More generally, if Z is a third manifold, $g : Z \to Y$ is of class C^1 and $z_0 \in Z$, $g(z_0) = y_0$, then $(f \circ g)_* = f_* \circ g_*$, $(f \circ g)^* = g^* \circ f^*$. When passing to sections, we see that if ω is a 1 form on X, then $f^*\omega$ is a well defined 1-form on Y (called the pull-back of ω by means of f). The corresponding push-forward $f_*\nu$ of a vector field ν can be defined when f is a diffeomorphism, but not in general. If $\gamma :]a,b[\to X$ is a C^1 curve and $t_0 \in]a,b[$, we recover the tangent vector $\gamma'(t_0)$ of γ at t_0 as $\gamma_*(\frac{\partial}{\partial t}(t_0))$.

The elementary theory of ordinary differential equations gives the following fact: if ν is a C^∞ vector field on X, then for every $x_0 \in X$, we can find $T_+(x_0)$, $T_-(x_0)$ in $]0, +\infty]$, such that we have a smooth (i.e. C^∞) curve:

$$] - T_-(x_0), T_+(x_0)[\ni t \mapsto \gamma(t) =: \exp(t\nu)(x_0) \in X$$
$$\text{with } \gamma(0) = x_0, \ \gamma'(t) = \nu(\gamma(t)).$$

If we choose $T_\pm$ maximal, we get lower semi-continuous functions $X \ni x \mapsto T_\pm(x)$ and a smooth map

$$\{(t, x) \in \mathbf{R} \times X; -T_-(x) < t < T_+(x)\} \ni (t, x) \mapsto \Phi(t, x) = \exp(t\nu)(x)$$

with $\Phi(0, x) = x$, $\frac{\partial}{\partial t}\Phi(t, x) = \nu(\Phi(t, x))$. When t is fixed, we can in general not define $\exp t\nu$ on all of X but only on the set of $x \in X$ for which

$-T_-(x) < t < T_+(x)$. We have

$$\exp t\nu(\exp s\nu(x)) = \exp(t+s)\nu(x),$$

for t, s, x such that the left member is defined.

The canonical 1 and 2 forms. Let $\pi : T^*X \to X$ be the natural projection (which is simply $(x, \xi) \mapsto x$ in canonical coordinates). For $\rho \in T^*X$, consider $\pi^* : T^*_{\pi(\rho)}X \to T^*_\rho(T^*X)$. Since $\rho \in T^*_{\pi(\rho)}X$, we can define the canonical 1 form $\omega_\rho \in T^*_\rho(T^*X)$, by $\omega_\rho = \pi^*(\rho)$. Varying ρ, we get a smooth 1 form ω on on T^*X, which in canonical coordinates has the expression: $\omega = \sum \xi_j dx_j$.

We next recall some facts about forms of higher degree. If L is a finite dimensional real vector space, and L^* the dual space, then we have a natural duality between the k fold exterior product spaces $\wedge^k L$ and $\wedge^k L^*$, given by

$$\langle u_1 \wedge \ldots \wedge u_k, v_1 \wedge \ldots \wedge v_k \rangle = \det(\langle u_j, v_k \rangle), \ u_j \in L, \ v_k \in L^*.$$

Without repeating the definition of exterior products and exterior product spaces, it may be useful to recall that exterior products $u_1 \wedge \ldots \wedge u_k$ are linear in each of their factors and change sign if we permute two neighboring factors. Moreover, if $e_1, \ldots, e_n$ form a basis for L, then a basis for $\wedge^k L$ is formed by the $e_{j_1} \wedge \ldots \wedge e_{j_k}$, for all $1 \leq j_1 < j_2 < \ldots < j_k \leq n$.

If M is a C^∞ manifold of dimension m, then a differential k form v is a section of the vector bundle $\wedge^k T^*M$. In local coordinates $x_1, \ldots, x_m$:

$$v = \sum_{|I|=k} v_I(x)dx^I, \tag{1.1}$$

where in general, $I = (i_1, \ldots, i_k) \in \{1, 2, \ldots, m\}^k$, $|I| = k$, $dx^I = dx_{i_1} \wedge \ldots \wedge dx_{i_k}$. The representation above becomes unique if we restrict the sum to the set of I with $i_1 < i_2 < \ldots < i_k$. If v is a k form of class C^1 locally given by (1.1), we define the $k+1$ form

$$dv = \sum_{|I|=k} dv_I \wedge dx^I. \tag{1.2}$$

dv is called the exterior differential of v and it can be shown that its definition does not depend on the choice of local coordinates or on the choice of the representation (1.1). We have the following facts:

$$d^2 = 0, \tag{1.3}$$

If ω is a $k+1$ form of class C^∞, which is closed in the sense that $d\omega = 0$, then in every open set in M which is diffeomorphic to a ball, we can find a smooth k form v, such that $dv = \omega$.   (1.4)

If $f : Y \to X$ is a smooth map between two smooth manifolds, then there is a unique way of extending the pull-back f^* from 1 forms to k forms by multilinearity. If v is a smooth k form on X, then $d(f^*v) = f^*(dv)$.   (1.5)

We now return to the canonical 1 form ω on T^*X, and define the canonical 2 form σ on T^*X as $\sigma = d\omega$. In canonical coordinates:

$$\sigma = \sum_1^n d\xi_j \wedge dx_j.$$ (1.6)

For $\rho \in T^*X$, $\sigma_\rho \in \wedge^2 T_\rho^*(T^*X)$ can be viewed as a linear form on $\wedge^2 T_\rho(T^*X)$ or equivalently as an antisymmetric bilinear form on $T_\rho(T^*X) \times T_\rho(T^*X)$. Mixing the two points of view, we write:

$$\sigma_\rho(t,s) = \langle \sigma_\rho, t \wedge s \rangle, \ t, s \in T_\rho(T^*X).$$

In canonical coordinates, and with the notation $t = (t_x, t_\xi)$ (so that $t = \sum(t_{x_j}\frac{\partial}{\partial x_j} + t_{\xi_j}\frac{\partial}{\partial \xi_j}))$, $s = (s_x, s_\xi)$, we get

$$\sigma_\rho(t,s) = \langle t_\xi, s_x \rangle - \langle s_\xi, t_x \rangle = \sum(t_{\xi_j} s_{x_j} - s_{\xi_j} t_{x_j}).$$

From this it is clear that σ_ρ is a non-degenerate bilinear form and consequently there is a bijection $H : T_\rho^*(T^*X) \to T_\rho(T^*X)$ determined by:

$$\sigma(s, Hu) = \langle s, u \rangle \ s \in T_\rho(T^*X), \ u \in T_\rho^*(T^*X).$$

In canonical coordinates, if $u = u_x dx + u_\xi d\xi = \sum(u_{x_j} dx_j + u_{\xi_j} d\xi_j)$, we get $Hu = u_\xi \frac{\partial}{\partial x} - u_x \frac{\partial}{\partial \xi}$. If $f(x,\xi)$ is of class C^1 on X (or on some open subset of X), we define the Hamilton field of f by $H_f = H(df)$. In canonical coordinates,

$$H_f = \sum_1^n \left(\frac{\partial f}{\partial \xi_j} \frac{\partial}{\partial x_j} - \frac{\partial f}{\partial x_j} \frac{\partial}{\partial \xi_j} \right).$$

If M is a manifold, $\rho \in M$, $t \in T_\rho M$, then we define the contraction $t \rfloor : \wedge^k T_\rho^* M \to \wedge^{k-1} T_\rho^* M$ as the adjoint of the left exterior multiplication $t \wedge : \wedge^{k-1} T_\rho M \to \wedge^k T_\rho M$. Then with $M = T^*X$, the Hamilton field is (equivalently) defined by the pointwise relation,

$$H_f \rfloor \sigma = -df.$$ (1.7)

If f, g are two C^1 functions defined on some open set in T^*X, we define their Poisson bracket as the continuous function

$$\{f, g\} = H_f(g) = \langle H_f, dg \rangle = \sigma(H_f, H_g),$$

where, in the second expression, we view H_f as a first order differential operator. In canonical coordinates,

$$\{f, g\} = \sum \left(\frac{\partial f}{\partial \xi_j} \frac{\partial g}{\partial x_j} - \frac{\partial f}{\partial x_j} \frac{\partial g}{\partial \xi_j} \right)$$

Notice that $\{f, g\} = -\{g, f\}$, and in particular that $\{f, f\} = 0$.

Lie derivatives. Let v be a smooth vector field on a manifold M and let ω be a smooth k form on M. Then the Lie derivative of ω along v is defined pointwise by

$$\mathcal{L}_v \omega = \left(\frac{d}{dt} \right)_{t=0} ((\exp tv)^* \omega).$$

If u is a another smooth vector field on M, we also define

$$\mathcal{L}_v u = \left(\frac{d}{dt} \right)_{t=0} ((\exp -tv)_* u).$$

Here we need of course to observe that the push-forward of a vector field by means of a local diffeomorphism can be defined locally. We have the following facts:

(1) When ω is a 0 form, i.e. a function, then $\mathcal{L}_v \omega = v(\omega)$.

(2) $\mathcal{L}_v u = [v, u] = vu - uv$, where u, v are viewed as first order differential operators in the last two expressions.

(3) $\mathcal{L}_v(d\omega) = d(\mathcal{L}_v \omega)$,

(4) $\mathcal{L}_v(\omega_1 \wedge \omega_2) = (\mathcal{L}_v \omega_1) \wedge \omega_2 + \omega_1 \wedge (\mathcal{L}_v \omega_2)$,

(5) $\mathcal{L}_v(u \rfloor \omega) = (\mathcal{L}_v u) \rfloor \omega + u \rfloor (\mathcal{L}_v \omega)$,

(6) $\mathcal{L}_v \omega = v \rfloor d\omega + d(v \rfloor \omega)$,

(7) $\mathcal{L}_{v_1 + v_2} = \mathcal{L}_{v_1} + \mathcal{L}_{v_2}$.

Lemma 1.1. *If f is a smooth function on some open subset in T^*X, then* $\mathcal{L}_{H_f} \sigma = 0$.

Proof. It suffices to make the calculation,

$$\mathcal{L}_{H_f}\sigma = H_f \rfloor d\sigma + d(H_f \rfloor \sigma) = H_f \rfloor d^2\omega - d^2 f = 0.$$

#

Locally, we can define the maps $\Phi_t = \exp tH_f$, when $|t|$ is sufficiently small and we have $\Phi_t^*\sigma = \sigma$. In fact, we have pointwise:

$$\frac{d}{dt}\Phi_t^*\sigma = (\frac{d}{ds})_{s=0}\Phi_t^*\Phi_s^*\sigma = \Phi_t^*\mathcal{L}_{H_f}\sigma = 0.$$

Lemma 1.2. *If f, g are two smooth functions defined on some open subset of T^*X, then $[H_f, H_g] = H_{\{f,g\}}$.*

Proof. We have to show that $[H_f, H_g]\rfloor\sigma = -d\{f, g\}$. This follows from the computation:

$$\begin{aligned}
-d\{f,g\} &= -d(\mathcal{L}_{H_f}g) = -(\mathcal{L}_{H_f}dg) = \mathcal{L}_{H_f}(H_g\rfloor\sigma)\\
&= [H_f, H_g]\rfloor\sigma + H_g\rfloor(\mathcal{L}_{H_f}\sigma) = [H_f, H_g]\rfloor\sigma.
\end{aligned}$$

#

Using the preceding lemma, it is easy to prove the Jacobi identity for three smooth functions,

$$\{f, \{g, h\}\} + \{g, \{h, f\}\} + \{h, \{f, g\}\} = 0.$$

Lagrangian manifolds. A submanifold $\Lambda \subset T^*X$ is called a Lagrangian manifold if $\dim \Lambda = \dim X$ and $\sigma_{|\Lambda} = 0$. In general, we define the restriction of a differential k form to a submanifold as the the pull-back of this form by means of the natural inclusion map, and there is a corresponding natural way of viewing the tangent space of a submanifold at some given point as a subspace of the tangent space of the ambient manifold at the same point. If Λ is a submanifold of T^*X and $\rho \in \Lambda$, then we define $T_\rho\Lambda^\perp \subset T_\rho(T^*X)$ as the orthogonal space with respect to σ of $T_\rho\Lambda \subset T_\rho(T^*X)$. The sum of the dimensions of $T_\rho\Lambda$ and $T_\rho\Lambda^\perp$ add up to the dimension of $T_\rho(T^*X)$, but there is no reason for $T_\rho\Lambda$ and $T_\rho\Lambda^\perp$ to have zero intersection. As a matter of fact, it is clear that a submanifold $\Lambda \subset T^*X$ is Lagrangian if and only if $T_\rho\Lambda = T_\rho\Lambda^\perp$ for every $\rho \in \Lambda$. That there are plenty of Lagrangian submanifolds follows from the following result.

Theorem 1.3. *Let $\Lambda \subset T^*X$ be a submanifold with $\dim \Lambda = \dim X$ and such that $\pi_{|_\Lambda} : \Lambda \to X$ is a local diffeomorphism (in the sense that every point ρ in Λ has a neighborhood in Λ which is mapped diffeomorphically by π onto a neighborhood of $\pi(\rho)$). Then Λ is Lagrangian if and only if for each point $\rho \in \Lambda$, we can find a (real) C^∞ function $\phi(x)$ defined near $\pi(\rho)$, such that Λ coincides near ρ with the manifold $\{(x, d\phi(x)); x \in$ some neighborhood of $\pi(\rho)\}$.*

Proof. If ω is the canonical 1 form, we notice that $d(\omega_{|_\Lambda}) = \sigma_{|_\Lambda}$. Therefore the following three statements are equivalent:

(1) Λ is Lagrangian.

(2) $\omega_{|_\Lambda}$ is closed (i.e. $d(\omega_{|_\Lambda}) = 0$).

(3) Locally on Λ, we can find a smooth function ϕ with $\omega_{|_\Lambda} = d\phi$.

If $x_1, \ldots, x_n$ are local coordinates on X, we can also view them (or rather their compositions with π) as local coordinates on Λ, and represent Λ by equations $\xi = \xi(x)$ in the corresponding canonical coordinates. Then (3) is equivalent to $\xi_j(x) = \frac{\partial \phi(x)}{\partial x_j}$, i.e. $\sum \xi_j(x) dx_j = d\phi$.   #

Hamilton–Jacobi equations. These equations are of the form $p(x, \phi'_x) = 0$, where p is a real-valued C^∞ function defined on some open subset of T^*X. Here we shall also assume that $dp(x, \xi) \neq 0$, when $p(x, \xi) = 0$. The basic idea in treating a Hamilton-Jacobi equation is to consider the Lagrangian manifold $\Lambda = \Lambda_\phi$ associated with ϕ as in the preceding theorem, and try to find such a manifold inside the hypersurface H defined by $p(x, \xi) = 0$. If $\rho \in \Lambda$, we shall then have $T_\rho\Lambda \subset T_\rho H$ (considering these tangent spaces as subspaces of $T_\rho T^*X$), and hence $T_\rho^\perp H \subset T_\rho\Lambda$, since $T_\rho\Lambda^\perp = T_\rho\Lambda$. It is easy to see that $T_\rho H^\perp = \mathbf{R}H_p$, so we must have $H_p \in T_\rho\Lambda$ at every point $\rho \in \Lambda$, or in other words H_p should be tangent to Λ at every point of Λ.

Proposition 1.4. *Let $\Lambda' \subset H$ be an isotropic submanifold (in the sense that $\sigma_{|_{\Lambda'}} = 0$) of dimension $n - 1$ passing through some given point $\rho_0 \in H$ and such that $H_p(\rho_0) \notin T_{\rho_0}\Lambda'$. Then in a neighborhood of ρ_0 we can find a Lagrangian manifold Λ such that $\Lambda' \subset \Lambda \subset H$ (in that neighborhood).*

Proof. According to the observation above it is natural to try

$$\Lambda = \{\exp(tH_p)(\rho); \; |t| < \epsilon, \; \rho \in \Lambda', \; |\rho - \rho_0| < \epsilon\}$$

for some sufficiently small $\epsilon > 0$. (Here $|\rho - \rho_0|$ is well-defined, if we choose some local coordinates.) Then $\Lambda' \subset \Lambda$ (near ρ_0) and since H_p is tangent to

H (by the relation $H_p p = 0$) we also have $\Lambda \subset H$. From the assumption $H_p(\rho_0) \notin T_{\rho_0}\Lambda'$ and the implicit function theorem, it also follows that Λ is a smooth manifold of dimension n.

In order to verify that Λ is Lagrangian, we first take $\rho \in \Lambda'$ (with $|\rho - \rho_0| < \epsilon$) and consider $T_\rho\Lambda = T_\rho\Lambda' \oplus \mathbf{R}H_p$. Then $\sigma_{\rho|_{T_\rho\Lambda \times T_\rho\Lambda}} = 0$ since $\sigma_{\rho|_{T_\rho\Lambda' \times T_\rho\Lambda'}} = 0$, $\sigma_\rho(H_p, H_p) = 0$, $\sigma_\rho(t, H_p) = \langle t, dp \rangle = 0$ for all $t \in T_\rho\Lambda' \subset T_\rho H$.

More generally, at the point $\rho_t = \exp(tH_p)(\rho)$, $\rho \in \Lambda'$, we have

$$T_{\rho_t}(\Lambda) = \exp(tH_p)_*(T_\rho\Lambda)$$

and for $u, v \in T_\rho\Lambda$ we get, using the fact that $\exp(tH_p)^*\sigma_{\rho_t} = \sigma_\rho$:

$$\sigma_{\rho_t}(\exp(tH_p)_* u, \exp(tH_p)_* v) = \sigma_\rho(u, v) = 0.$$

We have then verified that $\sigma_{|_\Lambda} = 0$, which suffices since Λ has the right dimension.   #

In the following we write $x = (x', x_n) \in \mathbf{R}^n$, $x' = (x_1, \ldots, x_{n-1}) \in \mathbf{R}^{n-1}$.

Theorem 1.5. *Let $p(x, \xi)$ be a real valued C^∞ function, defined in a neighborhood of some point $(0, \xi_0) \in T^*\mathbf{R}^n$, such that $p(0, \xi_0) = 0$, $\frac{\partial p}{\partial \xi_n}(0, \xi_0) \neq 0$. Let $\psi(x')$ be a real valued C^∞ function defined near 0 in $\mathbf{R}^{n-1}$ such that $\frac{\partial \psi}{\partial x'}(0) = \xi_0'$. Then there exists a real valued smooth function $\phi(x)$, defined in a neighborhood of $0 \in \mathbf{R}^n$, such that in that neighborhood:*

$$p(x, \phi'_x(x)) = 0, \quad \phi(x', 0) = \psi(x'), \quad \phi'_x(0) = \xi_0. \tag{1.8}$$

If $\widetilde{\phi}(x)$ is a second function with the same properties, then $\phi(x) = \widetilde{\phi}(x)$ in some neighborhood of 0.

Proof. In a suitable neighborhood of $(0, \xi_0) \in \mathbf{R}^{n-1} \times \mathbf{R}^n$ we have $p(x', 0, \xi) = 0$ if and only if $\xi_n = \lambda(x', \xi')$, where λ is a real valued C^∞ function, with $\lambda(0, \xi_0') = (\xi_0)_n$. Let

$$\Lambda' = \{(x, \xi); \ x_n = 0, \ \xi' = \frac{\partial \psi}{\partial x'}(x'), \ \xi_n = \lambda(x', \xi'), \ x' \in \text{neigh}(0)\},$$

where neigh (0) indicates some sufficiently small neighborhood of 0.

Then $\Lambda' \subset p^{-1}(0)$ is isotropic of dimension $n - 1$ and H_p is nowhere tangent to Λ' since H_p has a component $\frac{\partial p}{\partial \xi_n} \frac{\partial}{\partial x_n}$ with $\frac{\partial p}{\partial \xi_n} \neq 0$. Let $\Lambda \subset p^{-1}(0)$ be a Lagrangian manifold as in Proposition 1.4. The differential of $\pi_{|_\Lambda} : \Lambda \to \mathbf{R}^n$ is bijective at $(0, \xi_0)$, so if we restrict the attention to a sufficiently small neighborhood of that point, we can apply Theorem 1.3 and see that Λ is of the form $\xi = \phi'(x)$, $x \in$ neigh (0). We have then $p(x, \phi'(x)) = 0$, $\phi'(0) = \xi_0$. Since $\Lambda' \subset \Lambda$, we get $\frac{\partial \psi}{\partial x'}(x') = \frac{\partial \phi}{\partial x'}(x', 0)$, so modifying ϕ by a constant, we get $\phi(x', 0) = \psi(x')$. We leave the verification of the uniqueness statement as an exercise. #

We can view Λ as a union of integral curves of H_p, passing through Λ'. The projection of such an integral curve is an integral curve of the field $\nu = \sum_1^n \frac{\partial p}{\partial \xi_j}(x, \phi'_x(x))\frac{\partial}{\partial x_j}$, which can be identified with $H_{p|_\Lambda}$ via the projection $\pi_{|_\Lambda}$. If $q(x, \xi) = \sum_1^n \frac{\partial p}{\partial \xi_j}(x, \xi)\xi_j$, we have the trivial identity

$$\sum_1^n \frac{\partial p}{\partial \xi_j}(x, \phi'_x(x))\frac{\partial}{\partial x_j}\phi = q(x, \phi'_x(x)).$$

If $x = x(t)$ is an integral curve of ν with $x_n(0) = 0$, then we get $\phi(x(t)) = \psi(x'(0)) + \int_0^t q(x(s), \xi(s))ds$, where $\xi(s) = \phi'(x(s))$, so that $s \mapsto (x(s), \xi(s))$ is the integral curve of H_p with $x_n(0) = 0$, $\xi'(0) = \frac{\partial \psi}{\partial x'}(x'(0))$, $\xi_n(0) = \lambda(x'(0), \xi'(0))$.

If $\psi = \psi_\alpha$ depends smoothly on some parameters $\alpha \in \mathbf{R}^k$, then $\phi = \phi(x, \alpha)$ will be a smooth function of (x, α), and differentiating the equation $p(x, \phi'_x(x)) = 0$, we see that $\frac{\partial \phi}{\partial \alpha}$ is constant along the bicharacteristics curves, i.e. the x-space projections of the H_p integral curves in Λ_ϕ.

In order to recall the roots of symplectic geometry in classical mechanics, let us consider the case when $p = \frac{1}{2m}\xi^2 + V(x)$, where $V(x)$ is some smooth real potential and $m > 0$ is a constant. The equations for the H_p integral curves are $x'(t) = \frac{\xi(t)}{m} =: v$, $\xi'(t) = -V'(x(t))$, and if we eliminate $\xi(t)$, we get the differential equation for $x(t)$: $x''(t) = -\frac{V'(t)}{m}$. We can view this as the motion of a classical partical of mass m. ξ is the momentum, so that $\xi(t) = mv(t)$. $v(t)$ is the velocity. The total energy $\frac{1}{2m}\xi^2 + V(x)$ is a constant of the motion (i.e. constant on every integral curve). Finally $-\frac{V'(x)}{m}$ is the acceleration induced by the force $-V'(x)$.

Another motivation for working on the cotangent bundle comes from the general theory of partial differential equations. Consider a differential operator with smooth coefficients, $P : C^\infty(X) \to C^\infty(X)$, where X is a manifold. For every choice of smooth local coordinates, P takes the form $P = \sum_{|\alpha| \le m} a_\alpha(x)D_x^\alpha$, where $m \in \mathbf{N} = \{0, 1, 2, \ldots\}$ is the order of the operator and we use standard multi-index notation: $\alpha = (\alpha_1, \ldots, \alpha_n) \in \mathbf{N}^n$, $|\alpha| = \alpha_1 + \ldots + \alpha_n$, $D_x = (D_{x_1}, \ldots, D_{x_n})$, $D_{x_j} = \frac{1}{i}\frac{\partial}{\partial x_j}$, $D_x^\alpha = D_{x_1}^{\alpha_1} \ldots D_{x_n}^{\alpha_n}$. In the corresponding canonical coordinates (x, ξ), we define the principal symbol,

$$p(x, \xi) = \sum_{|\alpha|=m} a_\alpha(x) \xi^\alpha, \text{ where } \xi^\alpha = \xi_1^{\alpha_1} \dots \xi_n^{\alpha_n}.$$

Then one can check that $p(x, \xi)$ is a well-defined function in $C^\infty(T^*X)$. A quick way to see this is to notice that if a, $\phi \in C^\infty(X)$ and ϕ is real valued, then

$$P(a(x)e^{i\lambda\phi(x)}) = e^{i\lambda\phi(x)}(p(x, \phi'(x))\lambda^m a(x) + \mathcal{O}(\lambda^{m-1})), \text{ when } \lambda \to \infty.$$

Notes

The presentation here is close to that of [GrSj] and has been inspired by that of Duistermaat [Du]. The reader who wants to study the subject in depth could turn to the books of Sternberg [St] and Hofer–Zehnder [HoZe].

2. The WKB-method

This is a very general method for constructing rapidly oscillating approximate solutions to differential equations. We restrict our attention to the problem of constructing approximate solutions to the stationary Schrödinger equation

$$(-h^2\Delta + V(x) - E)u = 0 \tag{2.1}$$

in the semi-classical limit: $h \to 0$. Here V is a smooth real valued function, defined on some open set $X \subset \mathbf{R}^n$, and $\Delta = \sum_1^n \partial_{x_j}^2$ is the Laplace operator. With $-h^2\Delta + V(x)$ we associate the classical Hamiltonian

$$p(x, \xi) = \xi^2 + V(x), \tag{2.2}$$

and in some sense we may say that (2.1) is a quantum mechanical problem and that the corresponding problem of classical mechanics is to understand the nature of the trajectories of

$$H_p = 2\xi \cdot \partial_x - V'(x) \cdot \partial_\xi \tag{2.3}$$

in the energy surface $p(x, \xi) = E$. The general problem of so-called semi-classical analysis (or semi-classical approximation) is to relate the two problems in the limit when $h \searrow 0$. Both problems are difficult when considered globally, and easier when considered locally.

Let Λ be a Lagrangian submanifold of $p^{-1}(E)$, and assume, possibly after restricting our attention to some part of Λ, that Λ is of the form

$$\Lambda_\phi = \{(x, \phi'(x)); \ x \in \Omega\}, \ \Omega \text{ open subset of } X, \tag{2.4}$$

where $\phi \in C^\infty(\Omega; \mathbf{R})$, so that ϕ solves the characteristic equation:

$$(\phi'_x)^2 + V(x) - E = 0. \tag{2.5}$$

One way of producing such a manifold is the following. Let Γ be a hypersurface in $\mathbf{R}^n$ and let $\psi \in C^\infty(\Gamma; \mathbf{R})$ satisfy

$$(\psi'_x)^2 + V(x) - E < 0, \ x \in \Gamma, \tag{2.6}$$

where ψ'_x denotes the gradient of ψ when Γ is viewed as a Riemannian manifold with the induced metric, and $(\psi'_x)^2$ is the corresponding square of the norm. In fact, choose local coordinates $y'_1, \ldots, y'_n$ on Γ and corresponding local coordinates $y_1, \ldots, y_{n-1}, y_n$ on $\mathbf{R}^n$, such that $y_{j|_\Gamma} = y'_j$, $y_{n|_\Gamma} = 0$, and such that $\frac{\partial}{\partial y_n}$ is orthogonal to $T\Gamma$ at every point. In the corresponding canonical coordinates (y, η), the restriction of ξ^2 to Γ is equal to $q(y', \eta') +$

$a_{n,n}(y')\eta_n^2$, where q is a positive definite quadratic form in η' and $a_{n,n} > 0$, and the condition (2.6) becomes $q(y', \psi'(y')) + V(y') - E < 0$ on Γ. Theorem 1.5 can now be applied in a neighborhood of any fixed point $x_0 \in \Gamma$, and gives two solutions $\phi = \phi_\pm$ of the Hamilton–Jacobi problem

$$(\phi_x')^2 + V(x) - E = 0, \quad \phi_{|\Gamma} = \psi. \tag{2.7}$$

Moreover $\pi_x(H_{p|_{\Lambda_\phi}})$ is transversal to Γ.

Now return to the more general situation with $\Lambda = \Lambda_\phi$ of the form (2.4). We try to construct an approximate solution of (2.1) of the form

$$u(x; h) = e^{i\phi(x)/h} a(x; h). \tag{2.8}$$

We get

$$e^{-i\phi/h}(-h^2\Delta + V(x) - E)(e^{i\phi/h}a(x; h))$$
$$= (\sum(hD_{x_j} + \partial_{x_j}\phi)^2 + V(x) - E)a(x; h)$$
$$= ((\phi')^2 + V(x) - E)a + \sum h(D_{x_j} \circ \partial_{x_j}\phi + \partial_{x_j}\phi \circ D_{x_j})a - h^2\Delta a$$
$$= \frac{2h}{i}(\phi'(x) \cdot \partial_x + \frac{1}{2}\Delta\phi)a - h^2\Delta a, \tag{2.9}$$

where the last equality follows from the eikonal equation (2.5). We now look for $a(x; h)$ of the form

$$a(x; h) \sim a_0(x) + a_1(x)h + a_2(x)h^2 + \dots, \quad a_j \in C^\infty(\Omega), \tag{2.10}$$

in the sense that

$$|\partial_x^\alpha(a - \sum_0^N a_j(x)h^j)| \le C_{K,\alpha,N}h^{N+1}, \quad x \in K, \tag{2.11}$$

for all $K \subset\subset \Omega$, $\alpha \in \mathbf{N}^n$, $N \in \mathbf{N}$.

We remark here that in general if $a_j \in C^\infty(\Omega)$, $0 < h < 1$, then we can find $a \in C^\infty(\Omega)$ which satisfies (2.10). In fact, this follows from the classical Borel construction of a smooth function with a given formal Taylor series at some given point: Let $\chi \in C_0^\infty(\mathbf{R})$ be equal to 1 near 0 and put

$$a(x; h) = \sum_0^\infty a_j(x)h^j\chi(\lambda_j h) \tag{$*$}$$

for a suitable sequence $\lambda_j \to \infty$. By $C_0^\infty(\Omega)$ we denote the space of all $u \in C^\infty(\Omega)$ with compact support. Let $K_k \subset\subset \Omega$ be an increasing sequence of compact sets tending to Ω. Choosing λ_j sufficiently large, we can arrange that

$$\|h^j \chi(\lambda_j h) a_j(\cdot)\|_{C^k(K_k)} \leq h^{j-1} 2^{-j}, \; j \geq k, \; j = 1, 2, \ldots$$

Then $(*)$ converges in the C^k topology for every fixed k. Moreover,

$$a - \sum_0^N a_j(x) h^j = \sum_0^N a_j(x)(\chi(\lambda_j h) - 1) h^j + \sum_{N+1}^\infty a_j \chi(\lambda_j h) h^j,$$

and we check that the $C^k(K_k)$ norm of each of the sums to the right is $\mathcal{O}(h^{N+1})$. In this discussion, we can also replace h^j by h^{k_j} where $k_j \nearrow \infty$.

Substituting (2.10) into the last expression in (2.9) and requiring that each term in the resulting asymptotic expansion (of the same type as (2.10)) vanish, we get the sequence of transport equations:

$$(\phi'(x) \cdot \partial_x + \frac{1}{2}\Delta\phi)a_0(x) = 0, \tag{T_0}$$

$$(\phi'(x) \cdot \partial_x + \frac{1}{2}\Delta\phi)a_1(x) = \frac{i}{2}\Delta a_0, \tag{T_1}$$

$$(\phi'(x) \cdot \partial_x + \frac{1}{2}\Delta\phi)a_2(x) = \frac{i}{2}\Delta a_1, \tag{T_2}$$

$$\cdots\cdots\cdots\cdots\cdots\cdots\cdots$$

$$(\phi'(x) \cdot \partial_x + \frac{1}{2}\Delta\phi)a_n(x) = \frac{i}{2}\Delta a_{n-1}, \tag{T_n}$$

$$\cdots\cdots\cdots\cdots\cdots\cdots\cdots$$

Remark 2.1. In the one-dimensional case it is sometimes convenient to look for u of the form

$$\frac{b(x; h)}{(E - V(x))^{1/4}} e^{i\phi/h}, \text{ where now } (\phi')^2 + V(x) - E = 0, \tag{2.12}$$

so that for instance $\phi' = \sqrt{E - V(x)}$. The reason for this is that (T_0) becomes $(\phi'(x)\partial_x + \frac{1}{2}\phi''(x))a_0(x) = 0$ and this ordinary differential equation has the general solution

$$a_0(x) = C \exp\left(-\frac{1}{2}\int^x \frac{\phi''}{\phi'} dx\right) = C \exp\left(-\frac{1}{2}\log\phi'\right) = \frac{C}{\sqrt{\phi'}}, \tag{2.13}$$

so if $b(x; h) \sim b_0(x) + h b_1(x) + \ldots$ in (2.12), then we have $b_0 = \text{const}$. One can also look for solutions of the form $\psi^{-1/2} \exp\int^x \psi(y) dy$, with $\psi > 0$, leading to a non-linear 2nd order (Ricati) equation for ψ, which can be formally solved by an asymptotic series: $\psi(x; h) \sim \psi_0(x) + h^2\psi_2(x) + h^4\psi_4(x) + \ldots$, with $\psi_0(x) = \sqrt{E - V(x)}$.

Assuming that we have solved successively all the transport equations in Ω and that $a(x; h)$ is an asymptotic sum in the sense of (2.10), (2.11), we get

$$(-h^2\Delta + V(x) - E)(e^{i\phi(x)/h}a(x; h)) = e^{i\phi(x)/h}r(x; h), \qquad (2.14)$$

with $r \sim 0$:

$$|\partial_x^\alpha r(x; h)| \leq C_{K,\alpha,N}h^{N+1}, \quad x \in K \subset\subset \Omega, \; \alpha \in \mathbf{N}^n, \; N \in \mathbf{N}. \qquad (2.15)$$

We can solve the sequence of transport equations in the following situation. Assume that $\phi'(x) \neq 0$, and choose local coordinates (t, y) in a neighborhood of some point, so that $\phi'(x) \cdot \partial_x$ becomes ∂_t and so that the neighborhood becomes: $\Omega = \{(t, y) \in \mathbf{R} \times \mathbf{R}^{n-1}; |t| < \epsilon, |y| < \epsilon\}$ for some small $\epsilon > 0$. Then we get a unique solution of the transport equations, if we prescribe the restrictions to $t = 0$ of $a_0, a_1, \ldots$.

If we want to construct WKB-solutions globally (that is in some large given region) many difficulties may appear:

– ϕ may become singular in some region (caustics), in particular the construction above with a real valued phase ϕ will always be restricted to the classically allowed region: $V(x) - E < 0$.

– ϕ may be multivalued.

– The integral curves of $\phi' \cdot \partial_x$ may behave in such a way that we get problems with the transport equations.

The consideration of these difficulties (which are of a great mathematical and physical interest) by Keller [Ke] and Maslov [Ma1] are at the origin of various versions of the theory of Fourier integral operators, in particular the one by Hörmander [Höl].

In the classically forbidden region: $V(x) - E > 0$, we can construct local solutions of the form $e^{-\phi(x)/h}a(x; h)$ with ϕ real valued. Formally, we may write $-\phi = i(i\phi)$ so the earlier constructions work, replacing everywhere ϕ by $i\phi$. Hence we obtain the eikonal equation

$$-(\phi')^2 + V - E = 0,$$

and the sequence of transport equations

$$(\phi' \cdot \partial_x + \frac{1}{2}\Delta\phi)a_0 = 0, \qquad (T_0')$$

$$(\phi' \cdot \partial_x + \frac{1}{2}\Delta\phi)a_1 = \frac{1}{2}\Delta a_0, \tag{T$_1'$}$$

etc. When choosing an asymptotic sum of the resulting formal symbol a, we get the obvious analogue of (2.14), (2.15). The classical Hamiltonian is now $q(x, \xi) + E := -(p(x, i\xi) - E) = \xi^2 - V(x) + E$.

Notes

WKB-solutions are of great interest in spectral theory and can be used in proving existence results. Under suitable conditions exact eigenfunctions for some problems can be well approximated by the WKB-solutions. The reader is referred to Chapter 6 for further information in this context. There have been many generalizations and extensions of the WKB-method which cannot be described here, such as the behavior of WKB-solutions near caustics and the WKB-method in the analytic case. We refer the reader to [Ma1], [La], [Ke], [Gr], [Vo] and [CaNoPh].

3. The WKB-method for a potential minimum

We shall construct approximate eigenvalues and eigenfunctions for the Schrödinger operator in a multi-dimensional case. Consider

$$P = -\frac{1}{2}h^2\Delta + V(x) \tag{3.1}$$

in a neighborhood of $0 \in \mathbf{R}^n$ and assume that V is smooth, real valued with
a non-degenerate (local) minimum at 0:

$$V(0) = 0, \ V'(0) = 0, \ V''(0) > 0. \tag{3.2}$$

Here the last inequality is to be understood in the sense of symmetric
matrices. Put $p(x,\xi) = \frac{1}{2}\xi^2 + V(x)$, $q(x,\xi) = -p(x,i\xi) = \frac{1}{2}\xi^2 - V(x)$.
After a Euclidean change of variables x and the corresponding dual change
of variables ξ, and still denoting by q the function q expressed in the new
variables, we may assume that:

$$q(x,\xi) = \frac{1}{2}\xi^2 - \sum_1^n \frac{a_j}{2}x_j^2 + \mathcal{O}((x,\xi)^3), \tag{3.3}$$

where $a_j > 0$ are the eigenvalues of the Hessian of V at 0. Introduce the new
variables $y_j = y_j/\sqrt{b_j}$ and the corresponding dual variables $\eta_j = \xi_j\sqrt{b_j}$.
Then the function q becomes:

$$q(y,\eta) = \sum(\frac{1}{2b_j}\eta_j^2 - \frac{b_j a_j}{2}y_j^2) + \mathcal{O}((y,\eta)^3). \tag{3.4}$$

Choose b_j so that $\frac{1}{b_j} = b_j a_j$, i.e. $b_j = 1/\sqrt{a_j}$. Then we get

$$q(y,\eta) = \sum_1^n \frac{\lambda_j}{2}(\eta_j^2 - y_j^2) + \mathcal{O}((y,\eta)^3), \ \lambda_j = \sqrt{a_j}. \tag{3.5}$$

The Hamilton field becomes:

$$H_q = \sum_1^n \lambda_j(\eta_j\partial_{y_j} + y_j\partial_{\eta_j}) + \mathcal{O}((y,\eta)^2). \tag{3.6}$$

This field vanishes at $(0,0)$ and, consequently, we cannot apply the general
result about existence of solutions of the Hamilton–Jacobi equations in
Chapter 1.

In general, if $v(x, \partial_x) = \sum_1^m a_j(x)\partial_{x_j}$ is a smooth real vector field on $\mathbf{R}^m$ which vanishes at 0, then we consider the corresponding linearized vector field:

$$v_0(x, \partial_x) = \sum\sum \partial_{x_k} a_j(0) x_k \partial_{x_j} = \langle Ax, \partial_x \rangle, \qquad (3.7)$$

where $A = (\partial_{x_k} a_j(0))$ is called the matrix of the linearization. Using the fact that v vanishes at 0, we see that v_0 is an invariantly defined linear vector field on $T_0\mathbf{R}^m$ and that A is invariantly defined as a map: $T_0\mathbf{R}^m \to T_0\mathbf{R}^m$. In fact, A maps $\nu(0)$ to $\mu(0)$ if ν, μ are real vector fields, related by $\mu = \mathcal{L}_\nu v = [\nu, v]$.

The eigenvalues and eigenvectors of the linearization of a vector field are of importance for the properties of the integral curves near a stationary point (i.e. a point where the vector field vanishes). For instance, we have the so-called stable manifold theorem (see for instance Abraham–Marsden [AbMa], Abraham–Robbin [AbRo]) which exists in various versions. Here is one of them:

Theorem 3.1. *Let v be a C^∞ vector field defined near 0 in $\mathbf{R}^m$ which vanishes at 0. Let d be the number of eigenvalues with real part > 0, counted with their algebraic multiplicity, of the linearization $\partial v(0)/\partial x$ of v at 0. Then in a suitable neighborhood U of 0, there is a unique closed smooth connected d-dimensional submanifold $\Lambda \subset U$ containing 0 and with the following properties:*

(1) v is tangent to Λ at every point of Λ: $v_\rho \in T_\rho\Lambda$, $\forall \rho \in \Lambda$,

(2) The complexification $\mathbf{C} \otimes T_0\Lambda$ is the sum of the generalized eigenspaces corresponding to the eigenvalues with real part > 0.

Moreover, $\exists C > 0$ such that for all $x \in \Lambda$, we have $\|\exp(-tv)(x)\| \le Ce^{-t/C}\|x\|$, $t > 0$. Here $\|\cdot\|$ is the standard Euclidean norm. Replacing this norm by another suitable one, we may even arrange so that the prefactor 'C' to the right, in the last estimate, can be replaced by 1.

Returning to $v = H_q$ in (3.6), we get the linearization at $(0, 0)$:

$$F = \begin{pmatrix} 0 & \lambda_1 & & & & & \\ \lambda_1 & 0 & & & & & \\ & & 0 & \lambda_2 & & & \\ & & \lambda_2 & 0 & & & \\ & & & & \ddots & & \\ & & & & & 0 & \lambda_n \\ & & & & & \lambda_n & 0 \end{pmatrix} \qquad (3.8)$$

which has the structure of a block-matrix, when the coordinates are enumerated as y_1, η_1, y_2, η_2,..., and where the non-diagonal blocks are 0. The linearization of a Hamilton field is sometimes called a fundamental matrix.

In general, if F is a fundamental matrix, then $\exp{(tF)}$ is a symplectic matrix: $\sigma(\exp{(tF)}\nu, \exp{(tF)}\mu) = \sigma(\nu, \mu)$ and differentiating this and putting $t = 0$, we see that F is anti-symmetric with respect to the symplectic form:

$$\sigma(F\nu, \mu) + \sigma(\nu, F\mu) = 0. \tag{3.9}$$

Let

$$\mathbf{C}^{2n} = \bigoplus_{\mu \in \sigma(F)} E_\mu$$

be the Jordan decomposition of $\mathbf{C}^{2n}$ into generalized eigenspaces. Then we see that E_μ and E_λ are orthogonal with respect to σ iff $\mu + \lambda \neq 0$. In fact, assume that $\mu + \lambda \neq 0$ and let $x \in E_\mu$, $y \in E_\lambda$. Then $(\mu - F)^m \big|_{E_\mu} = 0$, for some $m > 0$, and similarly with μ replaced by λ, so

$$(\mu + F)^m \big|_{E_\lambda} = ((\mu + \lambda)I + (F - \lambda))^m \big|_{E_\lambda} : E_\lambda \to E_\lambda$$

is bijective. We can therefore write $y = (\mu + F)^m z$, $z \in E_\lambda$, and

$$\sigma(x, y) = \sigma(x, (\mu + F)^m z) = \sigma((\mu - F)^m x, z) = 0,$$

so E_μ and E_λ are orthogonal when $\mu + \lambda \neq 0$. Then clearly E_μ and $E_{-\mu}$ cannot be orthogonal, and σ even gives a natural way of identifying $E_{-\mu}$ with the dual space of E_μ. If we regroup the distinct eigenvalues into $\{0\}$, $\{\mu_1, -\mu_1\}, \ldots, \{\mu_d, -\mu_d\}$, then we get the decomposition into $d + 1$ spaces which are orthogonal for σ:

$$\mathbf{C}^{2n} = E_0 \oplus (E_{\mu_1} \oplus E_{-\mu_1}) \oplus \ldots \oplus (E_{\mu_d} \oplus E_{-\mu_d}).$$

Notice also that the restriction of the symplectic form to each of these subspaces is non-degenerate, that the dimension of E_0 is even, and that the dimension of E_μ is equal to the dimension of $E_{-\mu}$. Using also the fact that F is real, we see that the non-zero eigenvalues split into groups: λ_j, $-\lambda_j$, for $\lambda_j > 0$, $i\alpha_j$, $-i\alpha_j$ for $\alpha_j > 0$ and ζ_j, $\overline{\zeta_j}$, $-\zeta_j$, $-\overline{\zeta_j}$, where $\operatorname{Re}\zeta_j > 0$, $\operatorname{Im}\zeta_j > 0$.

Returning to F in (3.8), we get the real eigenvalues $\pm\lambda_j$, $j = 1, \ldots, n$. The corresponding eigenvectors are immediate to compute and the sum of the eigenspaces corresponding to positive eigenvalues is given by $\eta = y$, $y \in \mathbf{R}^n$.

Applying the stable manifold theorem, we see that in a suitable neighborhood of $(0, 0)$, we can find a closed n-dimensional submanifold Λ_+ such that

$(0,0) \in \Lambda_+$, $T_{(0,0)}\Lambda_+ = \{(y,y)\}$ and such that H_q is tangent to Λ_+ at every point.

Reversing the sign of q we also obtain a closed n-dimensional submanifold Λ_- with $(0,0) \in \Lambda_-$, $T_{(0,0)}\Lambda_- = \{(y,-y)\}$ such that H_q is tangent to Λ_-. We also have

$$\|\exp(tH_q)(\rho)\| \leq Ce^{-|t|/C}\|\rho\|, \text{ for } \rho \in \Lambda_\pm, \ \mp t > 0.$$

(We say that H_q is expansive on Λ_+ and contractive on Λ_-.) We call Λ_+ and Λ_- respectively the stable outgoing and the stable incoming manifolds. (A more standard terminology is to call Λ_+ the unstable manifold and to call Λ_- the stable manifold, however we use the slightly different terminology, thinking about the stability of the manifolds themselves under the H_q-flow.)

Lemma 3.2. $q_{|\Lambda_\pm} = 0.$

Proof. Since $q(\rho) = q(\exp(tH_q)(\rho))$, we get

$$q(\rho) = \lim_{t \to \mp \infty} q(\exp(tH_q)(\rho)) = q(0) = 0,$$

for $\rho \in \Lambda_\pm$. #

Lemma 3.3. $\Lambda_\pm$ *are Lagrangian manifolds.*

Proof. Using $x_1, \ldots, x_n$ as local coordinates on Λ_+, we have $v := H_{q|\Lambda_+} = \sum v_j(x)\partial_{x_j}$, and $\partial v/\partial x(0)$ has the eigenvalues $\lambda_1, \ldots, \lambda_n$. Recall that if $x(t)$ is an integral curve of v then by the stable manifold theorem, or by a direct argument, we have

$$\|x(t)\| \leq Ce^{-|t|/C}\|x(0)\|, \ t \to -\infty, \tag{3.10}$$

where we may even replace the first factor C on the right by 1. We look for the evolution of tangent vectors. Let $x(t,s)$ be an integral curve, depending smoothly on the additional parameter s. Differentiating the equation $\partial_t x = v(x)$, (viewing v as a vector depending on x) we get

$$\partial_t \partial_s x = \frac{\partial v}{\partial x}(x(t,s))\partial_s x.$$

Using (3.10) and the fact that $\partial v(0)/\partial x$ has only eigenvalues with strictly positive real part, we get

$$\|\partial_s x(t,s)\| \leq Ce^{-|t|/C}\|\partial_s x(0,s)\|, \ t \leq 0. \tag{3.11}$$

In other words, if

$$\rho(t) = \exp{(tH_q)}(\rho(0)), \ t \le 0, \delta(t) = (\exp{(tH_q)})_*\delta(0) \in T_{\rho(t)}\Lambda_+,$$

then

$$\|\delta(t)\| \le Ce^{-|t|/C}\|\delta(0)\|, \ t \le 0.$$

It is now easy to see that Λ_+ is Lagrangian. First we notice that the dimension is the right one, secondly that for $\nu, \mu \in T_\rho\Lambda_+$:

$$\sigma_\rho(\nu, \mu) = \sigma_{\exp{tH_q}(\rho)}((\exp{tH_q})_*\nu, (\exp{tH_q})_*\mu) \to 0, \ t \to -\infty,$$

so $\sigma_\rho(\nu, \mu) = 0.$ #

In the original coordinates we get a smooth real function $\phi(x)$ defined in a neighborhood of 0, such that Λ_+ is given by $\xi = \phi'(x)$ (and Λ_- is given by $\xi = -\phi'(x)$), $\phi(0) = 0$, $\phi'(0) = 0$, $\phi''(0) > 0$. (To get the last fact, just notice that in the y-coordinates, we have $\phi(y) = \frac{y^2}{2} + \mathcal{O}(y^3)$.) We also have of course, the eikonal equation: $q(x, \phi'_x) = 0$, or more explicitly: $(\phi'_x)^2 - V(x) = 0$.

We can now start the WKB-construction. We want to find $a(x; h) \sim a_0(x) + ha_1(x) + \ldots$ and $E \sim \sum_0^\infty E_j h^j$ such that

$$(P - hE)(ae^{-\phi/h}) = re^{-\phi(x)/h}, \ r = \mathcal{O}(h^N), \forall N \ge 0. \tag{3.12}$$

Here we notice in general that $P(ae^{-\phi/h}) = be^{-\phi/h}$, with $b = h\mathcal{L}(a) - h^2\Delta a/2$, where $\mathcal{L} = \nabla\phi \cdot \nabla + \Delta\phi/2$. Thus, if we try a with the asymptotic expansion above, it is enough to solve (in some fixed neighborhood of 0) the sequence of transport equations:

$$(\mathcal{L} - E_0)a_0 = 0, \tag{T_0}$$

$$(\mathcal{L} - E_0)a_1 = E_1 a_0 + \frac{1}{2}\Delta a_0, \tag{T_1}$$

$$(\mathcal{L} - E_0)a_2 = E_2 a_0 + E_1 a_1 + \frac{1}{2}\Delta a_1. \tag{T_2}$$

etc. Since the gradient of ϕ vanishes at 0 it is not obvious how to solve these equations and we have to study $\mathcal{L}$ more closely. Recall that $q = \frac{1}{2}\xi^2 - V(x)$, $V(x) = \sum \frac{a_j}{2}x_j^2 + \mathcal{O}(x^3)$. Denote by V_0 the leading quadratic part of the last expression, and write similarly, $\phi(x) = \phi_0(x) + \mathcal{O}(x^3)$. Then from $q(x, \phi'_x) = 0$, we get $(\phi'_0)^2 = V_0(x)$, and knowing also that ϕ_0 is positive definite, we get the unique solution $\phi_0(x) = \sum \frac{\lambda_j}{2}x_j^2$, and consequently, $\Delta\phi_0 = \sum \lambda_j$. We conclude that

$$\mathcal{L} = \sum(\lambda_j x_j + \mathcal{O}(x^2))\partial_{x_j} + \sum \frac{\lambda_j}{2} + \mathcal{O}(x).$$

Put

$$\mathcal{L}_0 = \sum \lambda_j x_j \partial_{x_j} + \sum \frac{\lambda_j}{2}.$$

We notice that if $\mathcal{P}^m_{\text{hom}}$ is the space of polynomials in n variables which are homogeneous of degree $m \in \mathbf{N}$, then $\mathcal{L}_0(\mathcal{P}^m_{\text{hom}}) \subset \mathcal{P}^m_{\text{hom}}$, and the monomials, x^α, $|\alpha| = m$, constitute a basis of eigenvectors of the restriction of $\mathcal{L}_0$ to $\mathcal{P}^m_{\text{hom}}$. The corresponding eigenvalues are $\sum \lambda_j(\alpha_j + \frac{1}{2})$, so $\mathcal{L}_0 - E_0$ is a bijection in $\mathcal{P}^m_{\text{hom}}$ iff $E_0 \notin \{\sum \lambda_j(\alpha_j + \frac{1}{2}); |\alpha| = m\}$.

Consider a homogeneous transport equation (like (T_0)):

$$(\mathcal{L} - E_0)f = 0, \tag{3.13}$$

where $f \in C^\infty$ does not vanish to infinite order at 0. Let $f = f_m(x) + \mathcal{O}(x^{m+1})$, $0 \neq f_m \in \mathcal{P}^m_{\text{hom}}$. Then we get $(\mathcal{L}_0 - E_0)f_m = 0$. We conclude that E_0 has to be an eigenvalue of $\mathcal{L}_0$: $E_0 = \sum(\alpha_j + \frac{1}{2})\lambda_j$, for some $\alpha \in \mathbf{N}^n$ of length m. We now assume for a given E_0:

There is precisely one $\alpha \in \mathbf{N}^n$ such that $E_0 = \sum \lambda_j(\frac{1}{2} + \alpha_j)$. $\qquad$ (H)

Let α_0 denote the unique α in the above assumption, and put $m_0 = |\alpha_0|$, $f_0 = x^{\alpha_0}$. We now work with formal power series at 0. Then we can construct a solution $f = \sum_{m_0}^\infty f_k$, $f_k \in \mathcal{P}^k_{\text{hom}}$, to (3.13) in the following way. Put $f_{m_0} = x^{\alpha_0}$. Then (in the sense of formal power series at 0): $(\mathcal{L} - E_0)f_{m_0} = \sum_{m_0+1}^\infty g_k$, where g_k is homogeneous of degree k. Let $\tilde{f}_{m_0+1}$ solve $(\mathcal{L}_0 - E_0)f_{m_0+1} = -g_{m_0+1}$. Then $(\mathcal{L} - E_0)(f_{m_0} + f_{m_0+1}) = \sum_{m_0+2}^\infty h_k$, etc.

Let $\tilde{f}_{m_0}$ be the (formal power series) solution of (3.13) that we have just constructed. More generally, inhomogeneous equations can be treated the same way, and we obtain

Proposition 3.4. *For every formal power series g at $x = 0$ there is a unique scalar $\lambda(g)$ such that $(\mathcal{L} - E_0)f = g - \lambda(g)\tilde{f}_{m_0}$ has a solution. The solution is unique up to a multiple of $\tilde{f}_{m_0}$.*

It is now clear how to construct formal power series solutions a_0, a_1, a_2, etc. to the sequence of transport equations, as well as a corresponding sequence $E_0, E_1, \ldots$. We leave as an exercise to the reader to show that the E_j are uniquely determined, once E_0 has been chosen (satisfying (H)). We also notice that if $2j < m_0$, then a_j vanishes to the order $m_0 - 2j$ at $x = 0$.

The final step in our construction is to pass from formal power series at $x = 0$ to actual C^∞-functions defined in some fixed j-independent neighborhood of

$x = 0$. If M is the open set where V and ϕ are defined, we let $\Omega \subset M$ be an open neighborhood of 0 and we say that Ω is *star-shaped* if the following statements hold:

(1) If $x \in \Omega$, then $\exp(t\nabla\phi \cdot \partial_x)(x)$ is well-defined and belongs to Ω for $t \leq 0$. Moreover $\exp(t\nabla\phi \cdot \partial_x)(x)$ converges to 0 when $t \to -\infty$.

(2) For every compact set $K \subset \Omega$, the set $\hat{K} = \{0\} \cup \{\exp(t\nabla\phi \cdot \partial_x)(x); x \in K, t \leq 0\}$ is a compact subset of Ω.

In view of (3.10) and the subsequent remark, we see that $B(0, r_0) = \{x \in \mathbf{R}^n; \|x\| < r_0\}$ is star-shaped if $r_0 > 0$ is sufficiently small. We now let Ω be star-shaped, and consider the equation

$$(\mathcal{L} - E_0)u = v$$

in the space of C^∞ functions on Ω which vanish to infinite order at 0. With $\nu = \nabla\phi \cdot \partial_x$ we only retain that this equation is of the form

$$(\nu + k(x))u = v, \tag{3.14}$$

where $k \in C^\infty(\Omega)$. Let $] - \infty, 0] \ni t \mapsto \gamma(t)$ be the ν-integral curve with $\gamma(0) = x$ for some given $x \in \Omega$. Along γ the differential equation takes the form:

$$(\frac{d}{dt} + k(\gamma(t)))u(\gamma(t)) = v(\gamma(t)).$$

Here $v(\gamma(t)) = \mathcal{O}(e^{-C|t|})$ for every $C > 0$ (since $\gamma(t)$ approaches 0 exponentially fast when $t \to -\infty$), and we have a unique solution with the same properties, given by

$$u(\gamma(t)) = \int_{-\infty}^{t} e^{-\int_s^t k(\gamma(\sigma))d\sigma} v(\gamma(s))ds. \tag{3.15}$$

Since $\gamma(0) = x$, we see that the only possible solution of (3.14) which vanishes to infinite order at 0 is given by:

$$u(x) = \int_{-\infty}^{0} e^{-\int_s^0 k(\exp \sigma\nu(x))d\sigma} v(\exp s\nu(x))ds, \tag{3.16}$$

and it only remains to show that this expression defines a function in the space we want. For x in a compact in Ω, there is a constant $C > 0$ such that

$$\|\exp s\nu(x)\| \leq Ce^{-|s|/C}\|x\|, \quad \|d_x\exp s\nu(x)\| \leq Ce^{-|s|/C},$$

and for the higher differentials we have estimates of the same type. Taking repeatedly higher and higher differentials of the expression (3.16) we verify the required properties, and we obtain:

Proposition 3.5. *Let Ω be star-shaped and let $v \in C^\infty(\Omega)$ vanish to infinite order at 0. Then the equation (3.14) has a unique solution with the same properties.*

We now consider the sequence of transport equations (T_j) in some star-shaped domain Ω. We recall that we already know how to solve these equations in the space of formal power series as $x = 0$, and we let $\tilde{a}_0, \tilde{a}_1, \tilde{a}_2, \ldots$ be smooth functions on Ω which represent a formal power series solution. Then we look for $a_j = \tilde{a}_j + b_j$, where $b_j \in C^\infty(\Omega)$ vanishes to infinite order at 0. Using the last proposition it is easy to see that there are such uniquely determined functions b_j. Now let $a(x; h) \sim \sum a_j(x) h^j$. We have proved:

Theorem 3.6. *Let E_0 satisfy (H) and let α_0 and m_0 be defined as after (H), and let Ω be star-shaped. Then we can find $a_j(x) \in C^\infty(\Omega)$ with $a_0(x) = x^{\alpha_0} + \mathcal{O}(|x|^{m_0+1})$, $a_j(x) = \mathcal{O}(|x|^{m_0-2j})$, $2j < m_0$, and uniquely determined real numbers E_1, E_2 such that if $E(h) \sim E_0 + E_1 h + \ldots$, then*

$$(P - hE(h))(e^{-\phi/h}a) = re^{-\phi/h},$$

where $|\partial_x^\alpha r(x; h)| \le C_{K,N,\alpha} h^N$, $x \in K$ for every $K \subset\subset \Omega$, $\alpha \in \mathbf{N}^n$, $N \in \mathbf{N}$.

This theorem can be generalized to the case of arbitrary values E_0 of the form $\sum \lambda_j(\alpha_j + \frac{1}{2})$, not necessarily satisfying (H), but the argument above becomes a little more complicated, and moreover one will in general get half-powers of h in the expansion of $E(h)$.

We end this chapter by recalling how the values E_0 can be viewed as eigenvalues of a certain harmonic oscillator. We start with the case of the standard harmonic oscillator on $\mathbf{R}$:

$$P = (-\frac{d^2}{dx^2} + x^2).$$

One eigenvalue is $\lambda_0 = 1$ and the corresponding normalized eigenfunction is

$$u_0 = \pi^{-\frac{1}{4}} e^{-x^2/2}.$$

To generate the other eigenvalues, introduce the annihilation operator $Z = \frac{d}{dx} + x$ and its adjoint, the creation operator $Z^* = -\frac{d}{dx} + x$. Then if we use the standard notation $[A, B] = AB - BA$ for the commutator of the operators A, B, we get $[Z, Z^*] = 2$, $P = ZZ^* - 1 = Z^*Z + 1$. Assuming we already found a function u_j with $Pu_j = \lambda_j u_j$, we try $u_{j+1} = Z^* u_j$. Then $PZ^* u_j = (Z^*Z + 1)Z^* u_j = Z^*(ZZ^* + 1)u_j = Z^*(P + 2)u_j = (\lambda_j + 2)Z^* u_j$. Hence u_{j+1} is an eigenfunction with eigenvalue $\lambda_{j+1} = \lambda_j + 2$. Thus we get the sequence of eigenfunctions $e_j = C_j(Z^*)^j(e^{-x^2/2}) = p_j(x)e^{-x^2/2}$, where

$C_j > 0$ is a normalization constant, determined by the requirement that the L^2-norm of e_j should be equal to 1. Here the polynomials p_j are called Hermite polynomials. It is clear that p_j is of degree j exactly, so an arbitrary polynomial is a linear combination of the p_j. Since the space of functions of the form $p(x)e^{-x^2/2}$, with p polynomial, is dense in $L^2(\mathbf{R})$, it follows that the orthonormal family $e_0, e_1, \ldots$ is an orthonormal basis. Anticipating a little on the general spectral theory for selfadjoint operators, that will be reviewed in chapter 4, we see that the spectrum of our one-dimensional harmonic oscillator is given by the eigenvalues $\lambda_j = 1 + 2j$ for $j = 0, 1, 2, \ldots$ and that each of these eigenvalues is simple.

In $\mathbf{R}^n$, we consider the generalized harmonic oscillator

$$P = -\frac{1}{2}\Delta + V_0(x)$$

where V_0 is a positive definite quadratic form. Then as we have seen in the beginning of this chapter, we can make a linear change of variables and reduce P to the operator

$$P = \sum \frac{\lambda_j}{2}(D_{y_j}^2 + y_j^2),$$

where $D = \frac{1}{i}\partial$. We then get the eigenvalues $\mu_\alpha = \sum \lambda_j(\alpha_j + \frac{1}{2})$, for $\alpha \in \mathbf{N}^n$, and a corresponding orthonormal basis of eigenfunctions:

$$u_\alpha(y) = C_\alpha(-\partial_{y_1} + y_1)^{\alpha_1} \cdot \ldots \cdot (-\partial_{y_n} + y_n)^{\alpha_n}(e^{-y^2/2}) = p_\alpha(y)e^{-y^2/2}.$$

If we consider the semi-classical harmonic oscillator

$$P_h = -\frac{h^2}{2}\Delta + V_0(x),$$

then the change of variables $x = \sqrt{h}\tilde{x}$, reduces P_h to hP_1, so we get the eigenvalues $h\sum\lambda_j(\alpha_j + \frac{1}{2})$, and the corresponding eigenfunctions expressed in the y coordinates: $h^{-n/4}p_\alpha(h^{-1/2}y)e^{-y^2/(2h)}$. In other words, the WKB-constructions earlier produce exact eigenvalues and eigenfunctions in the case of a (generalized) harmonic oscillator.

Notes

In this chapter we have followed Helffer-Sjöstrand [HeSj2].

4. Self-adjoint operators

In this chapter we review some of the standard theory, and apply it to a semi-classical Schrödinger operator with a potential well. Let $\mathcal{H}$ be a complex separable Hilbert space. The corresponding norm and inner product are denoted by $\| \cdot \|$, $(\cdot|\cdot)$. By definition, an unbounded operator $S : \mathcal{H} \to \mathcal{H}$ is given by a subspace $\mathcal{D}(S) \subset \mathcal{H}$, called the domain of S, and a linear operator $S : \mathcal{D}(S) \to \mathcal{H}$. (It might be better to speak about not necessarily bounded operators, since bounded operators are not excluded from the class of unbounded operators.) The graph of S is defined by:

$$\text{graph}\,(S) = \{(x, Sx); \, x \in \mathcal{D}(S)\}. \tag{4.1}$$

This is a subspace of $\mathcal{H} \times \mathcal{H}$ that we equip with the norm of $\mathcal{H} \times \mathcal{H}$: $\|(u, v)\|^2 = \|u\|^2 + \|v\|^2$. We say that S is closed if graph (S) is closed.

Proposition 4.1. *Let $S : \mathcal{H} \to \mathcal{H}$ be an unbounded operator with a dense domain $\mathcal{D}(S)$. Then there exists an unbounded operator $S^* : \mathcal{H} \to \mathcal{H}$ given by:*

$$\mathcal{D}(S^*) = \{v \in \mathcal{H}; \exists C(v) \geq 0 \text{ such that } |(Su|v)| \leq C(v)\|u\|, \, u \in \mathcal{D}(S)\}, \tag{4.2}$$

$$(Su|v) = (u|S^*v), \, \forall u \in \mathcal{D}(S), \, \forall v \in \mathcal{D}(S^*). \tag{4.3}$$

Proof. We define $\mathcal{D}(S^*)$ by (4.2). Then for $v \in \mathcal{D}(S^*)$, the linear form $\mathcal{D}(S) \ni u \mapsto (Su|v)$ has a unique continuous extension to $\mathcal{H}$, hence there exists a unique $w \in \mathcal{H}$ such that $(Su|v) = (u|w)$, $\forall u \in \mathcal{D}(S)$. The map $\mathcal{D}(S^*) \ni v \mapsto w$ is linear and we can define S^*v to be equal to w.　　　#

Notice that if we drop the assumption that $\mathcal{D}(S)$ is dense, then we can still define $\mathcal{D}(S^*)$ by (4.2), and at least one element $S^*v \in \mathcal{H}$ by (4.3). However, the vector S^*v is no longer uniquely defined, since it can be changed by addition of an arbitrary element of $\mathcal{D}(S)^\perp := \{u \in \mathcal{H}; \, (u|v) = 0, \, \forall v \in \mathcal{D}(S)\}$. On the other hand, if we assume that S is bounded (so that $\mathcal{D}(S) = \mathcal{H}$ and $\|Su\| \leq C\|u\|$, and $\|S\|$ denotes the smallest possible constant in the previous estimate) then S^* is bounded and $\|S^*\| = \|S\|$.

Let $J : \mathcal{H} \times \mathcal{H}$ be given by $J((u, v)) = (-v, u)$. Then $J^2 = -I$, where I denotes the identity operator. Viewing $\mathcal{H} \times \mathcal{H}$ as a Hilbert space with the scalar product $((u_1, u_2)|(v_1, v_2)) = (u_1|v_1) + (u_2|v_2)$, we see that $J^* = -J$, and that J is unitary. (We recall that a bounded operator $U : \mathcal{H} \to \mathcal{H}$ is unitary if it is isometric: $\|Uu\| = \|u\|$ and surjective. Equivalently, a unitary operator U is a bounded operator which satisfies: $U^*U = UU^* = I$. One can also define unitary operators between two different Hilbert spaces.) We have the following relation:

$$\text{graph}\,(S^*) = (J(\text{graph}\,(S)))^\perp = J((\text{graph}\,(S))^\perp).$$

In particular, S^* is always a closed operator. Notice also that $\overline{\text{graph}\,(S)} = (\text{graph}\,(S))^{\perp\perp} = J^2(\text{graph}\,(S)^{\perp\perp}) = (J((J\text{graph}\,(S))^\perp))^\perp$. Hence if $\mathcal{D}(S^*)$ is dense, then S^{**} is the closure of S; $S^{**} = \overline{S}$ in the sense that $\text{graph}\,(\overline{S}) := \overline{\text{graph}\,(S)} = \text{graph}\,(S^{**})$.

In general it is not obvious that the closure of the graph of S is the graph of an operator. If this is the case, we say that S is closable. The above discussion shows that if the domains of S and S^* are dense, then S is closable. Conversely, it is easy to see that if $\mathcal{D}(S)$ is dense and S is closable, then $\mathcal{D}(S^*)$ is dense and the closure of S is equal to S^{**}.

In general when S is densely defined we have the useful identity:

$$\text{Im}\,(S)^\perp = \text{Ker}\,(S^*).$$

Here we put $\text{Ker}\,(A) = \{u \in \mathcal{D}(A);\ Au = 0\}$ and $\text{Im}\,(A) = \{Au;\ u \in \mathcal{D}(A)\}$.

Definition 4.2. If $A, B : \mathcal{H} \to \mathcal{H}$ are unbounded operators, we say that $A \subset B$, if $\mathcal{D}(A) \subset \mathcal{D}(B)$ and $Bu = Au$ for all $u \in \mathcal{D}(A)$. Equivalently, A is contained in B if and only if we have the corresponding inclusion for the graphs.

If $\mathcal{D}(A)$ is dense and $A \subset B$, then $B^* \subset A^*$.

Definition 4.3. Let $A : \mathcal{H} \to \mathcal{H}$ be a densely defined unbounded operator. We say that A is symmetric if $A \subset A^*$ and selfadjoint if $A = A^*$.

Self-adjoint operators have very nice properties, so if A is a given symmetric operator we are interested in finding a selfadjoint operator B which contains A. We then say that B is a selfadjoint extension of A. Notice that every selfadjoint operator is closed, so if B is a selfadjoint extension of a symmetric operator A then, necessarily, $\overline{A} \subset B$. (Here we notice that every symmetric operator is closable.)

Definition 4.4. A symmetric operator A is called essentially selfadjoint if $\overline{A}$ is selfadjoint.

If A is essentially selfadjoint, then the closure of A is the only selfadjoint extension of A.

Theorem 4.5. *Let* $A : \mathcal{H} \to \mathcal{H}$ *be symmetric. Then (1), (2), (3), are equivalent, where:*

(1) A is selfadjoint,

(2) A is closed and $\mathrm{Ker}\,(A^* \pm i) = \{0\}$ *for the two signs,*

(3) $\mathrm{Im}\,(A \pm i) = \mathcal{H}$ *for the two signs.*

Proof. In order to show that (1) $\Rightarrow$ (2), it suffices to show that if $S : \mathcal{H} \to \mathcal{H}$ is symmetric, then $\mathrm{Ker}\,(S \pm i) = \{0\}$. If $u \in \mathrm{Ker}\,(S + i)$, then

$$0 = ((S + i)u|u) = (Su|u) + i\|u\|^2. \tag{4.4}$$

Since $(Su|u) = (u|Su) = \overline{(Su|u)}$, we see that $(Su|u)$ is real, and hence (4.4) implies that $u = 0$. Also notice that for u in the domain of S we have the inequality: $\|u\|^2 \leq \mathrm{Im}\,((S + i)u|u) \leq \|(S + i)u\|\|u\|$, and after dividing by $\|u\|$ we get the first half of

$$\|u\| \leq \|(S + i)u\|, \quad \|u\| \leq \|(S - i)u\|. \tag{4.5}$$

The second half is obtained in the same way. From (4.5) we deduce that if S is closed and symmetric then $\mathrm{Im}\,(S \pm i)$ are closed subspaces of $\mathcal{H}$.

We now show that (2) $\Rightarrow$ (3), so we assume that (2) holds. Since $\overline{\mathrm{Im}\,(A \mp i)} = (\mathrm{Ker}\,(A^* \pm i))^\perp = \mathcal{H}$, and since A is closed and symmetric, we have $\overline{\mathrm{Im}\,(A \mp i)} = \mathrm{Im}\,(A \mp i)$, and (3) follows.

We finally show that (3) $\Rightarrow$ (1). Assume (3) and let $x \in \mathcal{D}(A^*)$. Using (3) we know that there exists $y \in \mathcal{D}(A)$ such that $(A^* - i)x = (A - i)y$. Since $A \subset A^*$, we have $(A^* - i)(x - y) = 0$. Since $\mathrm{Ker}\,(A^* - i) = (\mathrm{Im}\,(A + i))^\perp = \mathcal{H}^\perp = \{0\}$, we have $\mathrm{Ker}\,(A^* - i) = \{0\}$, hence $x - y = 0$ and hence $\mathcal{D}(A^*) = \mathcal{D}(A)$. #

From the preceding result, it is easy to get the following characterization of essentially selfadjoint operators.

Corollary 4.6. *Let* $A : \mathcal{H} \to \mathcal{H}$ *be symmetric. Then (1) $\Leftrightarrow$ (2) $\Leftrightarrow$ (3), where:*

(1) A is essentially selfadjoint,

(2) $\mathrm{Ker}\,(A^* \pm i) = \{0\}$ *for both the signs,*

(3) $\mathrm{Im}\,(A \pm i)$ *is dense in* $\mathcal{H}$ *for both the signs.*

In both the results above, we can replace $\pm i$ by two arbitrary points z_+ and z_- with $\pm\mathrm{Im}\,z_\pm > 0$.

Let $A : \mathcal{H} \to \mathcal{H}$ be densely defined. We say that $z_0 \in \mathbf{C}$ belongs to the resolvent set, $\rho(A)$, if $(z_0 - A) : \mathcal{D}(A) \to \mathcal{H}$ is bijective and the inverse $(z_0 - A)^{-1}$ belongs to $\mathcal{L}(\mathcal{H}, \mathcal{H})$, the space of bounded operators: $\mathcal{H} \to \mathcal{H}$. We then put $R(z_0) = (z_0 - A)^{-1}$. For $z \in \mathbf{C}$, we have $(z - A)R(z_0) = I + (z - z_0)R(z_0)$, and since the norm of the last term is $|z - z_0| \, \|R(z_0)\|$, we see that for $|z - z_0| < 1/\|R(z_0)\|$, $(z - A)$ has a right inverse, $R(z_0)(I + (z - z_0)R(z_0))^{-1}$. Similarly we have the left inverse, $(I + (z - z_0)R(z_0))^{-1}R(z_0)$. We conclude that if $z_0 \in \rho(A)$, $|z - z_0| < 1/\|R(z_0)\|$, then $z \in \rho(A)$. This implies that $\rho(A)$ is an open set. By definition its complement in $\mathbf{C}$ (which is then a closed set) is called the spectrum of A. It is usually denoted by $\sigma(A)$ or by $\mathrm{sp}\,(A)$.

In general, if $z, w \in \rho(A)$, then the corresponding resolvents commute: $R(z)R(w) = R(w)R(z)$, and we have the important resolvent identity:

$$R(z) - R(w) = (w - z)R(z)R(w).$$

If A is selfadjoint, then $\sigma(A) \subset \mathbf{R}$, and essentially from the proof of (4.5) it follows that $\|R(z)\| \leq 1/|\mathrm{Im}\, z|$. Later we will see that we even have $\|R(z)\| = (\mathrm{dist}\,(z, \sigma(A)))^{-1}$ in the selfadjoint case. (Indeed, this will be immediate from the spectral theorem.)

Defect indices. The following is a general theorem on selfadjoint extensions of symmetric operators, that we state without proof.

Theorem 4.7. *Let $A : \mathcal{H} \to \mathcal{H}$ be a closed symmetric operator. Then:*

(1) $\dim \mathrm{Ker}\,(z - A^*)$ is constant for z in the open upper half-plane. The same holds for z in the open lower half-plane. We put $n_\pm = \dim \mathrm{Ker}\,(\pm i - A^*)$.

(2) One of the following holds: $\sigma(A)$ is the closed upper half-plane, the closed lower half-plane, the whole complex plane, or a subset of the real line.

(3) A is selfadjoint iff $\sigma(A) \subset \mathbf{R}$.

(4) A is selfadjoint iff $n_+ = n_- = 0$.

(5) A has a selfadjoint extension iff $n_+ = n_-$.

(6) If $n_+ = n_-$, then there is a bijection between the set of selfadjoint extensions of A and the set of unitary operators $U : \mathcal{H}_+ \to \mathcal{H}_-$, where $\mathcal{H}_\pm = \mathrm{Ker}\,(\pm i - A^*)$.

Remark. Let $V \in L^2_{\mathrm{loc}}(\mathbf{R}^n; \mathbf{R})$ and let $S = -\Delta + V(x)$, where $\Delta = \sum \partial^2_{x_j}$ is the Laplace operator, equipped with the domain. $\mathcal{D}(S) = C^\infty_0(\mathbf{R}^n)$. Then S is symmetric and the antilinear operator Γ of complex conjugation of functions.

$\Gamma u = \bar{u}$ commutes with S and with $S^* = A^*$, where A denotes the closure of S. Then $\Gamma(\mathcal{H}_\pm) = \mathcal{H}_\mp$, so we conclude that there exists at least one selfadjoint extension of S.

Friedrichs extensions. Let S be a symmetric operator such that $S \geq I$ in the sense that

$$(Su|u) \geq \|u\|^2, \ \forall u \in \mathcal{D}(S). \tag{4.6}$$

We can then associate with S the quadratic form:

$$q(u, u) = (Su|u) = |||u|||^2, \tag{4.7}$$

where the last equality defines a norm on $\mathcal{D}(S)$. Let $\mathcal{D}(q)$ be the completion of $\mathcal{D}(S)$ for this norm. (This completion is the abstract Hilbert space obtained as the set of equivalence classes of Cauchy sequences on $\mathcal{D}(q)$ for the norm $||| \cdot |||$, where the Cauchy sequences $(u_j)_1^\infty$ and $(v_j)_1^\infty$ are said to be equivalent if $|||u_j - v_j||| \to 0$.) Since every Cauchy sequence for the norm $||| \cdot |||$ is also a Cauchy sequence for the norm $\| \cdot \|$, we have a natural bounded linear map (of norm ≤ 1) $j : \mathcal{D}(q) \to \mathcal{H}$.

Lemma 4.8. *j is injective.*

Proof. Let $u \in \mathcal{D}(q)$, $j(u) = 0$. If $u_n \in \mathcal{D}(S)$, $|||u_n - u||| \to 0$, then $\|u_n\| \to 0$. We have

$$|||u|||^2 = \lim_{n \to \infty} \lim_{m \to \infty} q(u_n, u_m) = \lim_{n \to \infty} \lim_{m \to \infty} (Su_n|u_m) = \lim_{n \to \infty} 0 = 0.$$

$$\#$$

We can then view $\mathcal{D}(q)$ as a subspace of $\mathcal{H}$, and extend q to a quadratic form on $\mathcal{D}(q) \times \mathcal{D}(q)$. In general, we say that a quadratic form $q(u, u) \geq \|u\|^2$ defined on the subspace $\mathcal{D}(q) \times \mathcal{D}(q)$ is closed, if $\mathcal{D}(q)$ is complete for the norm $||| \cdot ||| = \sqrt{q(\cdot, \cdot)}$. We have just verified that the quadratic form q in (4.7) extended to $\mathcal{D}(q) \times \mathcal{D}(q)$ is closed. Moreover, it is densely defined in the sense that $\mathcal{D}(q)$ is dense.

Theorem 4.9. *Let $q(u, u) \geq \|u\|^2$ be a closed and densely defined quadratic form with domain $\mathcal{D}(q)$. Then there exists a unique selfadjoint operator $Q : \mathcal{H} \to \mathcal{H}$ with domain $\mathcal{D}(Q)$ contained in $\mathcal{D}(q)$ and with $(Qu|u) = q(u, u)$ for every $u \in \mathcal{D}(Q)$.*

Theorem 4.10. *Let $S : \mathcal{H} \to \mathcal{H}$ be a symmetric operator which is semi-bounded from below in the sense that $(Su|u) \geq -M\|u\|^2$, $u \in \mathcal{D}(S)$.*

Then there exists a unique selfadjoint extension A of S with the property that $\mathcal{D}(A) \subset \mathcal{D}(q)$, where q is the closed quadratic form associated with $S + (M + 1)I$, discussed prior to Theorem 4.9.

The operator A in the last result is called the Friedrichs extension of S.

Example. Let $\Omega \subset \mathbf{R}^n$ be open and let $0 \leq V \in C^\infty(\Omega)$. Then the symmetric operator $S = -\Delta + V$ with domain $C_0^\infty(\Omega)$ is semi-bounded from below, so we can define a corresponding Friedrichs extension. If we assume that Ω is bounded with a smooth (C^∞) boundary, and V is bounded and continuous on the closure of Ω, then the domain of the corresponding quadratic form is $H_0^1(\Omega)$, the closure of $C_0^\infty(\Omega)$ for the norm

$$\|u\|_{H^1}^2 = \sum_{|\alpha| \leq 1} \|D^\alpha u\|_{L^2}^2.$$

The domain of the corresponding Friedrichs extension then turns out to be $H_0^1(\Omega) \cap H^2(\Omega)$, and since all functions in this domain vanish on the boundary (in a sense defined by means of the Sobolev spaces), we may call the Friedrichs extension (in this case) the Dirichlet realization of $-\Delta + V$ on Ω.

Stronger results exist in the literature:

Theorem 4.11 (Kato [Ka1]). *If $0 \leq V \in L_{\text{loc}}^2(\mathbf{R}^n)$, $S = -\Delta + V$, with domain $\mathcal{D}(S) = C_0^\infty(\mathbf{R}^n)$, then S is essentially selfadjoint.*

Theorem 4.12 (Farris–Lavine [FaLa]). *Let $V = V_1 + V_2$ with $V_1 \in L_{\text{loc}}^2(\mathbf{R}^n)$, $V_1 \geq -C|x|^2 - D$ for some constants C, D, and with $V_2 \in L^p(\mathbf{R}^n)$, where we assume that $p \geq 2$ when $n \leq 3$, $p > 2$ when $n = 4$ and $p \geq n/2$ when $n \geq 5$. Then $-\Delta + V$ with domain $C_0^\infty(\mathbf{R}^n)$ is essentially selfadjoint.*

It follows from the last result that when $n \geq 3$ and $V \in L_{\text{loc}}^2$, $V \geq -C/|x| - D|x|^2 - E$, then $-\Delta + V$ with domain $\mathcal{D}(S)$ is essentially selfadjoint.

We next recall various forms of the spectral theorem for selfadjoint operators.

Theorem 4.13. *Let A be a selfadjoint operator on a separable Hilbert space $\mathcal{H}$. Then there exist a measure space $(M, \mathcal{M}, \mu)$, where μ is a finite measure, a measurable function $a : M \to \mathbf{R}$ and a unitary operator $U : \mathcal{H} \to L^2(d\mu)$ such that:*

(1) A vector $\psi \in \mathcal{H}$ belongs to $\mathcal{D}(A)$ iff $a \cdot U\psi \in L^2(d\mu)$.

(2) If $\psi \in \mathcal{D}(A)$, then $UA\psi = aU\psi$.

One may even arrange so that $L^2(d\mu)$ is the direct countable orthogonal sum of $L^2(\mathbf{R}; d\mu_j)$, for $j = 1, 2, \ldots$, where μ_j is a finite Borel measure on $\mathbf{R}$ and $a(x) = x$. More generally, in the situation described in the theorem, we have

$$\sigma(A) = \text{Im}_{\text{ess}}(a) = \{\lambda \in \mathbf{R}; \mu(a^{-1}([\lambda - \epsilon, \lambda + \epsilon])) > 0, \forall \epsilon > 0\}.$$

Functional calculus. If $h : \mathbf{R} \to \mathbf{C}$ is a bounded Borel function, and U and a are as in the preceding theorem, we define $h(A) = U^{-1}(h \circ a)U$. We then obtain the existence part of the following theorem:

Theorem 4.14. *Let $\mathcal{H}$ be a separable Hilbert space, and let $A : \mathcal{H} \to \mathcal{H}$ be a selfadjoint operator. Let*

$$\mathcal{B}_b(\mathbf{R}) = \{h : \mathbf{R} \to \mathbf{C}; \ h \ \text{is a bounded Borel function}\}.$$

Then there exists a unique map:

$$\mathcal{B}_b(\mathbf{R}) \ni h \mapsto h(A) \in \mathcal{L}(\mathcal{H}, \mathcal{H}),$$

with the properties (1)–(4):

(1) $h(A)^* = \overline{h}(A)$, $h_1(A) + h_2(A) = (h_1 + h_2)(A)$, $h_1(A)h_2(A) = (h_1 h_2)(A)$,

(2) $\|h(A)\|_{\mathcal{L}(\mathcal{H}, \mathcal{H})} \le \sup_{\mathbf{R}} |h|$ *(It is even enough to take the supremum over the spectrum of A.),*

(3) *If $h_n \in \mathcal{B}_b$, $h_n(x) \to x$, $n \to \infty$, $|h_n(x)| \le |x|$, then $h_n(A)\psi \to A\psi$ for every $\psi \in \mathcal{D}(A)$,*

(4) *If $h_n \to h$ pointwise and $\sup |h_n| \le C$, then $h_n(A) \to h(A)$ strongly, (This means that $h_n(A)\phi \to h(A)\phi$ for every $\phi \in \mathcal{H}$.)*

Moreover, we have

(5) *If $\psi \in \mathcal{D}(A)$, $\lambda \in \mathbf{R}$, $A\psi = \lambda\psi$, then $h(A)\psi = h(\lambda)\psi$,*

(6) *If $h \ge 0$, then $h(A) \ge 0$ (in the sense that $(h(A)\phi|\phi) \ge 0$ for all $\phi \in \mathcal{D}(A)$).*

Spectral measures. For every $\phi \in \mathcal{H}$ there is a unique Borel measure μ_ϕ of finite mass, such that for every bounded Borel function g: $(g(A)\phi|\phi) = \int g(\lambda)\mu_\phi(d\lambda)$. By polarization, we can also construct a measure $\mu_{\phi,\psi}$ such that $(g(A)\phi|\psi) = \int g(\lambda)\mu_{\phi,\psi}(d\lambda)$. These measures are called spectral measures. In the multiplicative representation of Theorem 4.13, we have $\mu_\phi =$

$a_*(|\phi|^2\mu)$ (i.e. direct image of the measure $|\phi|^2\mu$ under a). In fact, if g is a bounded Borel function, then

$$\int g(\lambda)(a_*(|\phi|^2\mu))(d\lambda) = \int g(a(m))|\phi(m)|^2\mu(dm)$$

$$= (g(A)\phi|\phi) = \int g(\lambda)\mu_\phi(d\lambda).$$

From measure theory, we now recall that every Borel measure on $\mathbf{R}$ has a unique decomposition: $\mu = \mu_{pp} + \mu_{ac} + \mu_{sc}$, where the three measures to the right are mutually singular, that is carried by disjoint sets, and where:

μ_{pp} is a pure point measure: $\mu_{pp} = \sum a_j\delta_{x_j}$, where the sum is countable or finite and δ_{x_j} denotes the Dirac measure at the point x_j,

μ_{ac} is an absolutely continuous measure, i.e. there exist a locally integrable function f with respect to the Lebesgue measure dx, such that $\mu_{ac} = f dx$,

μ_{sc} is singular continuous, that is μ_{sc} has no point masses ($\mu_{sc}(\{x\}) = 0$ for every $x \in \mathbf{R}$) and is carried by a set of Lebesgue measure 0.

Using this decomposition one can construct an orthogonal decomposition: $\mathcal{H} = \mathcal{H}_{pp} \oplus \mathcal{H}_{ac} \oplus \mathcal{H}_{sc}$ such that each of the closed subspaces to the right is invariant under $(A+i)^{-1}$, and such that if $\phi \in \mathcal{H}_{pp}$, $\in \mathcal{H}_{ac}$ or $\in \mathcal{H}_{sc}$, then μ_ϕ is pp, ac or sc respectively. There is an obvious way of defining the (selfadjoint) restriction of A to these subspaces, and $A|_{\mathcal{H}_{pp}}$ has an orthonormal (O.N.) basis of eigenvectors. Moreover, every eigenvector of A belongs to $\mathcal{H}_{pp}$. It is clear that the decomposition of $\mathcal{H}$ above is unique.

Let $\overline{\sigma_{pp}(A)}$, $\sigma_{ac}(A)$, $\sigma_{sc}(A)$ be the spectra of the restrictions of A to the corresponding three subspaces. Then $\overline{\sigma_{pp}(A)}$ is the closure of the the the set $\sigma_{pp}(A)$ of eigenvalues of A. (We recall that if $\phi \in \mathcal{D}(A)$, $\lambda \in \mathbf{R}$, then ϕ is called an eigenvector of A and λ is the corresponding eigenvalue.) It is easy to see that the spectrum of A is equal to $\overline{\sigma_{pp}(A)} \cup \sigma_{ac}(A) \cup \sigma_{sc}(A)$.

Spectral projectors. If Ω is a Borel set in $\mathbf{R}$, we put $P_\Omega = 1_\Omega(A)$, where 1_Ω denotes the characteristic function of Ω (equal to 1 on Ω and equal to 0 on $\mathbf{R} \setminus \Omega$). Then the P_Ω form a spectral family:

(a) P_Ω is an orthogonal projection,

(b) $P_\emptyset = 0$,

(c) If Ω_j, $j = 1, 2, \ldots$ is a sequence of Borel sets with $\Omega_j \nearrow \Omega$, when $j \to \infty$, then $P_{\Omega_j} \to P_\Omega$ strongly,

(d) $P_{\Omega_1}P_{\Omega_2} = P_{\Omega_1 \cap \Omega_2}$.

Put $P_\lambda = P_{]-\infty,\lambda]}$. If $\phi \in \mathcal{H}$, then $(\phi|P_\lambda\phi)$ is a bounded increasing function of λ and the corresponding Stieltjes measure is the spectral measure μ_ϕ:
$d(\phi|P_\lambda\phi) = \mu_\phi$.

If g is a complex valued Borel function on $\mathbf{R}$, we put

$$\mathcal{D}(g(A)) = \{\phi \in \mathcal{H}; \int |g(\lambda)|^2 d(\phi|P_\lambda\phi) < \infty\}.$$

This space is dense in $\mathcal{H}$ and we can define $g(A) : \mathcal{D}(g(A)) \to \mathcal{H}$ by

$$(g(A)\phi|\psi) = \int g(\lambda)d(P_\lambda\phi|\psi), \quad \phi,\psi \in \mathcal{D}(g(A)).$$

If g is real-valued, then $g(A)$ is selfadjoint. Formally, we write:

$$g(A) = \int g(\lambda)dP_\lambda.$$

In the situation of Theorem 4.13, if $g : \mathbf{R} \to \mathbf{C}$ is a Borel function, and $\phi \in \mathcal{H} \simeq L^2(\mu)$, then

$$\int |g(t)|^2 \mu_\phi(dt) = \int |g(t)|^2 a_*(|\phi|^2\mu) = \int |g(a(m))\phi(m)|^2 \mu(dm).$$

Hence, $\phi \in \mathcal{D}(g(A))$ iff $(g \circ a)\phi \in L^2$ and $g(A)$ can be identified with the operator of multiplication by $g \circ a$. In the special case when $g(\lambda) = \lambda$, we get $g(A) = A$, and we arrive at a second version of the spectral theorem:

$$A = \int \lambda dP_\lambda.$$

We also get:

Theorem 4.15. *If $f,g : \mathbf{R} \to \mathbf{C}$ are Borel functions, $\phi \in \mathcal{D}(g(A))$ and $g(A)\phi \in \mathcal{D}(f(A))$, then $\phi \in \mathcal{D}(fg(A))$ and $(fg)(A) = f(A)g(A)\phi$.*

We also notice that for $\lambda \in \mathbf{C} \setminus \sigma(A)$, we have $(A - \lambda)^{-1} = \int (t - \lambda)^{-1}dP_t$.

Stone's formula. For $\epsilon > 0$ and $-\infty < a < b < \infty$ consider

$$B_{\epsilon,a,b} = (2\pi i)^{-1} \int_a^b ((A - \lambda - i\epsilon)^{-1} - (A - \lambda + i\epsilon)^{-1})d\lambda,$$

where the integral can be defined as an operator valued Riemann integral. If $[a,b] \ni \lambda \mapsto g_\lambda \in C(\mathbf{R}) \cap L^\infty(\mathbf{R})$ is continuous, it is easy to see that:

$$\int_a^b g_\lambda(A)d\lambda = (\int_a^b g_\lambda(\cdot)d\lambda)(A),$$

and hence $B_{\epsilon,a,b} = f_\epsilon(A)$, where

$$f_\epsilon(t) = \frac{1}{2\pi i}\int_a^b ((t-\lambda-i\epsilon)^{-1} - (t-\lambda+i\epsilon)^{-1})d\lambda = \frac{1}{2\pi}\int_a^b \frac{2\epsilon}{(t-\lambda)^2 + \epsilon^2}d\lambda.$$

We see that $0 \le f_\epsilon(t) \le 1$ and that $f_\epsilon \to \frac{1}{2}(1_{[a,b]} + 1_{]a,b[})$, when ϵ tends to 0. We then obtain Stone's formula:

$$\frac{1}{2}(P_{[a,b]} + P_{]a,b[}) = \text{strong limit}_{\epsilon \searrow 0} B_{\epsilon,a,b},$$

with $B_{\epsilon,a,b}$ defined above.

Essential spectrum. Let $A : \mathcal{H} \to \mathcal{H}$ be selfadjoint. Then $\lambda \in \mathbf{R}$ belongs to $\sigma(A)$ iff $P_{]\lambda-\epsilon,\lambda+\epsilon[} \ne 0$ for every $\epsilon > 0$.

Definition 4.16. We say that $\lambda \in \mathbf{R}$ belongs to the essential spectrum $\sigma_{ess}(A)$ iff $P_{]\lambda-\epsilon,\lambda+\epsilon[}$ is of infinite rank for every $\epsilon > 0$.

It is easy to see that the essential spectrum is a closed set, contained in the spectrum. We also define the discrete spectrum $\sigma_{disc}(A) = \sigma(A) \setminus \sigma_{ess}(A)$. The discrete spectrum is then the union of all eigenvalues of A of finite multiplicity which are isolated from the rest of the spectrum. We have the following

Weyl criterion. Let $\lambda \in \mathbf{C}$. Then

(1) λ belongs to the spectrum of A iff $\exists$ a normalized sequence $\phi_n \in \mathcal{D}(A)$, $n = 1, 2, \ldots$ such that $(A - \lambda)\phi_n \to 0$,

(2) λ belongs to the essential spectrum of A iff there exists an infinite orthonormal sequence $\phi_n \in \mathcal{D}(A)$ such that $(A - \lambda)\phi_n \to 0$.

The following two results (which extend a classical theorem of H. Weyl) say roughly that the essential spectrum is unchanged under compact perturbations.

Theorem 4.17. *Let* $A, B : \mathcal{H} \to \mathcal{H}$ *be selfadjoint operators such that* $(A+i)^{-1} - (B+i)^{-1}$ *is compact. Then* $\sigma_{ess}(A) = \sigma_{ess}(B)$.

Definition 4.18. Let $A : \mathcal{H} \rightarrow \mathcal{H}$ be selfadjoint, $C : \mathcal{H} \rightarrow \mathcal{H}$, with $\mathcal{D}(A) \subset \mathcal{D}(C)$ and $C(A + i)^{-1}$ compact. Then we say that C is relatively compact with respect to A, or simply that C is A-compact.

Notice that if C is compact, then C is A-compact for any selfadjoint operator acting in the same Hilbert space.

Theorem 4.19. *Let A be selfadjoint and let C be symmetric and A-compact. Define $B = A + C$ with domain $\mathcal{D}(B) = \mathcal{D}(A)$. Then B is selfadjoint and $\sigma_{\mathrm{ess}}(A) = \sigma_{\mathrm{ess}}(B)$.*

Example. Let $V \in L^{\infty}(\mathbf{R}^n; \mathbf{R})$ and assume that $V(x) \rightarrow 0$, when $|x| \rightarrow \infty$. Let $A = -\Delta$ with $\mathcal{D}(A) = H^2(\mathbf{R}^n)$. Then $V(-\Delta + i)^{-1}$ is compact, so $\sigma_{\mathrm{ess}}(A + V) = \sigma_{\mathrm{ess}}(A) = [0, \infty[$. This means that $\sigma(A + V)$ is the union of $[0, \infty[$ and a finite or countable subset of $] - \infty, 0[$ with 0 as the only possible accumulation point, consisting of eigenvalues of finite multiplicity.

Example. More generally, let $V \in L^2_{\mathrm{loc}}(\mathbf{R}^n; \mathbf{R})$ be bounded from below and put $c = \underline{\lim}_{|x| \rightarrow \infty} V(x)$. Then $\inf \sigma_{\mathrm{ess}}(-\Delta + V) \geq c$. In fact, assume for instance that $c < +\infty$, put $A = -\Delta + \max(V, c)$, $B = V - \max(V, c)$ so that $B(x)$ is bounded and tends to 0 when $|x| \rightarrow \infty$. Since $\mathcal{D}(A) \subset H^1(\mathbf{R}^n)$, $(A + i)$ is bounded: $L^2 \rightarrow H^1$ and consequently $B(A + i)^{-1}$ is compact. Hence, $\sigma_{\mathrm{ess}}(A + B) = \sigma_{\mathrm{ess}}(A) \subset \sigma(A) \subset [c, \infty[$. When $\underline{\lim} V = +\infty$ the same arguments works if we replace c by any real number. In this case $-\Delta + V$ has a purely discrete spectrum, given by a sequence of eigenvalues (of finite multiplicity) which tends to $+\infty$. See also Persson [Pe].

The mini-max principle is a an important tool in the applications, and when applied to selfadjoint operators, bounded from below, it takes the form of a maxi-min principle. Let $A : \mathcal{H} \rightarrow \mathcal{H}$ be an unbounded selfadjoint operator which is semi-bounded from below in the sense that there exists a constant $C \in \mathbf{R}$ such that $A \geq -CI$. Assume that $\mathcal{H}$ is infinite dimensional. Let $\mu_1 \leq \mu_2 \leq \ldots$ be an increasing enumeration of all the eigenvalues of A with $\mu_j < \sigma_{\mathrm{ess}}(A)$, repeated according to their multiplicity. If the number N_0 of such eigenvalues is finite, we define $\mu_{N_0+1} = \mu_{N_0+2} = \ldots$ to be $\inf \sigma_{\mathrm{ess}}(A)$.

Let $1 \leq N < \infty$, let $u_1, \ldots, u_{N-1} \in \mathcal{H}$ be linearly independent and let $E = (u_1, \ldots, u_{N-1})^{\perp}$, where $(u_1, \ldots, u_{N-1})$ is the linear space spanned by $u_1, \ldots, u_{N-1}$.

Lemma 4.20. $\mathcal{D}(A) \cap E$ *is dense in* E.

Proof. Let $v_1, \ldots, v_{N-1} \in \mathcal{D}(A)$ be close to $u_1, \ldots, u_{N-1}$ in norm. Then the space $F = (v_1, \ldots, v_{N-1})$ is of dimension $N - 1$ and transversal to E, and there is a unique bounded projection $\Pi : \mathcal{H} \to E$ with $\text{Im}(\Pi) = E$, $\text{Ker}(\Pi) = F$. Then Π maps $\mathcal{D}(A)$ into $\mathcal{D}(A)$, and we also see that $\Pi(\mathcal{D}(A))$ is dense in E. #

Lemma 4.21. *We have*

$$\inf_{u \in \mathcal{D}(A) \cap E,\, \|u\|=1} (Au|u) \leq \mu_N(A).$$

Proof. We first assume that $\mu_N(A)$ is among the first N eigenvalues (repeated according to their multiplicities) strictly below $\sigma_{\text{ess}}(A)$. Let $e_1, \dots, e_N$ be a corresponding orthonormal family of eigenfunctions. Then $(e_1, \dots, e_N) \cap E \neq \{0\}$ and we let $u = \sum_1^N \lambda_j e_j$ be a normalized vector in the intersection, so that $\sum |\lambda_j|^2 = 1$. Clearly $u \in \mathcal{D}(A) \cap E$, and

$$(Au|u) = \sum_1^N \mu_j |\lambda_j|^2 \leq \mu_N \sum_1^N |\lambda_j|^2 = \mu_N.$$

In the case when $\mu_N = \inf \sigma_{\text{ess}}(A)$, the space $1_{]-\infty, \mu_N + \epsilon[}(A)(\mathcal{H}) \subset \mathcal{D}(A)$ is infinite dimensional for every $\epsilon > 0$ and has a non-trivial intersection with E. If u is a normalized vector in the intersection, we have $(Au|u) \leq \mu_N + \epsilon$, by the spectral theorem, and since $\epsilon > 0$ can be taken arbitrarily small, we obtain the lemma in this case also. #

Theorem 4.22. *Under the assumptions above, we have the maxi-min formula:*

$$\mu_N(A) = \sup_{\substack{u_1, \dots, u_{N-1} \in \mathcal{H}, \\ \text{linearly independent}}} \inf_{\substack{u \in (u_1, \dots, u_{N-1})^\perp \cap \mathcal{D}(A), \\ \|u\|=1}} (Au|u).$$

Proof. We have already seen that the RHS is $\leq \mu_N(A)$. When $\mu_N(A)$ is the Nth eigenvalue $< \inf \sigma_{\text{ess}}(A)$, we can take $u_j = e_j$, $1 \leq j \leq N-1$ with e_j as in the proof of the last lemma, and see that

$$\inf_{\substack{u \in (e_1, \dots, e_{N-1})^\perp \cap \mathcal{D}(A), \\ \|u\|=1}} (Au|u) = \mu_N(A).$$

The second case $\mu_N(A) = \inf \sigma_{\text{ess}}(A)$ can be treated similarly. #

We end this chapter with a rough determination of the low-lying eigenvalues of semi-classical Schrödinger operators of the type considered in Chapter 3. Further and deeper results in this direction will be given in Chapter 14. Let

M be either $\mathbf{R}^n$ or a closed bounded subset of $\mathbf{R}^n$ with smooth boundary and such that 0 belongs to the interior: $0 \in \text{int}(M)$. The arguments below will also work with only minor modifications in the case when M is a smooth compact Riemannian manifold, possibly with a boundary. Let $V \in C^\infty(M; [0, \infty[)$ with $V(x) = 0$ precisely for $x = 0$ and assume that $\liminf_{|x| \to \infty} V(x) > 0$, in the case when $M = \mathbf{R}^n$. Let

$$P = -h^2 \Delta + V, \qquad (4.8)$$

and denote by P also the corresponding Friedrichs extension, when starting from the symmetric operator (4.8) with domain $C_0^\infty(\text{int}(M))$. We know that $P \geq 0$ has purely discrete spectrum in $[0, c]$ for some $c > 0$, and we shall now determine the first approximation of the eigenvalues of P in any interval of the form $[0, C_0 h]$ in the limit $h \to 0$.

Let $V_0(x) = \frac{1}{2}\langle V''(0)x, x\rangle$ be the leading term in the Taylor expansion of V at 0, and introduce the harmonic oscillator,

$$P_0 = -h^2 \Delta + V_0(x), \quad x \in \mathbf{R}^n. \qquad (4.9)$$

This operator (realized through the Friedrichs extension) has a purely discrete spectrum, which we essentially computed in Chapter 3, and we saw there that the eigenvalues of $P_0 = P_0(h)$ are of the form $0 < E_0 h < E_1 h \leq E_2 h \leq \ldots$, where E_j are the eigenvalues of $P_0(1)$. A corresponding O.N. basis of eigenfunctions is given by $e_j(x; h) = h^{-n/4} e_j(h^{-1/2}x)$. Here $e_j(x) \in \mathcal{S}(\mathbf{R}^n)$ are Hermite functions up to a linear change of variables.

Let $0 < C_0 \notin \{E_0, E_1, \ldots\}$ and let N_0 be the number of E_js in $[0, C_0]$, so that $E_{N_0-1} < C_0 < E_{N_0}$.

Theorem 4.23. *Under the assumptions above, there exists $h_0 > 0$, such that for $0 < h \leq h_0$, P has precisely N_0 eigenvalues, $0 \leq \lambda_0(h) \leq \ldots \leq \lambda_{N_0-1}(h)$ in $[0, C_0 h]$. Moreover, $\lambda_j(h) = E_j h + \mathcal{O}(h^{3/2})$.*

Proof. Let $\chi \in C_0^\infty(\text{int}(M))$ be equal to 1 near 0. Since the L^2-norm of $(V(x) - V_2(x))\chi(x)e_j(x; h)$ is $\mathcal{O}(h^{3/2})$, we get

$$(P - E_j h)(\chi(x)e_j(x; h)) = r_j(x; h), \quad \|r_j\| = \mathcal{O}(h^{3/2}), \qquad (4.10)$$

and from this we conclude that for each $j \leq N_0 - 1$, there exists an eigenvalue μ_j of P with $\mu_j = E_j h + \mathcal{O}(h^{3/2})$, but to show that we get N_0 eigenvalues of P (counted with multiplicity) in this way, requires a little more work except when the E_j are all distinct.

Let $R \gg 1$ and choose a quadratic partition of unity,

$$\chi_0(x)^2 + \chi_1(x)^2 = 1, \qquad (4.11)$$

with $\chi_0, \chi_1 \in C^\infty(M; [0,1])$, $\chi_0 \in C_0^\infty(B(0, 2Rh^{1/2}))$, $\chi_0 = 1$ on $B(0, Rh^{1/2})$, $\partial^\alpha \chi_j = \mathcal{O}_\alpha((Rh^{1/2})^{-|\alpha|})$. Here $B(0,r)$ denotes the open ball in $\mathbf{R}^n$ of center 0 and radius r. Notice that

$$\chi_0[\Delta, \chi_0] + \chi_1[\Delta, \chi_1] = \chi_0 \Delta(\chi_0) + \chi_1 \Delta(\chi_1) = \mathcal{O}(\frac{1}{R^2 h}), \qquad (4.12)$$

which gives the so-called IMS localization formula,

$$(-\Delta u|u) = (-\Delta \chi_0 u|\chi_0 u)$$
$$+(-\Delta \chi_1 u|\chi_1 u) + ((\chi_0 \Delta(\chi_0) + \chi_1 \Delta(\chi_1))u|u), \ u \in \mathcal{D}(P). \ (4.13)$$

Combining (4.12), (4.13), we get for $u \in \mathcal{D}(P)$:

$$(Pu|u) = (P\chi_0 u|\chi_0 u) + (P\chi_1 u|\chi_1 u) + \mathcal{O}(\frac{h}{R^2})\|u\|^2. \qquad (4.14)$$

Here

$$(P\chi_1 u|\chi_1 u) \geq (V\chi_1 u|\chi_1 u) \geq E_{N_0} h\|\chi_1 u\|^2, \qquad (4.15)$$

if we choose first R large enough and then h small enough, so that

$$\inf_{|x| \geq Rh^{1/2}} V \geq E_{N_0} h.$$

On the other hand,

$$(P\chi_0 u|\chi_0 u) = (P_0 \chi_0 u|\chi_0 u) + \mathcal{O}(R^3 h^{3/2})\|\chi_0 u\|^2, \qquad (4.16)$$

when h is small enough, depending on R.

Assume that $\chi_0 u \perp e_j(\cdot; h)$, $0 \leq j \leq N_0 - 1$, so that $(P_0 \chi_0 u|\chi_0 u) \geq (E_{N_0} h)\|\chi_0 u\|^2$. Then from (4.14)–(4.16), we get

$$(Pu|u) \geq (E_{N_0} - \mathcal{O}(\frac{1}{R^2} + R^3 h^{1/2}))h\|u\|^2. \qquad (4.17)$$

From the maxi-min principle, it follows that the $(N_0 + 1)$st eigenvalue of P is $\geq (E_{N_0} - o(1))h$, when $h \to 0$.

It is easy to find a simple closed loop γ in $\{z \in \mathbf{C}; \mathrm{Re}\, z < C_0 h\}$ such that $\mathrm{dist}\,(z, \sigma(P) \cup \sigma(P_0(h))) \geq \epsilon_0 h$, $z \in \gamma$ and such that hE_j, $j \leq N_0 - 1$ are in the interior of γ. Here $\epsilon_0 > 0$ is some fixed number independent of h. Returning to (4.10), let $E \subset L^2(M)$ be the N_0-dimensional space spanned by $\chi e_j(\cdot; h)$, $j = 0, \ldots, N_0 - 1$, and observe that the functions χe_j form an almost O.N. basis in E in the sense that

$$(\chi e_j|\chi e_k) = \delta_{j,k} + \mathcal{O}(h^\infty). \qquad (4.18)$$

Rewrite (4.10) as

$$(z - P)(\chi e_j) = (z - E_j h)\chi e_j - r_j,$$

and apply $(z - P)^{-1}(z - E_j h)^{-1}$ for $z \in \gamma$:

$$(z - P)^{-1}\chi e_j = (z - E_j h)^{-1}\chi e_j + (z - P)^{-1}(z - E_j h)^{-1}r_j. \tag{4.19}$$

Let

$$\pi = \frac{1}{2\pi i}\int_\gamma (z - P)^{-1}dz$$

be the spectral projection associated with P, and the intersection of $\mathbf{R}$ and the interior of γ and let $F = \pi(L^2(M))$. From the conclusion after (4.17), we know that $\dim F \leq N_0$. From (4.19) we get

$$\pi(\chi e_j) = \chi e_j + k_j, \quad \|k_j\| = \mathcal{O}(h^{1/2}), \tag{4.20}$$

and it follows (when $h > 0$ is small enough) that the dimension of F is at least equal to N_0, so we have finally $\dim F = N_0$. Moreover, $f_j = \pi(\chi e_j)$, $0 \leq j \leq N_0 - 1$ form a basis in F, and if we use (4.10) again, we get $Pf_j = hE_j f_j + \mathcal{O}(h^{3/2})$. Let $f = (f_0, \ldots, f_{N_0-1})$ be the corresponding row vector and introduce the orthonormalized basis $g = f((f_j|f_k))^{-1/2}$. The matrix of $P_{|F}$ is then $\mathrm{diag}\,(hE_j) + \mathcal{O}(h^{3/2})$. If $\lambda_0 \leq \ldots \leq \lambda_{N_0-1}$ denote the eigenvalues of $P_{|F}$, it follows that $\lambda_j = hE_j + \mathcal{O}(h^{3/2})$. This completes the proof of the Theorem. #

Let E_j be one of the eigenvalues of the harmonic oscillator $P_0(1)$ with $j \leq N_0 - 1$, and assume that E_j is a simple eigenvalue. It follows from Theorem 3.6 that the corresponding eigenvalue $\lambda_j(h)$ has an asymptotic expansion $\sim h(a_0 + a_1 h + a_2 h^2 + \ldots)$, where $a_0 = E_j$. If we drop the assumption that E_j is simple then we still have an asymptotic expansion for $\lambda_j(h)/h$ with leading term E_j, provided that we allow for half powers of h. See Simon [Si1], Helffer–Sjöstrand [HeSj1].

Notes

Most of this chapter is a compilation of general and well-known facts for spectral theory. We have used [NaRi], [ReSi] and [CFKS]. The asymptotic behavior of the lowest eigenvalues of the Schrödinger operator in the semi-classical regime has been studied by many authors. See [Si1], [HeSj1]. In the one-dimensional case precise asymptotic expansions were computed by Combes, Duclos and Seiler [CDS]. The case of the asymptotic behavior when the minimum is degenerate has been studied by Martinez and Rouleux [MR].

5. The method of stationary phase

Let $X \subset \mathbf{R}^n$ be an open set, $\phi \in C^\infty(X; \mathbf{R})$ (i.e. a real valued smooth function) such that $d\phi \neq 0$ everywhere. If $u \in C_0^\infty(X)$, then the integral

$$I(\lambda) = \int e^{i\lambda\phi(x)} u(x) dx \qquad (5.1)$$

is rapidly decreasing when $\lambda \to +\infty$. This can be seen by repeated integrations by parts, using for instance the operator ${}^t L = \frac{1}{i\lambda|\phi'|^2} \sum \frac{\partial \phi}{\partial x_j} \frac{\partial}{\partial x_j}$ which satisfies ${}^t L(e^{i\lambda\phi}) = e^{i\lambda\phi}$. More precisely, we obtain:

For every compact $K \subset X$ and every $N \in \mathbf{N}$, there is

a constant $C = C_{K,\phi,N}$, such that $|I(\lambda)| \leq C(\sum_{|\alpha| \leq N} \sup |\partial^\alpha u(x)|)\lambda^{-N}$,

for $\lambda \geq 1$, $u \in C_0^\infty(X)$, $\operatorname{supp} u \subset K$. $\qquad (5.2)$

This means that if $\phi \in C^\infty(X; \mathbf{R})$, $u \in C_0^\infty(X)$, the asymptotic behaviour of $I(\lambda)$ when $\lambda \to +\infty$ is determined by ϕ, u in a neighborhood of the set of critical points of ϕ. (Recall that a critical point of a function is a point where the gradient of the function vanishes.) The most important (and most easy) case is the one of non-degenerate critical points. We say that the critical point $x_0 \in X$ of ϕ is non-degenerate if $\det \phi''(x_0) \neq 0$, where $\phi''(x_0) = (\frac{\partial^2 \phi}{\partial x_j \partial x_k})_{1 \leq j,k \leq n}$ is the Hessian of ϕ at x_0. Since $\phi''(x_0)$ is the differential of the map $X \ni x \mapsto \phi'(x) \in \mathbf{R}^n$ at x_0, it follows that x_0 is an isolated critical point, if it is a non-degenerate one. The Morse lemma gives local coordinates near a non-degenerate critical point x_0 of ϕ for which $\phi(x) - \phi(x_0)$ becomes a quadratic form.

Lemma 5.1. *Let $\phi \in C^\infty(X; \mathbf{R})$ and let $x_0 \in X$ be a non-degenerate critical point. Then there are neighborhoods U of $0 \in \mathbf{R}^n$ and V of x_0, and a C^∞ diffeomorphism $\kappa : V \to U$, such that*

$$\phi \circ \kappa^{-1}(x) = \phi(x_0) + \frac{1}{2}(x_1^2 + \ldots + x_r^2 - x_{r+1}^2 - \ldots - x_n^2).$$

Here r and $n - r$ are the numbers of positive and negative eigenvalues, respectively, of $\phi''(x_0)$.

Proof. After a translation and a linear change of coordinates, we may assume that $x_0 = 0$ and that

$$\phi(x) = \frac{1}{2}(x_1^2 + \ldots + x_r^2 - x_{r+1}^2 - \ldots - x_n^2) + \mathcal{O}(|x|^3) \quad x \to 0.$$

By Taylor's formula,

$$\phi(x) = \int_0^1 (1-t)\frac{\partial^2}{\partial t^2}(\phi(tx))dt = \frac{1}{2}\sum\sum q_{j,k}(x)x_jx_k = \frac{1}{2}\langle x, Q(x)x\rangle,$$

where $Q(x) = (q_{j,k}(x))$, $q_{j,k}(x) = 2\int_0^1(1-t)\frac{\partial^2\phi}{\partial x_j\partial x_k}(tx)dt$, so that $Q(0) = \phi''(0)$ is the diagonal matrix with diagonal elements equal to 1 at the first r places and with the remaining $n-r$ diagonal elements equal to -1. We look for κ of the form $\kappa(x) = A(x)x$, where the matrix $A(x)$ depends smoothly on x and satisfies $A(0) = I$. Then $A(x)$ should satisfy $\langle x, Q(x)x\rangle = \langle A(x)x, Q(0)A(x)x\rangle$, so it suffices to have $Q(x) = {}^tA(x)Q(0)A(x)$.

Let $\mathrm{Sym}(n, \mathbf{R})$ be the space of real symmetric $n \times n$ matrices and consider the map

$$\mathcal{F} : \mathrm{Mat}(n, \mathbf{R}) \ni A \mapsto {}^tAQ(0)A \in \mathrm{Sym}(n, \mathbf{R}),$$

where $\mathrm{Mat}(n, \mathbf{R})$ denotes the space of all real $n \times n$ matrices. The differential at the point $A = I$ is

$$d\mathcal{F} : \mathrm{Mat}(n, \mathbf{R}) \ni \delta A \mapsto {}^t(\delta A)Q(0) + Q(0)(\delta A) \in \mathrm{Sym}(n, \mathbf{R}).$$

$d\mathcal{F}$ is surjective with a right inverse given by: $\delta B \mapsto \frac{1}{2}Q(0)^{-1}\delta B$.

By the implicit function theorem, $\mathcal{F}$ has a local smooth right inverse $\mathcal{G}$, mapping a neighborhood of zero in $\mathrm{Sym}(n, \mathbf{R})$ into a neighborhood of 0 in $\mathrm{Mat}(n, \mathbf{R})$, with $\mathcal{F} \circ \mathcal{G} = \mathrm{id}$. We get $A(x)$ with the required properties by taking $A(x) = \mathcal{G}(Q(x))$. The map $\kappa(x) = A(x)x$ is a diffeomorphism from a neighborhood of 0 onto a neighborhood of 0, since $d\kappa(0) = A(0) = I$.   #

For $u \in L^1(\mathbf{R}^n)$, we define the Fourier transform $\mathcal{F}u(\xi) = \widehat{u}(\xi) = \int e^{-ix\cdot\xi}u(x)dx$ and we extend the definition to the case $u \in \mathcal{S}'(\mathbf{R}^n)$ in the usual way. If Q is a non-degenerate symmetric complex $n \times n$ matrix with $\mathrm{Im}\, Q \geq 0$ (in the sense of Hermitian matrices), then we know that

$$\mathcal{F} : e^{i\langle x, Qx\rangle/2} \mapsto (2\pi)^{n/2}(\det(\frac{1}{i}Q))^{-1/2}e^{-i\langle\xi, Q^{-1}\xi\rangle/2}.$$

Here we choose the continuous branch of the square root which is positive when $\frac{1}{i}Q$ is real and positive. In particular, when Q is real, we get

$$\mathcal{F} : e^{i\langle x, Qx\rangle/2} \mapsto (2\pi)^{n/2}e^{i\frac{\pi}{4}\mathrm{sgn}\,Q}|\det Q|^{-1/2}e^{-i\langle\xi, Q^{-1}\xi\rangle/2}, \qquad (5.3)$$

where we put $\mathrm{sgn}\, Q = r - (n - r)$, with r being the number of positive eigenvalues of Q.

For $u \in C_0^\infty(\mathbf{R}^n)$, we get from Parseval's formula:

$$\int e^{i\lambda\langle x,Qx\rangle/2}u(x)dx = (2\pi\lambda)^{-n/2}\frac{e^{i\frac{\pi}{4}\operatorname{sgn}Q}}{|\det Q|^{1/2}}\int e^{-i\langle\xi,Q^{-1}\xi\rangle/(2\lambda)}\widehat{u}(\xi)d\xi. \quad (5.4)$$

We now replace λ by $1/t$ and consider

$$\begin{aligned}I(t,u) &= \frac{|\det Q|^{\frac{1}{2}}e^{-i\frac{\pi}{4}\operatorname{sgn}Q}}{(2\pi t)^{\frac{n}{2}}}\int e^{i\langle x,Qx\rangle/(2t)}u(x)dx \\ &= \frac{1}{(2\pi)^n}\int e^{-it\langle\xi,Q^{-1}\xi\rangle/2}\widehat{u}(\xi)d\xi.\end{aligned}$$

For $u \in C_0^\infty$ this is a smooth function of $t \in [0,+\infty[$, and

$$\frac{\partial}{\partial t}I(t,u) = I(t,Pu),\ I(0,u) = u(0),\ \text{where } P = -\frac{i}{2}\langle D_x, Q^{-1}D_x\rangle.$$

Taylor's formula gives

$$I(t,u) = \sum_{k=0}^{N-1}\frac{I(0,P^k u)}{k!}t^k + \frac{t^N}{N!}R_N(u,t) = \sum_{k=0}^{N-1}\frac{(P^k u)(0)t^k}{k!} + \frac{t^N}{N!}R_N(u,t),$$

where

$$R_N(u,t) = N\int_0^1(1-s)^{N-1}I(st,P^N u)ds$$

so that

$$|R_N(u,t)| \le \frac{1}{(2\pi)^n}\|\mathcal{F}P^N u\|_{L^1} \le C_n\sum_{|\alpha|\le n+1}\|\partial^\alpha P^N u\|_{L^1}.$$

This gives

$$\int e^{i\lambda\langle Qx,x\rangle/2}u(x)dx$$
$$= \sum_{k=0}^{N-1}\frac{(2\pi)^{\frac{n}{2}}e^{i\frac{\pi}{4}\operatorname{sgn}Q}}{k!|\det Q|^{\frac{1}{2}}\lambda^{k+\frac{n}{2}}}(\frac{1}{2i}\langle D_x,Q^{-1}D_x\rangle)^k u(0) + \lambda^{-N-\frac{n}{2}}S_N(u,\lambda), \quad (5.5)$$

where

$$|S_N(u,\lambda)| \le \frac{1}{N!|\det Q|^{\frac{1}{2}}(2\pi)^n}\|\mathcal{F}(\frac{1}{2i}\langle D_x,Q^{-1}D_x\rangle)^N u\|_{L^1}$$
$$\le \frac{C_n}{N!|\det Q|^{\frac{1}{2}}}\sum_{|\alpha|\le n+1}\|\partial^\alpha(\frac{1}{2i}\langle D_x,Q^{-1}D_x\rangle)^N u\|_{L^1}. \quad (5.6)$$

Here the last sum can be replaced by $\|(\langle D_x, Q^{-1} D_x \rangle)^N u\|_{H^{\frac{n}{2}+\epsilon}}$ for any $\epsilon > 0$, provided that we allow C_n to depend on ϵ.

The following variant gives a sharper estimate on S_N and avoids the use of the Fourier transform. We start with the immediate estimate:

$$|I(t,u)| \leq C_n \frac{|\det Q|^{\frac{1}{2}}}{t^{n/2}} \|u\|_{L^1}.$$

With $f(t) = I(t,u)$, we Taylor expand at $t = 1$:

$$f(t) = \sum_{0}^{M-1} \frac{f^{(k)}(1)}{k!}(t-1)^k + \frac{(t-1)^M}{(M-1)!} \int_0^1 (1-s)^{M-1} f^{(M)}(1-s+st)\,ds.$$

Here for $0 \leq t \leq 1$:

$$\int_0^1 \frac{(1-s)^{M-1}}{(1-s+st)^{n/2}}\,ds = \mathcal{O}_{M,n}(1) A(M - \frac{n}{2}, t),$$

where $A(m,t)$ is 1 when $m > 0$, $1 + \log \frac{1}{t}$ when $m = 0$, and t^m when $-\frac{n}{2} < m < 0$. If we Taylor expand $I(t,u)$, we therefore get

$$|I(t,u)| \leq C_{n,M} |\det Q|^{\frac{1}{2}} \Big(\sum_{0}^{M-1} \|P^k u\|_{L^1} + \|P^M u\|_{L^1} A(M - \frac{n}{2}, t) \Big).$$

Taking $M = [\frac{n}{2}]$, the largest integer $\leq \frac{n}{2}$, we get

$$|I(t,u)| \leq C_n |\det Q|^{\frac{1}{2}} \Big(\sum_{0}^{[\frac{n}{2}]-1} \|P^k u\|_{L^1} + \|P^{[\frac{n}{2}]} u\|_{L^1} B_n(t) \Big),$$

where $B_n(t)$ is equal to $1 + \log \frac{1}{t}$ when n is even and $t^{-1/2}$ when n is odd.

Using this in the formula for R_N, with u replaced by $P^N u$, we get:

$$|R_N(u,t)| \leq C_{n,N} |\det Q|^{\frac{1}{2}} \Big(\sum_{0}^{[\frac{n}{2}]-1} \|P^{N+k} u\|_{L^1} + \|P^{[\frac{n}{2}]+N} u\|_{L^1} B_n(t) \Big).$$

It follows that

$$|S_N(u,\lambda)| \leq C_{n,N} \Big(\sum_{0}^{[\frac{n}{2}]-1} \|P^{N+k} u\|_{L^1} + \|P^{[\frac{n}{2}]+N} u\|_{L^1} B_n(\frac{1}{\lambda}) \Big). \qquad (5.7)$$

The following special case is one of the main ingredients in the theory of pseudodifferential operators. We replace n by $2n$ and x by $(x,y) \in \mathbf{R}^n$, and we put $Q = Q^{-1} = \begin{pmatrix} 0 & -I \\ -I & 0 \end{pmatrix}$. Then

$$\frac{1}{2}\langle (x,y), Q(x,y) \rangle = -x \cdot y, \quad \frac{1}{2i}\langle D_{(x,y)}, Q^{-1}D_{(x,y)} \rangle = \frac{1}{i}\sum \partial_{x_j}\partial_{y_j}.$$

Then (5.5), (5.6) become:

$$(\frac{\lambda}{2\pi})^n \iint e^{-i\lambda x \cdot y} u(x,y)\,dx\,dy = \sum_{k=0}^{N-1} \frac{1}{k!\,\lambda^k}((\frac{1}{i}\sum_1^n \partial_{x_j}\partial_{y_j})^k u)(0,0) + S_N(u,\lambda)$$

$$= \sum_{|\alpha| \leq N-1} \frac{1}{\lambda^{|\alpha|}\alpha!\,i^{|\alpha|}}(\partial_x^\alpha \partial_y^\alpha u)(0,0) + S_N(u,\lambda), \qquad (5.5')$$

where

$$|S_N(u,\lambda)| \leq \frac{C}{N!\,\lambda^N} \sum_{|\alpha+\beta| \leq 2n+1} \|\partial_x^\alpha \partial_y^\beta (\partial_x \cdot \partial_y)^N u\|_{L^1}, \qquad (5.6')$$

or where we could use the estimates resulting from (5.7).

We now return to the general situation. Combining the Morse lemma and (5.5), (5.6), we get

Proposition 5.2. *Let $\phi \in C^\infty(X;\mathbf{R})$ have the non-degenerate critical point $x_0 \in X$ and assume that $\phi'(x) \neq 0$ for $x \neq x_0$. Then there are differential operators $A_{2\nu}(D)$ of order $\leq 2\nu$, such that for every compact $K \subset X$ and every $N \in \mathbf{N}$, there is a constant $C = C_{K,N}$ such that for every $u \in C^\infty(X) \cap \mathcal{E}'(K)$*

$$\left| \int e^{i\lambda\phi(x)} u(x)\,dx - (\sum_0^{N-1}(A_{2\nu}(D_x)u)(x_0)\lambda^{-\nu-\frac{n}{2}})e^{i\lambda\phi(x_0)} \right|$$

$$\leq C\lambda^{-N-\frac{n}{2}} \sum_{|\alpha| \leq 2N+n+1} \sup |\partial^\alpha u(x)|, \quad \lambda \geq 1.$$

Here $A_0 = \dfrac{(2\pi)^{\frac{n}{2}} e^{i\frac{\pi}{4}\operatorname{sgn}\phi''(x_0)}}{|\det\phi''(x_0)|^{\frac{1}{2}}}.$

Recall that a differential operator of order $\leq m$ on an open set $X \subset \mathbf{R}^n$ is an operator of the form $P(x, D_x) = \sum_{\alpha \in \mathbf{N}^n, |\alpha| \leq m} a_\alpha(x)D_x^\alpha.$

Proof. Let V be a neighborhood of x_0 as in Lemma 5.1, and let $\chi \in C_0^\infty(V)$ be equal to 1 near x_0. Then $\int e^{i\lambda\phi}\chi u\,dx$ can be reduced to an integral of the form (5.4) (with a new u), by means of the Morse lemma. The value of the Jacobian of the change of variables at the point 0 can be determined from the relation $\phi''(x_0) = {}^t d\kappa(0) \circ Q(0) \circ d\kappa(0)$ (in the notations of Lemma 5.1 and its proof). The integral $\int e^{i\lambda\phi(x)}(1 - \chi(x))u(x)dx$ can be treated by means of repeated integrations by parts.   #

There are many variants of the situation discussed above. The phase and amplitude may depend smoothly on additional parameters, in which case, one has to check that the arguments above go through, with smooth parameter dependence everywhere. The phase may also be critical on a submanifold instead of at isolated points, and using Fubini's theorem this case can be reduced to the parameter dependent case just mentioned.

Notes

The main reference for this chapter is [Hö1], See also [Hö4]. We have followed closely [GrSj], while (5.7) may be new.

6. Tunnel effect and interaction matrix

In the study of semi-classical approximation one is often interested in the Schrödinger equation, either in its stationary version $Pu = Eu$, or in its time dependent version, $\frac{1}{i}\frac{\partial}{\partial t}u = Pu$. Here $P = -h^2\Delta + V(x)$ and the goal is to compare the (semi-classical) limit $h \to 0$ for these equations, with properties of the corresponding classical dynamical system generated by the Hamiltonian $p(x,\xi) = \xi^2 + V(x)$. We then consider the energy surface $p(x,\xi) = E$ and the corresponding classical particles are the Hamilton trajectories in $p^{-1}(E)$:

$$\frac{d}{dt}(x(t),\xi(t)) = H_p(x(t),\xi(t)),$$

where H_p is the Hamilton vector field:

$$H_p = \sum_1^n (\frac{\partial p}{\partial \xi_j}\frac{\partial}{\partial x_j} - \frac{\partial p}{\partial x_j}\frac{\partial}{\partial \xi_j}),$$

and we are working over some open set in $\mathbf{R}^n$. For the stationary Schrödinger equation we expect the solutions to be concentrated to the energy surface $p^{-1}(E)$; and for the time dependent Schrödinger equation we expect the solutions to decompose into wave packets that move along the classical trajectories.

In this chapter, we shall study phenomena where the expected correspondence between classical and quantum mechanics is not so clear, specifically we shall study what happens in the classically forbidden region, $\{x; V(x) > E\}$, where no classical particles of energy E can exist. It turns out that the corresponding eigenfunctions have to be exponentially small in this region. However, they are not identically zero, and this is at the origin of the interesting tunnel effect which we shall discuss.

a. Lithner–Agmon estimates.

Proposition 6.1. *Let $\Omega \subset \mathbf{R}^n$ be bounded with C^2-boundary. Let $V \in C(\overline{\Omega};\mathbf{R})$, $\Phi \in \mathrm{Lip}(\overline{\Omega};\mathbf{R})$ (the space of real valued Lipschitz continuous functions). Then the gradient $\nabla\Phi$ is well defined in $L^\infty(\Omega)$ as the almost everywhere limit of $\nabla(\chi_\epsilon * \Phi) = \chi_\epsilon * \nabla\Phi$, when $\epsilon \to 0$, where $\chi_\epsilon(x) = \frac{1}{\epsilon^n}\chi(\frac{x}{\epsilon}) \in C_0^\infty(B(0,\epsilon))$ is a standard mollifier. For every $u \in C^2(\overline{\Omega})$ satisfying $u_{|\partial\Omega} = 0$, we have*

$$h^2\int_\Omega \|\nabla(e^{\Phi/h}u)\|^2 dx + \int_\Omega (V(x) - |\nabla\Phi|^2)e^{2\Phi/h}|u|^2 dx$$

$$= \mathrm{Re}\int_\Omega e^{2\Phi/h}Pu(x)\overline{u(x)}dx. \tag{6.1}$$

Proof. Let first $\Phi \in C^2(\overline{\Omega})$. Then by Green's formula:

$$h^2 \int \|\nabla(e^{\Phi/h}u)\|^2 dx = -\int h^2 \Delta(e^{\Phi/h}u) \cdot e^{\Phi/h}\overline{u}dx$$

$$= -\int e^{2\Phi/h} h^2 \Delta(u)\overline{u}dx - 2\int e^{2\Phi/h} \sum_1^n \frac{\partial\Phi}{\partial x_j} h \frac{\partial u}{\partial x_j}\overline{u}dx$$

$$-\int h^2 \Delta(e^{\Phi/h})ue^{\Phi/h}\overline{u}dx$$

$$= \mathrm{I} + \mathrm{II} + \mathrm{III}.$$

Here

$$\mathrm{Re\,(II)} = -\sum\int e^{2\Phi/h}\frac{\partial\Phi}{\partial x_j}h\frac{\partial}{\partial x_j}|u|^2 dx = \sum\int h\frac{\partial}{\partial x_j}(e^{2\Phi/h}\frac{\partial\Phi}{\partial x_j})|u|^2 dx$$

$$= \int e^{2\Phi/h}(2\|\nabla\Phi\|^2 + h\Delta\Phi)|u|^2 dx.$$

For III, we use: $-h^2 \Delta(e^{\Phi/h}) = -e^{\Phi/h}(\|\nabla\Phi\|^2 + h\Delta\Phi)$. Then

$$\mathrm{I} + \mathrm{II} + \mathrm{III} = -\mathrm{Re}\int e^{2\Phi/h}h^2\Delta(u)\overline{u}dx + \int e^{2\Phi/h}\|\nabla\Phi\|^2|u|^2 dx,$$

and (6.1) follows in this case.

Now let $\Phi \in \mathrm{Lip}\,(\overline{\Omega})$, and denote by the same letter some Lipschitz extension of this function to $\mathbf{R}^n$. Put $\Phi_\epsilon = \chi_\epsilon * \Phi$. Then $\nabla\Phi_\epsilon$ is bounded in L^∞ and $\nabla\Phi \in L^\infty(\Omega)$, where $\nabla\Phi$ is defined in the sense of distributions. For the last statement, we let $\phi \in C_0^\infty(\mathbf{R}^n; \mathbf{R}^n)$:

$$\left|\int \nabla\Phi \cdot \phi dx\right| = \left|\lim_{\epsilon \to 0}\int \nabla\Phi_\epsilon \cdot \phi dx\right| \leq C\|\phi\|_{L^1},$$

and recall that L^∞ is the dual of L^1. By integration theory, we have $\nabla\Phi = \lim_{\epsilon \to \infty} \nabla\Phi_\epsilon$ a.e. since $\nabla\Phi_\epsilon = \chi_\epsilon * \nabla\Phi$. To get (6.1) in general, we first write (6.1) with Φ replaced by Φ_ϵ, let $\epsilon \to 0$ and use the dominated convergence theorem.   #

We indicate a shorter proof of (6.1) for smooth Φ, using the natural L^2-norms for differential 1 forms; $\|\sum u_j(x)dx_j\|^2 = \sum\|u_j\|^2$ and letting $(d\Phi^\wedge)u(x) = u(x)d\Phi(x)$, $d\Phi^{\lrcorner}(\sum u_j(x)dx_j) = \sum u_j(x)\frac{\partial\Phi}{\partial x_j}(x)$, $du(x) = \sum\frac{\partial u}{\partial x_j}dx_j$, $d^*(\sum u_j(x)dx_j) = -\sum\frac{\partial u_j}{\partial x_j}(x)$. This proof is easy to extend to the case of Riemannian manifolds. Notice that if $v = e^{\Phi/h}u$, then for $u \in C^2(\overline{\Omega})$ vanishing on $\partial\Omega$:

$$\mathrm{Re\,}(e^{2\Phi/h}(-h^2\Delta)u|u)$$
$$= \mathrm{Re\,}(e^{\Phi/h}(-h^2\Delta)e^{-\Phi/h}v|v) = \mathrm{Re\,}(e^{-\Phi/h}(-h^2\Delta)e^{\Phi/h}v|v).$$

Here

$$e^{-\Phi/h}(-h^2\Delta)e^{\Phi/h} = (e^{-\Phi/h}(hd)^* e^{\Phi/h})(e^{-\Phi/h}(hd)e^{\Phi/h})$$
$$= ((hd)^* - d\Phi^\lrcorner)(hd + d\Phi^\wedge)$$
$$= ((hd)^*(hd) - \|d\Phi\|^2) + ((hd)^*(d\Phi^\wedge) - (d\Phi^\lrcorner)(hd)).$$

The last term in the last member is anti-symmetric, while the first term in the last member is symmetric, and we get

$$\mathrm{Re}\,(e^{\Phi/h}(-h^2\Delta)e^{-\Phi/h}v|v) = \|hdv\|^2 - (\|d\Phi\|^2 v|v),$$

which gives (6.1).

Proposition 6.2. *Under the assumptions of Proposition 6.1, let* $F_+, F_- \in L^\infty(\Omega)$, $F_\pm \geq 0$ *satisfy*

$$V(x) - (\nabla\Phi(x))^2 = F_+(x)^2 - F_-(x)^2 \ a.e.$$

Then

$$h^2\|\nabla(e^{\Phi/h}u)\|^2 + \frac{1}{2}\|F_+ e^{\Phi/h}u\|^2 \leq \|\frac{1}{F_+ + F_-}e^{\Phi/h}Pu\|^2 + \frac{3}{2}\|F_- e^{\Phi/h}u\|^2.$$
$$(6.2)$$

Proof. From (6.1) we get

$$h^2\|\nabla(e^{\Phi/h}u)\|^2 + \|F_+ e^{\Phi/h}u\|^2$$
$$\leq \|\frac{1}{F_+ + F_-}e^{\Phi/h}Pu\|\,\|(F_+ + F_-)e^{\Phi/h}u\| + \|F_- u e^{\Phi/h}\|^2$$
$$\leq \|\frac{1}{F_+ + F_-}e^{\Phi/h}Pu\|^2 + \frac{1}{4}\|e^{\Phi/h}F_+ u + e^{\Phi/h}F_- u\|^2 + \|F_- u^{\Phi/h}\|^2$$
$$\leq \|\frac{1}{F_+ + F_-}e^{\Phi/h}Pu\|^2 + \frac{1}{2}\|e^{\Phi/h}F_+ u\|^2 + \frac{3}{2}\|e^{\Phi/h}F_- u\|^2,$$

which implies (6.2).   #

The preceding results have simple extensions to the case when we replace $\mathbf{R}^n$ by an n-dimensional complete Riemannian manifold M. In that case we let Δ be the Laplace–Beltrami operator, and we choose the natural norms and scalar products for gradients according to the recipes of Riemannian geometry and adapt the alternative proof of (6.1) indicated above.

In Chapter 4 we obtained a lower bound for the essential spectrum of a Schrödinger operator on $\mathbf{R}^n$ with a lower semi-bounded potential. We shall now reexamine such operators using Lithner–Agmon estimates. Let $V \in C^0(\mathbf{R}^n; \mathbf{R})$ be bounded from below: $\inf_{\mathbf{R}^n} V > E_0$ for some $E_0 \in \mathbf{R}$. Let P denote the Friedrichs extension of $-\Delta + V(x)$ for $C_0^\infty(\mathbf{R}^n)$; $P = -\Delta + V$, so that the domain of P is contained in the closure of $C_0^\infty(\mathbf{R}^n)$ for the quadratic form

$$q(u) = \int (\|\nabla u\|^2 + (V - E_0)|u|^2) dx.$$

Let $\chi \in C_0^\infty(\mathbf{R}^n)$ be equal to 1 near 0 and let $\Phi \in \mathrm{Lip}\,(\mathbf{R}^n)$ be constant for large $|x|$. If $u \in \mathcal{D}(P)$, then (6.1) is valid if we replace u by $u_R(x) = \chi(\frac{x}{R})u(x)$. Moreover, when $R \to \infty$, we have with convergence in L^2: $u_R \to u$, $|V - |\nabla\Phi|^2|^{1/2}u_R \to |V - |\nabla\Phi|^2|^{1/2}u$, $Pu_R \to Pu$, so, passing to the limit, we obtain (6.1) also for $u \in \mathcal{D}(P)$. Here we get the last convergence and the fact that $u_R \in \mathcal{D}(P)$ by considering the commutator $[P, \chi(\frac{x}{R})]$. Hence we also get (6.2) for $u \in \mathcal{D}(P)$, when Φ is constant outside a compact set.

Now assume that $V(x) \geq 2\alpha > 0$ for all $x \in \mathbf{R}^n$. Then $\inf \sigma(P) \geq 2\alpha$. Let $v \in L^2$ have compact support in $B(0, R_0)$. Let $u \in \mathcal{D}(P)$ be the solution of $Pu = v$. Put

$$\Phi_R(x) = \sqrt{\alpha}1_{R_0 \leq |x| \leq R}(x)(|x| - R_0) + \sqrt{\alpha}1_{|x| \geq R}(x)(R - R_0).$$

Then $(\nabla\Phi_R)^2 \leq \alpha$ a.e., so $V - (\nabla\Phi_R)^2 \geq \alpha$ a.e. We choose $F_- = 0$, $F_+(x) = \sqrt{V(x) - (\nabla\Phi_R(x))^2} \geq \sqrt{\alpha}$ and get from (6.2)

$$\|\nabla(e^{\Phi_R}u)\|^2 + \frac{\alpha}{2}\|e^{\Phi_R}u\|^2 \leq \frac{1}{\alpha}\|e^{\Phi_R}v\|^2.$$

Combining this with

$$\|e^{\Phi_R}\nabla u\| \leq \|\nabla(e^{\Phi_R}u)\| + \sqrt{\alpha}\|e^{\Phi_R}u\|,$$

we get

$$\sqrt{\alpha}\|e^{\Phi_R}\nabla u\| + \alpha\|e^{\Phi_R}u\| \leq 2\|e^{\Phi_R}v\|.$$

Put $\Phi = \Phi_\infty$. Letting $R \to \infty$, we get

Proposition 6.3. *Assume* $V \geq 2\alpha > 0$ *and let* $v \in L^2$ *have support in* $|x| \leq R_0$. *If* $u \in \mathcal{D}(P)$ *is the unique solution of* $Pu = v$, *then*

$$\sqrt{\alpha}\|e^\Phi\nabla u\| + \alpha\|e^\Phi u\| \leq 2\|e^\Phi v\|,$$

where $\Phi(x) = \sqrt{\alpha}1_{\{|x| \geq R_0\}}(x)(|x| - R_0)$.

Using this result, it is easy to verify more directly that

$$\inf \sigma_{\mathrm{ess}}(P) \geq \liminf_{|x| \to \infty} V(x),$$

when V is continuous and bounded from below.

b. The Lithner–Agmon metric and decay of eigenfunctions.

Let M denote either a compact connected Riemannian manifold of dimension n, or $\mathbf{R}^n$. Let $V \in C^\infty(M; \mathbf{R})$ and assume in the second case that $E_0 := \liminf_{|x|\to\infty} V(x) > -\infty$. Let $E \in \mathbf{R}$, with $E < E_0$ in the second case, and introduce the Lithner–Agmon (LA) metric:

$$(V(x) - E)_+ dx^2, \tag{6.3}$$

where $a_+ = \max(a, 0)$ and dx^2 denotes the the Riemannian metric on M. For a piecewise C^1 curve γ, we can define its length $|\gamma|$ in the LA-metric, and if $x, y \in M$, we define the LA distance $d(x, y)$ between x and y as inf $|\gamma|$ over all piecewise C^1 curves γ joining y to x. This distance may be degenerate in the sense that we may have $d(x, y) = 0$ when $x \neq y$. We have, however, standard properties such as:

$$d(x, y) = d(y, x), \; d(x, z) \leq d(x, y) + d(y, z), \tag{6.4}$$

$$|d(x, z) - d(x, y)| \leq d(y, z). \tag{6.5}$$

Moreover, $y \mapsto d(x, y)$ is a locally Lipschitz function and

$$|d(x, z) - d(x, y)| \leq ((V(y) - E)_+ + o(1))^{1/2} \|z - y\|_y, \tag{6.6}$$

when $z \to y$ and where $\| \cdot \|_y$ denotes the natural norm induced from the natural norm on the tangent space $T_y M$ via the standard identification of a neighborhood of y in M and a neighborhood of 0 in $T_y M$. It follows that for every x,

$$\|\nabla_y d(x, y)\| \leq (V(y) - E)_+^{1/2}, \text{ for a.a. } y, \tag{6.7}$$

and for all y:

$$\|\nabla_x d(x, y)\| \leq (V(x) - E)_+^{1/2}, \text{ for a.a. } x. \tag{6.8}$$

If $U \subset M$, we put

$$d(x, U) = \inf_{y \in U} d(x, y).$$

Again, we have $|d(x, U) - d(y, U)| \leq d(x, y)$, so that $\|\nabla_x d(x, U)\| \leq (V(x) - E)_+^{1/2}$ a.e. on M.

Let $E = 0$ for simplicity, and assume consequently that $\liminf_{|x|\to\infty} V(x) > 0$, in the case when $M = \mathbf{R}^n$. Let $U = \{x \in M; V(x) \leq 0\}$. Then U is compact. Let $u = u(x; h)$ be a normalized eigenfunction of the Schrödinger operator $P = -h^2 \Delta + V(x)$, with eigenvalue $\lambda(h)$ which tends to zero, when h tends to zero. Here, we assume that h varies in some subset of $]0, h_0]$, $h_0 > 0$ which has 0 as an accumulation point.

Proposition 6.4. *If* $\widetilde{\Phi}(x) = d(U, x)$, *then for every* $\epsilon > 0$, *we have for* $h > 0$ *small enough depending on* ϵ:

$$\|\nabla(e^{(1-\epsilon)\widetilde{\Phi}/h}u)\| + \|e^{(1-\epsilon)\widetilde{\Phi}/h}u\| \leq C_\epsilon e^{\epsilon/h}, \text{ when } M = \mathbf{R}^n,$$

$$\|\nabla(e^{\widetilde{\Phi}/h}u)\| + \|e^{\widetilde{\Phi}/h}u\| \leq C_\epsilon e^{\epsilon/h}, \text{ when } M \text{ is compact}.$$

Proof. We give the proof in the case when M is a compact manifold, and then we simply mention the additional argument that is needed in order to treat the $\mathbf{R}^n$ case. Apply Proposition 6.2 with $\Phi(x) = (1-\epsilon)\widetilde{\Phi}(x)$ and with V replaced by $V - \lambda(h)$. Then

$$V - \lambda(h) - (\nabla\Phi)^2 = V - \lambda - (1-\epsilon)^2\|\nabla\widetilde{\Phi}\|^2 \geq V - \lambda - (1-\epsilon)^2 V_+.$$

In the complement of U, we get

$$V - \lambda(h) - (\nabla\Phi)^2 \geq (1 - (1-\epsilon)^2)V - \lambda(h) = (2\epsilon - \epsilon^2)V - \lambda.$$

Let $U_\epsilon = \{x \in M; V(x) \leq \epsilon\}$. Then outside U_ϵ:

$$V - \lambda - (\nabla\Phi)^2 \geq 2\epsilon^2 - \epsilon^3 - \lambda(h) \geq \epsilon^2 - \lambda(h) \geq \frac{1}{2}\epsilon^2,$$

assuming that $\epsilon \leq 1$ and that h is sufficiently small depending on ϵ. Now take $F_\pm$ such that $F_+ = \sqrt{V - \lambda(h) - (\nabla\Phi)^2}$ and $F_- = 0$ outside U_ϵ. Then (6.2) gives

$$\|h\nabla(e^{(1-\epsilon)\widetilde{\Phi}/h}u)\|^2 + \|e^{(1-\epsilon)\widetilde{\Phi}/h}u\|^2 \leq C_\epsilon\|e^{(1-\epsilon)\widetilde{\Phi}/h}u\|^2_{U_\epsilon},$$

which gives with a new constant C_ϵ:

$$\|e^{(1-\epsilon)\widetilde{\Phi}/h}h\nabla u\|^2 + \|e^{(1-\epsilon)\widetilde{\Phi}/h}u\|^2 \leq C_\epsilon\|e^{(1-\epsilon)\widetilde{\Phi}/h}u\|^2_{U_\epsilon}.$$

Let K denote the maximum of $\widetilde{\Phi}$ on M and let $\delta(\epsilon) = \sup_{U_\epsilon} \widetilde{\Phi}$, so that $\delta(\epsilon) \to 0$ when $\epsilon \to 0$. Then

$$\|e^{\widetilde{\Phi}/h}h\nabla u\|^2 + \|e^{\widetilde{\Phi}/h}u\|^2 \leq Ce^{2\epsilon K/h + 2\delta(\epsilon)/h}\|u\|^2_{U_\epsilon} \leq Ce^{2(\epsilon K + \delta(\epsilon))/h}.$$

This implies the desired estimate with a new C and with a new ϵ, which can be chosen arbitrarily small. When $M = \mathbf{R}^n$ we take $\Phi_R(x) = (1-\epsilon)\chi_R(\widetilde{\Phi}(x))$, where $0 < R$ is a constant and $\chi_R(t) = t1_{[0,R]}(t) + R1_{\{t>R\}}(t)$. Since Φ_R is a constant for large x, we can apply Proposition 6.2. Using the same argument as above we get:

$$h^2\|\nabla(e^{\Phi_R/h}u)\|^2 + \int_{(2\epsilon-\epsilon^2)V(x)-\lambda\geq0}((2\epsilon-\epsilon^2)V(x)-\lambda)e^{2\Phi_R/h}|u|^2dx$$

$$\leq C\int_{(2\epsilon-\epsilon^2)V(x)-\lambda\leq0}e^{2(1-\epsilon)\widetilde{\Phi}(x)/h}|u|^2dx.$$

For $\lambda \leq a(\epsilon)$ ($a(\epsilon) \to 0$ when $\epsilon \to 0$), the right hand side can be estimated by $\widetilde{C}(\epsilon)e^{\epsilon/h}$. Letting $R \to \infty$, and using the fact that $(2\epsilon - \epsilon^2)V(x) - \lambda \geq C(\epsilon) > 0$ when $\widetilde{\Phi}(x) > \epsilon$, we get the desired estimate.   #

From this one can obtain pointwise estimates on the eigenfunctions, using classical a priori estimates for the Laplace operator. We will return to this question and the problem of finding more refined estimates later on.

Remark. Proposition 6.4 remains valid if M is a compact Riemannian C^2 manifold with boundary, and P the corresponding Dirichlet realization of $-h^2\Delta + V$.

c. The interaction matrix.

We give the discussion in the case when M is a compact Riemannian manifold. The case $M = \mathbf{R}^n$ can be treated in the same way with only minor changes under the assumption that $\liminf_{|x|\to\infty} V(x) > 0$.

Assume that
$$\{x \in M; V(x) \leq 0\} = U_1 \cup \ldots \cup U_N,$$

where U_j are compact disjoint sets (the 'potential wells'). Let d be the LA distance at energy zero. Assume that

$$\operatorname{diam}_d(U_j) = 0, \forall j, \tag{6.9}$$

where diam_d denotes the diameter with respect to d. For a small value of $\eta > 0$, we introduce $B(U_j, \eta) = \{x \in M; d(x, U_j) < \eta\}$. We shall assume without loss of generality that the boundary of $B(U_j, \eta)$ is smooth, since otherwise it is easy to make small changes in the arguments below. Consider the operator P_{M_j}, defined to be the Dirichlet realization of P on

$$M_j = M \setminus \cup_{k\neq j} B(U_k, \eta). \tag{6.10}$$

(Equivalently P_{M_j} is the Friedrichs extension of P from $C_0^\infty(\operatorname{int}(M_j))$. It is easy to show that the domain of P_{M_j} is equal to $H^2(M_j) \cap H_0^1(M_j)$, where

$H^k(M_j)$ denotes the classical Sobolev space of order k and H_0^1 is the closure of $C_0^\infty(\text{int}\,(M_j))$ in $H^1(M_j)$. It is also well-known that P_{M_j} has a purely discrete spectrum. Let $N_{P_{M_j}}(\lambda)$ denote the number of eigenvalues $\leq \lambda$. If $V \geq -C_0$, then the mini-max principle shows that

$$N_{P_{M_j}}(\lambda) \leq N_{-h^2\Delta-C_0}(\lambda) = \#\{\text{ eigenvalues of } -h^2\Delta_{M_j}, \leq \lambda + C_0\}$$
$$= \#\{\text{eigenvalues of } -\Delta_{M_j}, \leq h^{-2}(\lambda + C_0)\}.$$

Here Δ_{M_j} denotes the Dirichlet realization of the Laplace operator on M_j. Now it is a classical result that $N_{-\Delta_{M_j}}(\lambda) \leq C(1 + \lambda)^{n/2}$, so we get

$$N_{P_{M_j}}(\lambda) \leq C(\lambda + C_0)^{n/2}h^{-n}.$$

We shall use only that the number of eigenvalues of P_{M_j} in a fixed interval grows at most as a polynomial in $1/h$.

Let $I(h) = [\alpha(h), \beta(h)]$ be an interval and let $a(h) > 0$ be a function defined for $h \in J \subset]0,1]$ with $0 \in \overline{J}$. We assume

$$I(h) \to \{0\}, \quad h \to 0, \tag{6.11}$$

$$a(h) \geq \frac{1}{C_\epsilon}e^{-\epsilon/h} \text{ for every } \epsilon > 0, \tag{6.12}$$

$\forall j$, P_{M_j} has no eigenvalues in $]\alpha(h) - 2a(h), \alpha(h)[\cup]\beta(h), \beta(h) + 2a(h)[$.
$$\tag{6.13}$$

The purpose is to study the spectrum of P near $I(h)$ in terms of the spectral information that we may have about the P_{M_j}. Let us first consider the resolvent of P.

Definition. Let $A = A_h$, $h \in J$ be a family of operators $L^2(M) \to H^1(M)$ and let $f \in C^0(M \times M; \mathbf{R})$. We say that the kernel $A(x,y)$ of A (using the same letter for an operator and its distribution kernel) is $\widehat{\mathcal{O}}(e^{-f(x,y)/h})$ if for all $x_0, y_0 \in M$ and $\epsilon > 0$, there exist neighborhoods $V, U \subset M$ of x_0 and y_0 and a constant C, such that

$$\|Au\|_{H^1(V)} \leq Ce^{-(f(x_0,y_0)-\epsilon)/h}\|u\|_{L^2(U)},$$

for all $u \in L^2(M)$, with support in U. By $\widehat{\mathcal{O}}(e^{-f_1/h} + \ldots + e^{-f_k/h})$, we mean $\widehat{\mathcal{O}}(e^{-\min(f_j)/h})$, when $f_1, \ldots, f_k$ are finitely many functions with the same properties as f.

Here we make two observations.

(1) If $A(x,y) = \hat{\mathcal{O}}(e^{-f/h})$, $B(x,y) = \hat{\mathcal{O}}(e^{-g/h})$, then $A \circ B(x,y) = \hat{\mathcal{O}}(e^{-k/h})$, where $k(x,y) = \min_{z \in M}(f(x,z) + g(z,y))$.

(2) Let $A(x,y) = \hat{\mathcal{O}}(e^{-f(x,y)/h})$ and let $\phi, \psi \in C^0(M)$ satisfy: $\phi(x) \geq -f(x,y) + \psi(y)$, for all $x, y \in M$. Then $\|e^{-\phi/h} Au\| \leq C_\epsilon e^{\epsilon/h} \|e^{-\psi/h} u\|$, for all $u \in L^2(M)$.

As an example, let $V > 0$ be a positive potential on M. Then $(P - z)^{-1}$ exists for $z \in K(h)$ when $h > 0$ is small enough, if $K(h)$ is some bounded subset of $\mathbf{C}$ which tends to $\{0\}$ when $h \to 0$, and we have uniformly w.r.t. z

$$(P - z)^{-1}(x,y) = \hat{\mathcal{O}}(e^{-d(x,y)/h}).$$

This still holds when M is a compact manifold with boundary.

To see this, let v have its support in a small neighborhood of y_0. It then suffices to estimate the solution of u of the equation $(P - z)u = v$ by means of LA estimates.

Next we consider the case of one potential well, $U \subset M$, which is assumed to have diameter 0 for the LA distance (at energy 0). Let $K(h) \subset \mathbf{C}$ tend to $\{0\}$, when $h \to 0$ and assume that for every $\epsilon > 0$,

$$\text{dist}\,(K(h), \sigma(P)) \geq \frac{1}{C_\epsilon} e^{-\epsilon/h}.$$

Proposition 6.5. *Under the above assumptions, we have*

$$(P - z)^{-1}(x,y) = \hat{\mathcal{O}}(e^{-d(x,y)/h}),$$

uniformly, w.r.t. $z \in K(h)$.

Proof. Let $M_0 = M \setminus B(U, \eta)$, and let $\theta \in C_0^\infty(B(U, 2\eta))$ be equal to 1 near $\overline{B(U, \eta)}$. Let $\hat{\theta} \in C_0^\infty(B(U, 3\eta))$ be equal to 1 near $\overline{B(U, 2\eta)}$. Using the resolvent identity, we get the following representation:

$$(P_M - z)^{-1} = (1 - \theta)(P_{M_0} - z)^{-1}(1 - \hat{\theta}) + (P - z)^{-1}\hat{\theta}$$
$$+ (P - z)^{-1}\hat{\theta}[P, \theta](P_{M_0} - z)^{-1}(1 - \hat{\theta}). \quad (6.14)$$

From the decay estimates and the fact that $\|(P - z)^{-1}\|_{\mathcal{L}(L^2, L^2)} \leq C_\epsilon e^{\epsilon/h}$ for every $\epsilon > 0$, we see that

$$((P - z)^{-1}\hat{\theta})(x,y) = \hat{\mathcal{O}}(e^{\frac{1}{h}(-d(x,U)+3\eta)}).$$

then (6.14) implies that

$$(P - z)^{-1}(x, y) = \widehat{\mathcal{O}}(e^{\frac{1}{h}(-d(x,y)+6\eta)}).$$

Here we can choose $\eta > 0$ arbitrarily small, and the proposition follows.   #

We now return to the general situation with N potential wells of diameter 0 for the LA distance. We assume that $K(h) \to \{0\}$ when $h \to 0$ and that for arbitrarily small $\eta > 0$ and for all $\epsilon > 0$, there are constants $C_\epsilon = C_{\epsilon,\eta}$ such that

$$\text{dist}\,(K(h), \sigma(P_{M_j})) \geq \frac{1}{C_\epsilon} e^{-\epsilon/h},$$

when h is small enough depending on η.

Proposition 6.6. *Under the above assumptions and for h small enough, we have $K(h) \cap \sigma(P) = \emptyset$ and $(P - z)^{-1}(x, y) = \widehat{\mathcal{O}}(e^{-d(x,y)/h})$, uniformly for $z \in K(h)$.*

Proof. Let $\theta_j \in C_0^\infty(B(U_j, 2\eta))$, $\theta_j = 1$ near $\overline{B(U_j, \eta)}$ and put $\chi_j = 1 - \sum_{k \neq j} \theta_k$. Also choose $\widetilde{\chi}_j \in C_0^\infty(\text{int}\,(M_j))$, $= 1$ near $B(U_j, C_0)$, $C_0 \gg \eta$ independent of η, s.t. $\sum \widetilde{\chi}_j = 1$. In order to construct the resolvent of P, we put

$$R_0(z) = \sum_1^N \chi_j (P_{M_j} - z)^{-1} \widetilde{\chi}_j.$$

Notice that $R_0(z)(x, y) = \widehat{\mathcal{O}}(e^{-d(x,y)/h})$ and that χ_j is equal to 1 near the support of $\widetilde{\chi}_j$ (when η is small enough). We have

$$(P - z)R_0 = I - K,$$

$$K = \sum_{k \neq j} \sum [P, \theta_k](P_{M_j} - z)^{-1} \widetilde{\chi}_j.$$

Since $(P_{M_j} - z)^{-1} = \widehat{\mathcal{O}}(e^{-d(x,y)/h})$, it is clear that

$$\|K\|_{\mathcal{L}(L^2, L^2)} \leq C_\epsilon e^{-(C_0-\epsilon)/h}, \ \forall \epsilon > 0,$$

where $C_0 = \min_{j \neq k} d(\text{supp}\,\widetilde{\chi}_j, B(U_k, 2\eta)) > 0$. Hence $(1 - K)^{-1}$ exists for h sufficiently small, and then $(P - z)^{-1}$ exists and is given by $(P - z)^{-1} = R_0(z)(I - K)^{-1}$. Let D be the diameter of M for the LA-distance. Then for N_0 sufficiently large, we have

$$(I - K)^{-1} = \sum_0^{N_0} K^j + \mathcal{O}(e^{-D/h}), \text{ in } \mathcal{L}(L^2, L^2).$$

Denoting the remainder in the last equation by L, it is clear that

$$(R_0 \circ L)(x, y) = \widehat{\mathcal{O}}(e^{-d(x,y)/h}).$$

If we relax the definition of $\widehat{\mathcal{O}}$ by replacing H^1 by L^2, then $K(x, y) = \widehat{\mathcal{O}}(e^{-d(x,y)/h})$. Hence $K^j(x, y) = \widehat{\mathcal{O}}(e^{-d(x,y)/h})$, and consequently

$$R_0 \circ K^j(x, y) = \widehat{\mathcal{O}}(e^{-d(x,y)/h}),$$

where we again use the original definition of $\widehat{\mathcal{O}}$. This gives the Proposition.
#

Let $\mu_{j,1}, \ldots, \mu_{j,m_j}$ be the eigenvalues of P_{M_j} in $I(h)$ and let $\phi_{j,1}, \ldots, \phi_{j,m_j} \in L^2(M_j)$ be a corresponding O.N. system. It will be convenient to write $\alpha = (j, k)$, $1 \le k \le m_j$, and put $j(\alpha) = j$. (These quantities also depend on η, but the discussion below will also imply that this dependence is very weak, so we shall not insist too heavily on this dependence.)

Let $\theta_j \in C_0^\infty(B(U_j, 2\eta))$ be equal to 1 near $\overline{B(U_j, \eta)}$ and put $\chi_j = 1 - \sum_{k \ne j} \theta_k$,

$$\psi_\alpha = \chi_{j(\alpha)} \phi_\alpha.$$

Then for every $\epsilon > 0$:

$$\phi_\alpha, \psi_\alpha = \mathcal{O}(e^{-(d(U_{j(\alpha)}, x) - \epsilon)/h}) \text{ uniformly on } M_j,$$

and similarly for all derivatives of ϕ_α, ψ_α. (Actually, we only need $\|e^{d(U_{j(\alpha)}, x)/h} \phi_\alpha\| = \mathcal{O}(e^{\epsilon/h})$ for all $\epsilon > 0$, and similarly for ψ_α, $\nabla \phi_\alpha$, $\nabla \psi_\alpha$.)
Then

$$(P - \mu_\alpha)\psi_\alpha = r_\alpha,$$

where $\operatorname{supp} r_\alpha \subset \cup_{k \ne j(\alpha)} B(U_k, 2\eta)$, $r_\alpha = \mathcal{O}(e^{(\epsilon - d(U_{j(\alpha)}, x))/h})$ for all $\epsilon > 0$, and similarly for all derivatives.

Let $F \subset L^2(M)$ be the space associated with $\sigma(P) \cap (I(h) + [-\frac{a(h)}{2}, \frac{a(h)}{2}])$, and let $E \subset L^2(M)$ be the space spanned by the ψ_α.

Proposition 6.7. $\dim E = \dim F$.

Before the proof we make some general considerations. If E_1, E_2 are closed subspaces of a Hilbert space $\mathcal{H}$, we introduce the non-symmetric distance

$$\overrightarrow{d}(E_1, E_2) = \sup_{x \in E_1, \|x\|=1} d(x, E_2) = \|(1 - \pi_{E_2})_{|E_1}\|$$
$$= \|\pi_{E_1} - \pi_{E_2}\pi_{E_1}\| = \|\pi_{E_1} - \pi_{E_1}\pi_{E_2}\|,$$

where d denotes the natural distance in $\mathcal{H}$ and π_{E_j} is the orthogonal projection onto E_j. Notice that we have the 'oriented' triangle inequality:

$$\overrightarrow{d}(E_1, E_3) \leq \overrightarrow{d}(E_1, E_2) + \overrightarrow{d}(E_2, E_3).$$

Lemma 6.8. *If* $\overrightarrow{d}(E_1, E_2) < 1$, *then*

(a) $\pi_{E_2|E_1} : E_1 \to E_2$ *is injective,*

and

(b) $\pi_{E_1|E_2} : E_2 \to E_1$ *is surjective.*

In particular $\dim E_1 \leq \dim E_2$.

Proof. Let us first prove (a). If $\pi_{E_2|E_1}$ is not injective, then there exists $x_1 \in E_1$ with $\|x_1\| = 1$, such that $\pi_{E_2} x_1 = 0$. Then $d(x_1, E_2) = \|x_1 - \pi_{E_2} x_1\| = \|x_1\| = 1$, so $\overrightarrow{d}(E_1, E_2) = 1$, contrary to the assumption in the lemma.

To show (b) it is enough to show the surjectivity of

$$\pi_{E_1} \pi_{E_2|E_1} = 1_{E_1} - (1_{E_1} - \pi_{E_1} \pi_{E_2})_{|E_1} = I - K.$$

But here the assumptions of the lemma imply that $\|K\| < 1$.   #

Put $A_{j,k} = \pi_{E_j|E_k} = E_k \to E_j$, for $j \neq k \in \{1, 2\}$. Notice that $A_{2,1}$ and $A_{1,2}$ are adjoints to each other.

Lemma 6.9. *If* $\overrightarrow{d}(E_1, E_2)$ *and* $\overrightarrow{d}(E_2, E_1)$ *are both* < 1, *then they are equal.*

Proof. We have

$$\overrightarrow{d}(E_1, E_2)^2 = \sup_{\|x_1\|=1} 1 - \|A_{2,1} x_1\|^2,$$

so

$$\inf_{\|x_1\|=1} \|A_{2,1} x_1\|^2 = 1 - \overrightarrow{d}(E_1, E_2)^2,$$

which implies that $A_{2,1}$ is injective with a bounded left inverse. Similarly, $A_{1,2}$ is injective with a bounded left inverse. Since these two operators are adjoints to each other, it follows that they are bijective and that their inverses have the same norm. The last identity now shows that

$$\|A_{2,1}^{-1}\| = (1 - \vec{d}\,(E_1, E_2)^2)^{-1/2},$$

and since the norm to the right is equal to the norm of the inverse of $A_{1,2}$, which has an analogous expression, we conclude that $\vec{d}\,(E_1, E_2) = \vec{d}\,(E_2, E_1)$.   #

Proof of Proposition 6.7. If $j(\alpha) \neq j(\beta)$, then

$$|(\psi_\alpha|\psi_\beta)| \leq C_\epsilon e^{(\epsilon - d(U_{j(\alpha)}, U_{j(\beta)}))/h},$$

for every $\epsilon > 0$. If $j(\alpha) = j(\beta)$, then

$$(\psi_\alpha|\psi_\beta) = \delta_{\alpha,\beta} + \mathcal{O}_\epsilon(e^{\frac{1}{h}(\epsilon - 2\min_{k \neq j(\alpha)} d(U_{j(\alpha)}, U_k) + 4\eta)})$$
$$= \delta_{\alpha,\beta} + \widetilde{\mathcal{O}}(e^{-\frac{2}{h}\min_{k \neq j(\alpha)} d(U_{j(\alpha)}, U_k)}).$$

Here we recall that most of our quantities depend on the small parameter $\eta > 0$, and write $\widetilde{\mathcal{O}}(e^{-f/h})$ for $\mathcal{O}_{\epsilon,\eta}(e^{(\epsilon + \epsilon(\eta) - f)/h})$, where $\epsilon(\eta) \to 0$ when $\eta \to 0$. Let $\mathcal{D}'$ be the $N \times N$ matrix with diagonal elements 0 and with the off-diagonal element equal to $e^{-d(U_j, U_k)/h}$. The above inequalities can be resumed by:

$$((\psi_\alpha|\psi_\beta)) = I + \widetilde{\mathcal{O}}(\mathcal{D}' + \mathcal{D}'^2),$$

where the estimates for matrices are to be understood elementwise, uniformly with respect to the row and column indices, so that $(\phi_\alpha|\phi_\beta) - \delta_{\alpha,\beta} = \widetilde{\mathcal{O}}((\mathcal{D}' + \mathcal{D}'^2)_{j(\alpha),j(\beta)})$.

Let $S_0 = \min_{j \neq k} d(U_j, U_k)$. Then

$$\Psi := ((\psi_\alpha|\psi_\beta)) = I + \widetilde{\mathcal{O}}(e^{-S_0/h}).$$

Here we get the corresponding estimate in the ordinary matrix norm, since the size of the matrix is $\mathcal{O}(h^{-n})$. It follows that $\Psi^{-1/2} = I + \widetilde{\mathcal{O}}(e^{-S_0/h})$. Let ψ denote the row vector of all the ψ_α. Then we get an O.N. basis in E:

$$\widetilde{\psi} = \psi \circ \Psi^{-1/2}.$$

We introduce $v_\alpha = \pi_F \psi_\alpha$, where

$$\pi_F = \frac{1}{2\pi i} \int_\gamma (z - P)^{-1} dz,$$

and γ is the simple positively oriented loop around $I(h)$ given by the set of complex points in $\mathbf{C}$ at a distance $a(h)$ from $I(h)$. Since $(P - \mu_{j(\alpha)})\psi_\alpha = r_\alpha$, we have

$$(z - P)\psi_\alpha = (z - \mu_{j(\alpha)})\psi_\alpha - r_\alpha,$$

so that

$$(z - P)^{-1}\psi_\alpha = (z - \mu_{j(\alpha)})^{-1}\psi_\alpha + (z - P)^{-1}(z - \mu_{j(\alpha)})^{-1}r_\alpha,$$

and

$$v_\alpha = \psi_\alpha + \frac{1}{2\pi i}\int_\gamma (z - P)^{-1}(z - \mu_{j(\alpha)})^{-1}r_\alpha dz.$$

From the estimates on r_α and on the resolvent, we get $v_\alpha - \psi_\alpha = \widetilde{\mathcal{O}}(e^{-\delta_{j(\alpha)}(x)/h}) = \widetilde{\mathcal{O}}(e^{-S_0/h})$ in the L^2 and in the H^1 sense, where

$$\delta_j(x) = \min_{k \neq j} d(U_j, U_k) + d(U_k, x).$$

Using also that $\dim E = \mathcal{O}(h^{-n})$, we deduce that $\overrightarrow{d}(E, F) = \widetilde{\mathcal{O}}(e^{-S_0/h})$.

It remains to estimate $\overrightarrow{d}(F, E)$. Let $u \in \mathcal{D}(P)$ be a normalized eigenfunction: $(P - \lambda)u = 0$, $\lambda \in I(h) + [-\frac{a}{2}, \frac{a}{2}]$. We return to the formula for $R_0(z)$ in the proof of Proposition 6.6. We have

$$(P_{M_j} - z)(\widetilde{\chi}_j u) = (\lambda - z)\widetilde{\chi}_j u + \mathcal{O}(e^{-C_0/h}),$$

and it follows as above that

$$\frac{1}{2\pi i}\int_\gamma (z - P_{M_j})^{-1}dz\widetilde{\chi}_j u = \widetilde{\chi}_j u + \widetilde{\mathcal{O}}(e^{-C_0/h}).$$

Hence if we put $\pi_0 = -\frac{1}{2\pi i}\int R_0(z)dz$, which has its range in E, we get

$$\pi_0 u = u + \widetilde{\mathcal{O}}(e^{-C_0/h}).$$

Again, using that $\dim F = \mathcal{O}(h^{-n})$, we deduce that

$$\overrightarrow{d}(F, E) = \widetilde{\mathcal{O}}(e^{-C_0/h}).$$

Hence,

$$\overrightarrow{d}(E, F) = \overrightarrow{d}(F, E) < 1,$$

which concludes the proof of Proposition 6.7, in view of the lemmas 6.8, 6.9.
#

The next problem is to determine the matrix of $P_{|_F}$ for some suitable basis. We shall orthonormalize the basis v_α. Recall that

$$v_\alpha - \psi_\alpha = \widetilde{\mathcal{O}}(e^{-\delta_{j(\alpha)}(x)/h}). \tag{6.15}$$

Since $v_\alpha - \psi_\alpha$ and $v_\beta - \psi_\beta$ are orthogonal to F, we have

$$(v_\alpha|v_\beta) - (\psi_\alpha|\psi_\beta) = (v_\alpha|v_\beta - \psi_\beta) + (v_\alpha - \psi_\alpha|\psi_\beta) = -(v_\alpha - \psi_\beta|v_\beta - \psi_\beta).$$

Using (6.15), we get

$$(v_\alpha|v_\beta) - (\psi_\alpha|\psi_\beta) = \widetilde{\mathcal{O}}(e^{-\tau_{\alpha,\beta}/h}),$$

where

$$\tau_{\alpha,\beta} = \min_{j \neq j(\alpha), k \neq j(\beta)} d(U_{j(\alpha)}, U_j) + d(U_j, U_k) + d(U_k, U_{j(\beta)}).$$

This can also be written in matrix form:

$$((v_\alpha|v_\beta)) - ((\psi_\alpha|\psi_\beta)) = \widetilde{\mathcal{O}}(\mathcal{D}'(I + \mathcal{D}')\mathcal{D}') = \widetilde{\mathcal{O}}(\mathcal{D}'^2 + \mathcal{D}'^3).$$

Let us introduce $T = (t_{\alpha,\beta})$, where $t_{\alpha,\beta} = 0$ when $j(\alpha) = j(\beta)$, and $t_{\alpha,\beta} = (\psi_\alpha|\psi_\beta)$ when $j(\alpha) \neq j(\beta)$. Then

$$((\psi_\alpha|\psi_\beta)) = I + T + \widetilde{\mathcal{O}}(\mathcal{D}'^2 + \mathcal{D}'^3),$$

so

$$((v_\alpha|v_\beta)) = I + T + \widetilde{\mathcal{O}}(\mathcal{D}'^2 + \mathcal{D}'^3). \tag{6.16}$$

We do the same calculations for $(Pv_\alpha|v_\beta)$. Since we control the exponential decrease of $v_\alpha - \psi_\alpha$ in H^1, we obtain similarly

$$((Pv_\alpha|v_\beta)) = ((P\psi_\alpha|\psi_\beta)) + \widetilde{\mathcal{O}}(\mathcal{D}'^2 + \mathcal{D}'^3). \tag{6.17}$$

Here

$$(P\psi_\alpha|\psi_\beta) = \frac{1}{2}(P\psi_\alpha|\psi_\beta) + \frac{1}{2}(\psi_\alpha|P\psi_\beta)$$
$$= \frac{1}{2}(\mu_\alpha + \mu_\beta)(\psi_\alpha|\psi_\beta) + \frac{1}{2}((r_\alpha|\psi_\beta) + (\psi_\alpha|r_\beta)).$$

Using the fact that $((\psi_\alpha|\psi_\beta)) = I + T + \widetilde{\mathcal{O}}(\mathcal{D}'^2)$, we can write this in matrix form,

$$((P\psi_\alpha|\psi_\beta)) = \text{diag}\,(\mu_\alpha) + \frac{1}{2}T\,\text{diag}\,(\mu_\alpha) + \frac{1}{2}\text{diag}\,(\mu_\alpha)\,T$$
$$+ \frac{1}{2}((r_\alpha|\psi_\beta) + (\psi_\alpha|r_\beta)) + \widetilde{\mathcal{O}}(\mathcal{D}'^2 + \mathcal{D}'^3). \tag{6.18}$$

Here

$$(\psi_\alpha | r_\beta) = -h^2 \int (\psi_\alpha 2 \nabla \chi_{j(\beta)} \cdot \nabla \phi_\beta + \psi_\alpha (\Delta \chi_{j(\beta)}) \phi_\beta) dx,$$

and

$$-h^2 \int \psi_\alpha (\Delta \chi_{j(\beta)}) \phi_\beta dx$$

$$= h^2 \int \chi_{j(\alpha)} \nabla \phi_\alpha \cdot \nabla \chi_{j(\beta)} \phi_\beta dx + h^2 \int \chi_{j(\alpha)} \phi_\alpha \nabla \chi_{j(\beta)} \cdot \nabla \phi_\beta dx + \widetilde{\mathcal{O}}(\mathcal{D}'^2),$$

so

$$(\psi_\alpha | r_\beta) = h^2 \int \chi_{j(\alpha)} (\phi_\beta \nabla \chi_{j(\beta)} \cdot \nabla \phi_\alpha - \phi_\alpha \nabla \chi_{j(\beta)} \cdot \nabla \phi_\beta) dx + \widetilde{\mathcal{O}}(\mathcal{D}'^2).$$

$$(6.19)$$

(Here we made the computations as if M were equal to $\mathbf{R}^n$ at least locally, but it is easy to see that (6.19) holds in the more general Riemannian case.) If $j(\alpha) = j(\beta)$, this quantity is $\mathcal{O}(\mathcal{D}'^2)$. Let $w_{\alpha,\beta}$ be 0 when $\alpha = \beta$ and $w_{\alpha,\beta} = h^2 \int \chi_{j(\alpha)} (\phi_\beta \nabla \phi_\alpha - \phi_\alpha \nabla \phi_\beta) \cdot \nabla \chi_{j(\beta)} dx$ otherwise. We also introduce the symmetrized quantity, $\widehat{w}_{\alpha,\beta} = \frac{1}{2}(w_{\alpha,\beta} + w_{\beta,\alpha})$ and let $\widehat{W}$ be the corresponding matrix. Then by (6.17–19):

$$(Pv_\alpha | v_\beta) = \operatorname{diag}(\mu_\alpha) + \frac{1}{2} T \operatorname{diag}(\mu_\alpha) + \frac{1}{2} \operatorname{diag}(\mu_\alpha) T$$

$$+ (\widehat{w}_{\alpha,\beta}) + \widetilde{\mathcal{O}}(\mathcal{D}'^2 + \mathcal{D}'^3). \quad (6.20)$$

Put $V = ((v_\alpha | v_\beta))$. Then if v denotes the row vector (v_α), $vV^{-1/2}$ is an O.N. basis in F. The matrix of $P_{|F}$ with respect to this basis is

$$V^{-1/2}((Pv_\alpha | v_\beta))V^{-1/2}. \quad (6.21)$$

From (6.16) we get

$$V^{-1/2} = I - \frac{1}{2} T + \widetilde{\mathcal{O}}(\mathcal{D}'^2 + \mathcal{D}'^3), \quad (6.22)$$

and combining this with (6.20), we see that the matrix (6.21) becomes

$$(I - \frac{1}{2} T + \widetilde{\mathcal{O}}(\mathcal{D}'^2 + \mathcal{D}'^3))$$

$$\times (\operatorname{diag}(\mu_\alpha) + \frac{1}{2} T \operatorname{diag}(\mu_\alpha) + \frac{1}{2} \operatorname{diag}(\mu_\alpha) T + \widehat{W} + \widetilde{\mathcal{O}}(\mathcal{D}'^2 + \mathcal{D}'^3))$$

$$\times (I - \frac{1}{2} T + \widetilde{\mathcal{O}}(\mathcal{D}'^2 + \mathcal{D}'^3)).$$

Now $\widehat{W}, T = \widetilde{\mathcal{O}}(\mathcal{D}')$, so the matrix (6.21) is equal to

$$\text{diag}\,(\mu_\alpha) + \widehat{W} + \widetilde{\mathcal{O}}(\mathcal{D}'^2 + \mathcal{D}'^3).$$

We have proved:

Theorem 6.10. *Under the assumptions above, the matrix of* $P_{|F}$ *w.r.t. the O.N. basis* $vV^{-1/2}$ *is of the form*

$$\text{diag}\,(\mu_\alpha) + \widehat{W} + \widetilde{\mathcal{O}}(\mathcal{D}'^2 + \mathcal{D}'^3).$$

We now discuss the problem of 'computing' the interaction coefficients appearing in the matrix $\widehat{W}$. Assume that $j(\alpha) \neq j(\beta)$ and that there is $a \in]0, d(U_{j(\alpha)}, U_{j(\beta)}) - 2\eta[$ such that $\mu_\alpha - \mu_\beta = \mathcal{O}(e^{-(a-\epsilon)/h})$ for every fixed $\epsilon > 0$. By integration by parts, we have

$$w_{\alpha,\beta} = (\mu_\alpha - \mu_\beta)\int \psi_\alpha \psi_\beta dx + w_{\beta,\alpha},$$

so

$$w_{\alpha,\beta} - w_{\beta,\alpha} = \mathcal{O}(e^{-(d(U_{j(\alpha)}, U_{j(\beta)})+a-\epsilon)/h}),$$

for every $\epsilon > 0$. Let E be the 'ellipsoid' defined by

$$d(U_{j(\alpha)}, x) + d(U_{j(\beta)}, x) \leq d(U_{j(\alpha)}, U_{j(\beta)}) + a.$$

Assume that E does not intersect any of the $B(U_k, 2\eta)$ with $k \notin \{j(\alpha), j(\beta)\}$. Let $\Omega \subset M$ be open with smooth boundary such that

$$\Omega \supset \overline{B(U_{j(\alpha)}, 2\eta)}, \quad M \setminus \Omega \supset \overline{B(U_{j(\beta)}, 2\eta)}.$$

Let $\Gamma = \partial\Omega \cap E$, and let n denote the unit normal on Γ oriented into Ω. Using Green's formula, one can then prove that

$$w_{\alpha,\beta} = h^2 \int_\Gamma (\phi_\alpha \frac{\partial\phi_\beta}{\partial n} - \phi_\beta \frac{\partial\phi_\alpha}{\partial n})dS + \mathcal{O}(e^{-(d(U_{j(\alpha)}, U_{j(\beta)})+a-\epsilon)/h}). \quad (6.23)$$

In view of the weighted estimates we have for ϕ_α, ϕ_β, we expect the main contribution to the integral in (6.23) to come from the intersection between Γ and the union of minimal LA-geodesics joining $U_{j(\alpha)}$ and $U_{j(\beta)}$. We refer to Remark 2.13 in Helffer–Sjöstrand [HeSj2] for further details.

Under additional assumptions it is possible to determine the order of magnitude of the interaction coefficients $w_{\alpha,\beta}$ and even to give asymptotic expansions. Let us describe one such situation. Assume that all the wells are reduced to single points: $U_j = x_j$, and that V has a non-degenerate minimum in each of these points. Let us further assume that the localized harmonic oscillators for each of these points have the same lowest eigenvalue E_0, and choose the interval $I(h)$ to be of the form $[h(E_0 - \epsilon_0), h(E_0 + \epsilon_0)]$ for some fixed sufficiently small ϵ_0. Then we have precisely one eigenvalue $\mu_\alpha = \mu_{j(\alpha)}$ for each j, and $\mu_\alpha = h(E_0 + \mathcal{O}(h))$, according to the discussion at the end of Chapter 4. In the appendix of the present chapter, we show (also for some excited eigenvalues) that the eigenfunction ϕ_α is well approximated by the WKB-eigenfunction constructed in Chapter 3, provided that we restrict the attention to a star-shaped subdomain of M_α, (where we write $M_\alpha = M_{j(\alpha)}$). Since we assumed E_0 to be the ground state of the local harmonic oscillator for each well, we also know that the amplitude for each eigenfunction is an elliptic symbol (i.e. with a non-vanishing leading term in its asymptotic expansion) of order $n/4$. For given α, β as in (6.23), let us assume that there is no other U_γ (so that $\gamma \notin \{\alpha, \beta\}$) with $d(U_\alpha, U_\beta) = d(U_\alpha, U_\gamma) + d(U_\gamma, U_\beta)$. If we choose η (in the definition of the M_j) sufficiently small, then the minimal geodesics from U_α to U_β are contained in the intersection of M_α and M_β, except in $B(U_\alpha, \eta) \cup B(U_\beta, \eta)$. In particular, the intersection of Γ with the union of minimal geodesics from U_α to U_β is contained in the region where both ϕ_α and ϕ_β are well-approximated by the corresponding WKB-functions constructed in Chapter 3. In the appendix, we established this approximation only in L^2 and H^1 norms, but further easy arguments permit us to extend this approximation to the uniform norms and for all derivatives. Alternatively, we can reexamine how (6.23) is obtained by means of the Green formula and use only the approximation result of the appendix. Both methods permit us to replace ϕ_α, ϕ_β by their WKB-approximations in (6.23). After those substitutions, we see that the integrand in (6.23) is of the order of magnitude $h^{-1-n/2}e^{-(d(U_\alpha, x) + d(x, U_\beta))/h}$, and we get an estimate,

$$h^{1/2}e^{-d(U_\alpha, \beta)/h} \leq -w_{\alpha, \beta} \leq Ch^{1-n/2}e^{-d(U_\alpha, \beta)/h}. \qquad (6.24)$$

If we do not want to assume the exponential smallness of $\mu_\alpha - \mu_\beta$, leading to (6.23), we can still obtain (6.24) directly from the definition of $w_{\alpha, \beta}$ after (6.19). Now add the assumption that there are only finitely many LA-geodesics γ joining U_α to U_β and that they are all non-degenerate in the following sense. Let H be a hypersurface intersecting γ transversally at some point x_0, strictly between U_α and U_β, then for $x \in H$ close to x_0 we have

$$d(U_\alpha, x) + d(U_\beta, x) - d(U_\alpha, U_\beta) \sim |x - x_0|^2. \qquad (6.25)$$

It follows from general facts in Riemannian geometry that this definition is independent of the choice of the hypersurface H. Under this additional assumption, we can apply the method of stationary phase (see Chapter 5) to the integral in (6.23) (after replacing ϕ_α, ϕ_β by their WKB-approximations), and get

$$-w_{\alpha, \beta} = h^{1/2}a(h)e^{-d(U_\alpha, U_\beta)/h}, \quad a(h) \sim a_0 + ha_1 + \dots, \quad a_0 > 0. \qquad (6.26)$$

The leading coefficient a_0 can be computed, and we refer to Wilkinson [Wi], Dobrokotov–Kolokol'tsov–Maslov [DoKoMa] for further details.

In the following, we shall discuss some examples where some symmetry group implies that the eigenvalues associated with the different wells are all equal. We will assume that we are in the situation where we only have to consider one eigenvalue for each well, and that the interaction coefficients have the expected order of magnitude, as in the special situation discussed above.

Example 1. We consider the classical case of a symmetric double well potential. Let $N = 2$ and assume that there is an isometry $\iota : M \to M$ such that $\iota(U_1) = U_2$, $\iota(U_2) = U_1$, $V \circ \iota = V$. Then P_{M_1} and P_{M_2} have the same eigenvalues. Let μ be an isolated eigenvalue of P_{M_1} such that $]\mu - 2a(h), \mu + 2a(h)[$ contains no other eigenvalues, where $a(h) \geq \frac{1}{C_\epsilon} e^{-\epsilon/h}$, $\forall \epsilon > 0$. If $S_0 = d(U_1, U_2)$, then the matrix of $P_{|_F}$ in a suitable O.N. basis is given by

$$\begin{pmatrix} \mu & w \\ w & \mu \end{pmatrix} + \tilde{\mathcal{O}}(e^{-2S_0/h}).$$

As indicated above, we may assume that

$$\frac{1}{C_\epsilon} e^{(-S_0-\epsilon)/h} \leq |w| \leq C_\epsilon e^{(-S_0+\epsilon)/h},$$

for every $\epsilon > 0$. In order to handle the remainder appearing in the interaction matrices, we shall use the following simple consequence of the mini-max principle:

Lemma 6.11. *Let A, B be two Hermitian $N \times N$ matrices with eigenvalues $\lambda_1 \leq \lambda_2 \leq \ldots \leq \lambda_N$ and $\mu_1 \leq \mu_2 \leq \ldots \leq \mu_N$ respectively. Then $|\lambda_j - \mu_j| \leq \|A - B\|$, for every j.*

Applying this, we conclude that the eigenvalues of P in $]\mu - a, \mu + a[$ are of the form

$$\lambda_\pm = \mu \pm w + \tilde{\mathcal{O}}(e^{-2S_0/h}).$$

In particular, we observe a splitting:

$$\lambda_+ - \lambda_- = 2w + \tilde{\mathcal{O}}(e^{-2S_0/h}).$$

Example 2. Let V be a $\frac{2\pi}{N}$-periodic potential on $S^1 = M$ with N consecutive potential wells: $U_1, \ldots, U_N$. If $\iota : M \to M$ denotes rotation by $\frac{2\pi}{N}$, then $\iota(U_j) = U_{j+1}$ (and we put $U_{N+1} = U_1$, $\iota^*(V) = V$).

More generally, we assume that we have N wells and an isometry $\iota :$ $M \to M$ leaving M invariant and permuting $U_1, \ldots, U_N$ cyclically. In the general case, we also assume that $S_0 = d(U_1, U_2)$ is strictly less than $S_1 = \min_{j-k \neq -1,0,1} d(U_j, U_k)$, where we consider that $j, k \in \mathbf{Z}/N\mathbf{Z}$. Since $S_1 \leq 2S_0$, we get the matrix for $P_{|_F}$:

$$\mu I + \frac{1}{2} \begin{pmatrix} 0 & w & 0 & \ldots & w \\ w & 0 & w & \ldots & 0 \\ 0 & w & 0 & \ldots & 0 \\ \cdot & \cdot & \cdot & \ldots & \cdot \\ w & 0 & \ldots & w & 0 \end{pmatrix} + \widetilde{\mathcal{O}}(e^{-S_1/h}).$$

Let us compute the eigenvalues of

$$\frac{1}{2} \begin{pmatrix} 0 & 1 & 0 & \ldots & 1 \\ 1 & 0 & 1 & \ldots & 0 \\ 0 & 1 & 0 & \ldots & 0 \\ \cdot & \cdot & \cdot & \ldots & \cdot \\ 1 & 0 & \ldots & 1 & 0 \end{pmatrix},$$

which can be viewed as $\frac{1}{2}(\tau_{-1} + \tau_{+1})$ on $\ell^2(\mathbf{Z}/N\mathbf{Z})$, where $\tau_{\pm 1} u(k) = u(k \mp 1)$. Try an eigenvector of the form $u(k) = e^{i\theta k}$. The obvious periodicity condition implies that $\theta N = 2\pi j$, so $j \in \mathbf{Z}$, $\theta = 2\pi \frac{j}{N}$, for $j = 0, 1, \ldots, N-1$. We get the N linearly independent eigenvectors $u_j(k) = e^{i2\pi jk/N}$, with the corresponding eigenvalues, $\cos(2\pi j/N)$, $j = 0, 1, \ldots, N-1$.

The conclusion is that P has N eigenvalues in $[\mu - a, \mu + a]$, and they are of the form: $\mu + w \cos(2\pi j/N) + \widetilde{\mathcal{O}}(e^{-S_1/h})$ for $j = 0, 1, \ldots, N-1$. We see that the value $\mu + w + \widetilde{\mathcal{O}}(e^{-S_1/h})$ (for $j = 0$) and when N is even, the value $\mu - w + \widetilde{\mathcal{O}}(e^{-S_1/h})$ (for $j = N/2$), are simple. The other eigenvalues seem to be double and we may ask whether this is really the case.

Usually, multiple eigenvalues are due to invariance of the operator under group actions and the question ending the last example can be answered by finding such an invariance. Let us consider such invariances more generally, and let us restrict our attention to the actions of finite groups. Let G be a finite group of isometries $M \to M$ leaving V invariant. If $g \in G$, $u \in L^2(M)$, we put $g_* u = u(g^{-1}x)$. Then g_* is unitary and commutes with P. If F is the spectral subspace defined before, then $g_* : F \to F$, and $[g_*, \pi_F] = 0$ for all $g \in G$.

Lemma 6.12. *We can choose the θ_j above such that $g_* \theta_j = \theta_{\sigma_g(j)}$, where σ_g is the permutation of $\{1, \ldots, N\}$ given by $g(U_j) = U_{\sigma_g(j)}$.*

The functions θ_j are constructed by a procedure of averaging over the elements of the group, and we leave the simple details to the reader. Now take ϕ_j, χ_j, $\psi_j = \chi_j\phi_j$, $v_j = \pi_F\psi_j$ as before. Then $g_*\psi_j = \pm\psi_{\sigma_g(j)}$, and we assume for simplicity that only the plus sign appears here. (This will be the case if we work with the lowest eigenvalue for each P_{M_j}.) Then $g_*v_j = v_{\sigma_g(j)}$. Recall the construction of the O.N. basis $e_1, \ldots, e_N$ in F: We look for a positive selfadjoint matrix $(s_{j,k})$, such that $v_k = \sum s_{j,k}e_j$, with $e_1, \ldots, e_N$ orthonormal. Then

$$(v_j|v_k) = (\sum_\nu s_{j,\nu}e_\nu | \sum_\mu s_{k,\mu}e_\mu),$$

so $((v_j|v_k)) = S \circ ((e_\nu|e_\mu)) \circ S$. Thus

$$S^2 = ((v_j|v_k)), \tag{6.27}$$

and S is the unique positive selfadjoint matrix satisfying (6.27). (Uniqueness follows if we consider the spectral decomposition of S.) To sum up, there is a unique positive selfadjoint matrix S (now viewed as an operator $F \to F$) such that $v_k = Se_k$, with $e_1, \ldots, e_N$ orthonormal. Then

$$Se_{\sigma_g(k)} = v_{\sigma_g(k)} = g_*v_k = g_*Se_k = (g_*Sg_*^{-1})g_*e_k,$$

and the uniqueness of S implies that $S = g_*Sg_*^{-1}$ and $e_{\sigma_g(k)} = g_*e_k$.

In other words, the action of G on F is induced by the action of the corresponding well permutation group on the basis vectors $e_1, \ldots, e_N$. In practice, we combine this with the fact that every eigenspace of $P_{|F}$ is invariant under the G action.

Back to Example 1. Let the full matrix of $P_{|F}$ with respect to e_1, e_2 be $M = \begin{pmatrix} a & b \\ b & c \end{pmatrix}$, and use the fact that this matrix must commute with the matrix induced by the permutation of e_1 and e_2; $\begin{pmatrix} 0 & 1 \\ 1 & 0 \end{pmatrix}$. It follows that $a = c$ and the eigenvalues of M are then $a \pm b$.

Back to Example 2. Let M be the matrix of $P_{|F}$ with respect to the basis $e_1, \ldots, e_N$. We know that M commutes with the matrix of cyclic permutations:

$$\pi = \begin{pmatrix} 0 & 1 & 0 & 0 & \ldots & 0 \\ 0 & 0 & 1 & 0 & \ldots & 0 \\ 0 & 0 & 0 & 1 & \ldots & 0 \\ 0 & 0 & 0 & 0 & \ldots & 0 \\ \cdot & \cdot & \cdot & \cdot & \ldots & \cdot \\ 1 & 0 & 0 & 0 & \ldots & 0 \end{pmatrix}.$$

The eigenvalues of π are simple and of the form $\lambda = e^{2\pi i j/N}$, $j = 0, 1, \ldots, N-1$, and the corresponding eigenvectors are of the form

$$
\begin{pmatrix}
1 \\
\lambda \\
\lambda^2 \\
\cdot \\
\cdot \\
\lambda^{N-1}
\end{pmatrix}. \tag{6.28}
$$

Because of the commutativity, the corresponding one-dimensional eigenspaces are also eigenspaces of M, or in other words, the eigenvectors of π are also eigenvectors of M, so the earlier computation of the eigenvectors remains valid when the remainder terms are taken into account. The eigenvectors (6.28) of M are non-proportional to real vectors, except for $j = 0$ and when N is even for $j = N/2$, and since M is a real matrix, this implies that the eigenvalues of M associated with the cases $j \neq 0, N/2$ are at least double. The earlier computation then also shows that they are exactly double when h is small enough.

Example 3. Assume that we have N potential wells and a symmetry group G with the property that for all (i,j), $(\widetilde{i}, \widetilde{j})$ in $\{1, \ldots, N\}^2$ with $i \neq j$, $\widetilde{i} \neq \widetilde{j}$, there exists a $g \in G$ such that $(\sigma_g(i), \sigma_g(j))$ is equal to $(\widetilde{i}, \widetilde{j})$ or $(\widetilde{j}, \widetilde{i})$. Let $M = (m_{i,j})$ be the real symmetric matrix of $P_{|F}$. Then

$$
m_{i,j} = (e_i | M e_j) = (g_* e_i | M g_* e_j) = \ldots = m_{\widetilde{i}, \widetilde{j}}.
$$

Also assume that for every $(i, \widetilde{i}) \in \{1, \ldots, N\}^2$, there exists a $g \in G$ such that $\sigma_g(i) = \widetilde{i}$. Then we obtain $m_{i,i} = m_{\widetilde{i}, \widetilde{i}}$. We conclude that

$$
M = \mu I +
\begin{pmatrix}
0 & w & w & w & \ldots & w \\
w & 0 & w & w & \ldots & w \\
w & w & 0 & w & \ldots & w \\
w & w & w & 0 & \ldots & w \\
\cdot & \cdot & \cdot & \cdot & \ldots & \cdot \\
w & w & w & w & \ldots & 0
\end{pmatrix}.
$$

The eigenvalues of the rank 1 matrix with the entry 1 everywhere are: N with multiplicity 1 and 0 with multiplicity $N-1$. It follows that the eigenvalues of M (assuming that $w \neq 0$) are $\mu - w$ with multiplicity $N-1$ and $\mu + (N-1)w$ with multiplicity 1. As a special case of this example, we may think of the case when $n = 3$ and we have four wells at the summits of a regular tetrahedron.

In more general situations, it is useful to use representation theory for groups, and in particular Shur's lemma. We refer to Helffer–Sjöstrand [HeSj1–2] and Serre [Se] for further details.

Schrödinger operators with potential wells have turned out to be useful in topology, and questions concerning the multiplicity of eigenvalues for the Laplacian in Riemannian geometry, and there is now a rather considerable amount of articles on these subjects. See Bismut [Bis], Helffer–Sjöstrand [HeSj3], Witten [W], Kappeler [Kap], Colin de Verdière [CdV1-2]. See also [Si2].

Appendix

In this appendix, we shall show that the exact eigenfunctions for the one well problem are well approximated in suitable star-shaped regions by the WKB-expressions in Chapter 3. Let M be a bounded closed connected subset of $\mathbf{R}^n$ with smooth boundary. Everything works with only minor modifications in the case when M is a smooth compact Riemannian manifold, possibly with boundary, or when $M = \mathbf{R}^n$. Let $x_0 \in \text{int } M$ and let $0 \leq V \in C^\infty(M)$ vanish at x_0 precisely and have a non-degenerate minimum there, so that $V''(x_0) > 0$. We shall study $P = -h^2 \Delta + V$ defined as before by means of the Friedrichs extension.

Let $\phi(x)$ be the phase function constructed in Chapter 3 in some open neighborhood of x_0, and let $d(x) = d(x, x_0)$ denote the LA-distance from x_0 to x. Recall that $\phi(x)$ is a solution of the eikonal equation $q(x, \phi_x') = 0$, where $q(x, \xi) = \xi^2 - V(x)$.

Proposition A.1. *In a sufficiently small neighborhood of x_0, we have* $\phi(x) = d(x)$.

Proof. Let $\gamma : [0, 1] \to M$ be a a C^1 curve. Then

$$|\phi(\gamma(1)) - \phi(\gamma(0))| = |\int_0^1 \nabla\phi(\gamma(t)) \cdot \gamma'(t)dt| \leq \int_0^1 \sqrt{V(\gamma(t))}|\gamma'(t)|dt = |\gamma|, \tag{A.1}$$

where by definition the last expression is the length of γ for the LA-metric. Taking the infimum over all C^1 curves from x to y, we get,

$$|\phi(x) - \phi(y)| \leq d(x, y),$$

and in particular for $y = x_0$:

$$0 \leq \phi(x) \leq d(x).$$

If γ is an integral curve of $\nabla\phi$ up to reparametrization, then we have equality in (A.1):

$$|\phi(\gamma(1)) - \phi(\gamma(0))| = |\gamma| \geq d(x, y).$$

Hence $|\phi(x) - \phi(y)| \leq d(x,y)$, with equality if there is $\nabla\phi$-integral curve from x to y. But for any x near 0, there is a unique ∇_ϕ integral curve from x to 0, and, consequently, $\phi(x) = d(x)$, asserted.   #

Let $x_1 \in \operatorname{int} M$ and let $\gamma :] - \infty, 0] \to \operatorname{int} M$ be a minimal geodesic for the LA-metric from x_0 to x_1. If we write $d(y, x_0) = d(y, B(x_0, \epsilon)) + \epsilon$, when $y \notin B(x_0, \epsilon)$, and where we let $B(x_0, r)$ denote the open ball of radius x_0 and center r with respect to the LA-metric, and notice that $B(x_0, \epsilon)$ has smooth boundary, it follows from general facts about Riemannian manifolds, see Milnor [Mi], that there is a neighborhood Ω of $x_0 \cup \gamma(] - \infty, 0[)$, such that

(1) $\phi(x) = d(x, x_0) \in C^\infty(\Omega)$,

(2) Ω is star-shaped in the sense of Chapter 3. (The backward integral curve of $\nabla\phi$ through a given point $x \in \Omega$ then gives (up to reparametrization) the unique LA geodesic joining x to x_0.)

Let $u \in C^\infty(\Omega)$ be an approximate eigenfunction of P constructed as in Chapter 3, with the associated eigenvalue $hE(h)$, where $E(h) \sim E_0 + hE_1 + \dots$, and where E_0 is a simple eigenvalue of the localized harmonic oscillator $-\Delta + \frac{1}{2}\langle V''(x_0)x, x \rangle$. As we saw in the end of Chapter 4, there is a corresponding exact eigenvalue of the form $hE(h) + \mathcal{O}(h^\infty)$, and this is the only eigenvalue in some interval $[hE(h) - \epsilon_0 h, hE(h) + \epsilon_0 h]$, for some $\epsilon_0 > 0$. Without loss of generality we may assume that the exact eigenvalue is precisely $hE(h)$. Let $v \in C^\infty(M)$ be the corresponding exact normalized eigenfunction. If $\chi \in C_0^\infty(\Omega)$ is equal to 1 near x_0, we know that

$$\|\chi u - v\| = \mathcal{O}(h^\infty).$$

Here we write $g(h) = \mathcal{O}(h^\infty)$ if $|g(h)| \leq C_N h^N$ for every $N \geq 0$. We shall first establish a sharpened LA-estimate for v, where we shall not yet use the assumption that E_0 is a simple eigenvalue of the localized harmonic oscillator.

Proposition A.2. *There exist constants $C, \widetilde{C}$, such that*

$$h^2\|(1 + \frac{d}{h})^{-C} e^{d/h}\nabla v\|^2 + h\|(1 + \frac{d}{h})^{-C} e^{d/h} v\|^2 \leq \widetilde{C}h,$$

where $d = d(x, x_0)$.

Proof. Put

$$\Phi(x) = d(x) - Ch \log\max(\frac{d(x)}{h}, C).$$

Then for $d(x) \leq Ch$, we have $V - {\Phi'_x}^2 = 0$, and for $d(x) \geq Ch$:

$$\nabla\Phi(x) = (1 - \frac{Ch}{d(x)})\nabla d(x), \quad \text{a.e.,}$$

in view of the following general observation. Let $f \in C^1$ and let ϕ be a Lipschitz function defined in suitable domains. Let $\phi_\epsilon = \phi * \chi_\epsilon$, where χ_ϵ is a standard regularizer. Then the Lipschitz function $f \circ \phi$ is the limit of $f \circ \phi_\epsilon$ in the sense of distributions when $\epsilon \to 0$, and $\nabla(f \circ \phi_\epsilon) = (f' \circ \phi_\epsilon)\nabla\phi_\epsilon \to (f' \circ \phi)\nabla\phi$ a.e. Consequently $\nabla(f \circ \phi) = (f' \circ \phi)\nabla\phi$ a.e.

It follows that for $d(x) \geq Ch$:

$$V(x) - \nabla\Phi^2 \geq (1 - (1 - \frac{Ch}{d})^2)V(x) = V(x)(\frac{2Ch}{d} - (\frac{Ch}{d})^2) \geq Ch\frac{V(x)}{d(x)}.$$

Now $\frac{1}{C_0} \leq \frac{V(x)}{d(x)} \leq C_0$, for some constant $C_0 > 1$, so

$$V(x) - (\nabla\Phi)^2 \geq \widetilde{C}h \text{ a.e. in the region } d(x) \geq Ch.$$

Here we can assume that $\widetilde{C}$ is as big as we like by choosing C sufficiently large. In the following we shall assume that we have chosen $\widetilde{C} > E_0$.

Choose F_+, F_- as in Proposition 6.2 with V replaced by $V(x) - hE(h)$,

$$F_-(x) = 0 \text{ and } F_+(x) = \sqrt{V - hE(h) - (\nabla\Phi)^2} \text{ for } d(x) \geq Ch,$$

$$F_+ + F_- \sim \sqrt{h}, \text{ for } d(x) \leq Ch.$$

Then

$$h^2\|\nabla(e^{\Phi/h}v)\|^2 + \frac{h}{\widetilde{C}}\|e^{\Phi/h}v\|^2 \leq \widetilde{C}h\|e^{\Phi/h}v\|^2_{B(0,Ch)} \leq \widehat{C}h.$$

Here we notice that

$$e^{\Phi/h} \sim (1 + \frac{d}{h})^{-C}e^{d/h}.$$

Moreover,

$$\|h\nabla(e^{\Phi/h}v)\| \leq \|e^{\Phi/h}h\nabla v\| + \|(\nabla\Phi)e^{\Phi/h}v\|,$$

and $|\nabla\Phi| \leq V^{1/2} \leq Ch^{1/2}\frac{V^{1/2}}{h^{1/2}} \leq \widetilde{C}h^{1/2}(\frac{d}{h})^{1/2}$, so we get Proposition A.2. #

An immediate consequence of Proposition A.2 is that there is a number N_0 (depending on E_0) such that

$$\|e^{d/h}\nabla v\| + \|e^{d/h}v\| = \mathcal{O}(h^{-N_0}).$$

We now come to the main result of this appendix, where we recall that u is the asymptotic eigenfunction, constructed in Chapter 3, and that Ω is a star-shaped neighborhood of a minimal LA-geodesic.

Theorem A.3. *For every compact set $K \subset \Omega$ and every $N \in \mathbf{N}$, we have*

$$\|e^{d/h}\nabla(u - v)\|_{L^2(K)} + \|e^{d/h}(u - v)\|_{L^2(K)} = \mathcal{O}(h^N).$$

Proof. For a given K, let $\widehat{K}$ be the union of all minimal LA-geodesics from K to x_0. Choose $\chi \in C_0^\infty(\Omega)$, equal to 1 in a neighborhood of $\widehat{K}$. For some sufficiently small $\epsilon > 0$, put

$$\Phi_\epsilon(x) = \min(\Phi(x), F_\epsilon(x)),$$

where

$$F_\epsilon(x) = (1 - \epsilon) \inf_{y \in \operatorname{supp} \nabla\chi} (d(0, y) + d(y, x)).$$

Then:

(1) There exists $h_0 > 0$ such that for $0 < h \le h_0$, $\Phi_\epsilon(x)$ is equal to $\Phi(x)$ in a neighborhood of $\widehat{K}$, and equal to $(1 - \epsilon)d(x)$ on $\operatorname{supp} \nabla\chi$.

(2) We have $V - |\nabla F_\epsilon(x)|^2 \ge (1 - (1 - \epsilon)^2)V(x) \ge \epsilon V(x)$, so $V(x) - hE(h) - (\nabla\Phi_\epsilon)^2$ is $\ge \frac{Ch}{C_0}$ for $d(x) \ge Ch$ and $= \mathcal{O}(h)$ for $d(x) \le Ch$.

(3) Put $w = u - v$. Applying the weighted L^2-estimates to χw, we get:

$$h^2\|\nabla(e^{\Phi/h}w)\|^2_{L^2(K)} + h\|e^{\Phi/h}w\|^2_{L^2(K)}$$
$$\le C(\frac{1}{h}\|e^{\Phi_\epsilon/h}(\chi(P - hE)w + [P, \chi]w)\|^2 + h\|e^{\Phi_\epsilon/h}w\|_{L^2(\{d(x)\le Ch\})})$$
$$= \mathcal{O}(h^\infty).$$

The proof is complete.   #

Notes

Proposition 6.2 is a semi-classical version of results of Lithner [Li] and Agmon [Ag]. In the one dimensional case, sharp estimates can be obtained by using techniques of ordinary differential equations, see Gérard–Grigis [GeGr], Harrel [Ha1], Kirsch–Simon [KiSi1] and Nakamura [Na]. In higher dimensions we mention in addition to the work of [HeSj1,2] on which this chapter is based, the papers of Simon [Si1,2], Kirsch–Simon [KiSi2], Martinez [M1,2]. The case of Schrödinger operators with magnetic field has been studied by B.Helffer–Sjöstrand [HeSj8].

Results concerning the tunnel effect in periodic semi-classical situations are due to Harrel [Ha2], Simon [Si3], A.Outassourt [Ou], and the more general case of infinitely many potential wells has been treated by Carlsson [Car]. A broad survey of the tunnel effect and interaction matrices can be found in the book of Helffer [He].

7. h-pseudodifferential operators

Let $V \simeq \mathbf{R}^n$ be an n-dimensional vector space and let V' be its dual. The pseudo-differential operators (from now on pseudors for short) that we consider will be operators from $\mathcal{S}(V)$ to $\mathcal{S}'(V)$ (and most of the time with better continuity properties), associated with a symbol $a(x, \xi)$ defined on $V \times V'$ by means of a rule of quantization, depending on a fixed parameter $t \in [0, 1]$ and on a "Planck's constant" $h \in]0, 1]$. Formally the operator associated with a is

$$Au(x) = \mathrm{Op}_{h,t}(a)u(x) = \frac{1}{(2\pi h)^n} \iint e^{i(x-y)\cdot\eta/h} a(tx + (1 - t)y, \eta)u(y)dyd\eta.$$

(7.1)

Here the case $t = 1/2$ corresponds to the Weyl quantization, which has the most pleasant general properties, and the the case $t = 1$ corresponds to the standard quantization, which is traditionally used in the theory of partial differential equations. In the formula (7.1) dy denotes the Lebesgue measure associated with some basis in V, and $d\eta$ denotes the Lebesgue measure associated with the corresponding dual basis. Then $dyd\eta$ is invariantly defined and is indeed the symplectic volume element associated with the invariantly defined symplectic 2 form $\sigma(x, \xi; y, \eta) = \xi \cdot y - x \cdot \eta$ on $(V \times V')^2$. Here we write $\xi \cdot y = \xi y = \langle \xi, y \rangle = \sum \xi_i y_i$.

Proposition 7.1. *If $a \in \mathcal{S}(V \times V')$, then $\mathrm{Op}_{h,t}(a)$ is continuous $\mathcal{S}' \to \mathcal{S}$.*

Proof. If we work with the h-Fourier transform, $\mathcal{F}_h u(\xi) = \mathcal{F}u(\xi/h) = \int e^{-ix\cdot\xi/h} u(x)dx$, then the distribution kernel of $A = \mathrm{Op}_{h,t}(a)$ is

$$K_A(x, y) = \check{a}(tx + (1 - t)y, x - y),$$

(7.2)

where $\check{a}(x, y) = (\mathcal{F}_h^{-1})_{\eta \to y}(a(x, \eta))$ denotes the partial inverse Fourier transform in the second variable. Hence $\check{a} \in \mathcal{S}$, so $K_A \in \mathcal{S}$ and the proposition follows.   #

Notice that if we define the adjoint $A^* : \mathcal{S}' \to \mathcal{S}$ by $(Au|v) = (u|A^*v)$, $u, v \in \mathcal{S}$, $(u|v) = \int u\bar{v}dx$, then

$$A^* = \mathrm{Op}_{h,1-t}(\bar{a}) \text{ if } A = \mathrm{Op}_{h,t}(a).$$

(7.3)

Proposition 7.2. *If $a \in \mathcal{S}'(V \times V')$, then $A = \mathrm{Op}_{h,t}(a)$ is well-defined and continuous: $\mathcal{S} \to \mathcal{S}'$ in the sense that A maps a convergent sequence in $\mathcal{S}$ to a weakly convergent sequence in $\mathcal{S}'$.*

Proof. In this case we can define the distribution kernel $K_A \in S'$ by (7.2) and the corresponding operator has then the required continuity property.#

It is also easy to see that if a_j is a (weakly) convergent sequence in S', then $\mathrm{Op}_{h,t}(a_j)$ is convergent in the space of continuous operators $S \to S'$. Using this remark, many of the arguments below can be justified, by approximation of a given symbol in S' by a sequence of symbols in S.

As a converse to the last proposition, one can show that if $A : S(V) \to S'(V)$ is continuous, then the corresponding distribution kernel K_A belongs to $S'(V \times V)$, and for given h, t as above we have a unique $\breve{a}$ satisfying (7.2), given by

$$\breve{a}(x,y) = K_A(x + (1-t)y, x - ty), \qquad (7.4)$$

so $A = \mathrm{Op}_{h,t}(a)$ for a uniquely determined $a \in S'(V \times V')$, given by

$$a(x,\eta) = \int e^{-iy\cdot\eta/h} K_A(x + (1-t)y, x - ty)dy. \qquad (7.5)$$

Notice also that $\mathrm{Op}_{h,t}(a) = \mathrm{Op}_{1,t}(a(\cdot, h(\cdot\cdot)))$. In the case of the Weyl quantization, we shall use the simplified notation $\mathrm{Op}_{h,\frac{1}{2}}(a) = \mathrm{Op}_h(a) = a^w(x, hD_x)$. Similarly for the standard quantization ($t = 1$) we sometimes write $\mathrm{Op}_{h,1}(a) = a(x, hD_x)$.

In order to give some motivation for the Weyl quantization, we consider the real linear form $L(x,\xi) = x^* \cdot x + \xi^* \cdot \xi$, $x^* \in V'$, $\xi^* \in V$. Then $\mathrm{Op}_{h,t}(L)$ is independent of t and equals $L(x, hD) = x^* \cdot x + \xi^* \cdot hD_x$. This operator is symmetric, when equipped with the domain $S(V)$. We claim that L is essentially selfadjoint. In fact, let $u \in D(L^*)$, and let $v = L^*u \in L^2$. Then we first see that $Lu = v$ in the sense of distributions. Let $\chi \in S(V)$, $\chi(0) = 1$, and put $u_\epsilon = \chi(\epsilon x)\chi(\epsilon D)u \in S(V)$, for $\epsilon > 0$. Then, if $\|\cdot\|$ denotes the norm in L^2, we get

$$\|u - u_\epsilon\| \leq \|\chi(\epsilon x)(\chi(\epsilon D) - 1)u\| + \|(\chi(\epsilon x) - 1)u\| \to 0,$$

$$\|Lu_\epsilon - v\|$$
$$\leq \|[L, \chi(\epsilon x)]\chi(\epsilon D)u\| + \|\chi(\epsilon x)[L, \chi(\epsilon D)]u\| + \|\chi(\epsilon x)\chi(\epsilon D)v - v\| \to 0,$$

when $\epsilon \to 0$, using Parseval's formula to see that $\|[L, \chi(\epsilon D)]u\| \to 0$. It follows that L^* is the closure of L, so L is essentially selfadjoint, as claimed.

Consider now the problem of constructing the unitary group:

$$U_t = e^{-itL(x,hD)/h}.$$

In other words, for $u \in S$ we want to find $v(t,x)$, C^1 in t with values in S such that $hD_tv(t,x) + L(x,hD)v = 0$, $v(0,x) = u(x)$ where u is an arbitrary element of S. Indeed, by the spectral theorem, if $u \in \mathcal{D}(L)$, then $v(t) = U_tu$ is the unique solution in $C^0(\mathbf{R}; \mathcal{D}(L)) \cap C^1(\mathbf{R}; L^2)$ of the initial value problem above. We try $v(t,x) = \mathrm{Op}_h(e^{-itL(x,\xi)/h})u =: \widetilde{U_t}u$. Then $v(0,x) = u(x)$, and if we write $a_t = \exp(-itL(x,\xi)/h)$, we have

$$hD_t\widetilde{U_t}u = \mathrm{Op}_h(hD_t(a_t))u, \ hD_t(a_t) = -L(x,\xi)e^{-itL(x,\xi)/h}. \tag{7.6}$$

Moreover, we may apply $L(x,hD)$ inside the sign of integration and obtain

$$L(x,hD)\widetilde{U_t}u(x)$$
$$= \frac{1}{(2\pi h)^n} \iint e^{i(x-y)\cdot\eta/h} L(x,\eta+hD)(a_t(\frac{x+y}{2},\eta))u(y)dyd\eta.$$

Here

$$L(x,\eta+hD)(a_t(\frac{x+y}{2},\eta)) =$$
$$L(\frac{x+y}{2},\eta)a_t(\frac{x+y}{2},\eta) + \frac{1}{2}x^* \cdot (x-y)a_t(\frac{x+y}{2},\eta) + \frac{h}{2i}(\xi^* \cdot \partial_x a_t)(\frac{x+y}{2},\eta).$$

The second term contributes to the integral by

$$\frac{1}{(2\pi h)^n} \iint \frac{1}{2}(x^* \cdot hD_\eta)(e^{i(x-y)\cdot\eta/h})a_t(\frac{x+y}{2},\eta)u(y)dyd\eta$$
$$= \frac{1}{(2\pi h)^n} \iint e^{i(x-y)\cdot\eta/h}(-\frac{1}{2}(x^* \cdot hD_\eta)a_t)(\frac{x+y}{2},\eta)u(y)dyd\eta.$$

More generally we have proved that for any $a \in S'$:

$$L(x,hD)\mathrm{Op}_h(a) = \mathrm{Op}_h(b), \text{ where } b(x,\xi) = L(x,\xi)a(x,\xi) + \frac{h}{2i}\{L,a\}. \tag{7.7}$$

For the special symbol a_t (and more generally for any symbol of the form $f(L(x,\xi))$), we have $\{L,a_t\} = 0$, so

$$L(x,hD)\mathrm{Op}_h(a_t) = \mathrm{Op}_h(La_t). \tag{7.8}$$

Combining this with (7.6), we get $(hD_t + L(x,hD))v = 0$, $v(0,x) = u$. It will follow from Lemma 7.8 below that v is a C^∞ function of t with values in $S \subset \mathcal{D}(L)$, so we obtain

$$U_tu = \mathrm{Op}_h(e^{-itL/h})u, \ u \in S. \tag{7.9}$$

When $L(x,\xi) = x^* \cdot x$, we have $e^{-itL(x,hD)/h} = e^{-itx^*\cdot x/h}$ (multiplication operator) and when $L(x,\xi) = \xi^* \cdot \xi$, we get $e^{-itL(x,hD)/h} = e^{-t\xi^*\cdot\partial_x} = \tau_{t\xi^*}$, where $\tau_{t\xi^*}u(x) = u(x - t\xi^*)$ is the operator of translation by the vector $t\xi^*$.

From (7.9) we get the formula:

$$e^{-iL(x,hD)/h} = e^{-ix^* \cdot x/2h} T_{\xi^*} e^{-ix^* \cdot x/2h}, \qquad (7.10)$$

and it follows that if $M(x,\xi) = y^* \cdot x + \eta^* \cdot \xi$ is a second real linear form, then

$$e^{-iL(x,hD)/h} e^{-iM(x,hD)/h} = e^{i\{L,M\}/2h} e^{-i(L+M)(x,hD)/h}. \qquad (7.11)$$

Heuristically, we associate $e^{-iL(x,hD)/h}$ with the mapping $(x,\xi) \mapsto (x,\xi) + H_L$, where $H_L = (\xi^*, -x^*)$ is the Hamilton vector of L. Notice also that $L(x,\xi) = \sigma(x,\xi; H_L)$.

Using (7.9), we shall decompose a general pseudor. Assume that $a(x,\xi) \in \mathcal{S}(V \times V')$ and start by writing Fourier's inversion formula:

$$a(x,\xi) = \frac{1}{(2\pi h)^{2n}} \iint e^{i(x \cdot x^* + \xi \cdot \xi^*)/h} \, \widehat{a}(x^*,\xi^*) dx^* d\xi^*, \qquad (7.12)$$

where

$$\widehat{a}(x^*,\xi^*) = \iint e^{-i(x \cdot x^* + \xi \cdot \xi^*)/h} a(x,\xi) dx d\xi$$

is the h-Fourier transform. Taking the Weyl quantization, we get

$$a^w(x,hD) = \frac{1}{(2\pi h)^{2n}} \iint \widehat{a}(x^*,\xi^*) e^{\frac{i}{h}(x^* \cdot x + \xi^* \cdot hD)} dx^* d\xi^*$$

$$= \frac{1}{(2\pi h)^{2n}} \int \widehat{a}(\ell) e^{\frac{i}{h}\ell(x,hD)} d\ell, \qquad (7.13)$$

with uniform convergence in the space $\mathcal{L}(L^2, L^2)$ since the operators $e^{i\ell(x,hD)/h}$ are unitary.

Composition of symbols. Let $a, b \in \mathcal{S}(V \times V')$. Using the preceding representation, we get

$$a^w \circ b^w = \frac{1}{(2\pi h)^{4n}} \iint \widehat{a}(\ell) \widehat{b}(m) e^{\frac{i}{h}\ell(x,hD)} \circ e^{\frac{i}{h}m(x,hD)} d\ell dm$$

$$= \frac{1}{(2\pi h)^{4n}} \iint \widehat{a}(\ell) \widehat{b}(m) e^{\frac{i}{2h}\{\ell,m\}} e^{\frac{i}{h}(\ell+m)(x,hD)} d\ell dm$$

$$= \frac{1}{(2\pi h)^{2n}} \int \widehat{c}(r) e^{\frac{i}{h}r(x,hD)} dr = c^w,$$

where

$$\widehat{c}(r) = \frac{1}{(2\pi h)^{2n}} \int_{\ell+m=r} \widehat{a}(\ell) \widehat{b}(m) e^{\frac{i}{2h}\{\ell,m\}} d\ell.$$

We shall show that $c = \tilde{c}$, where

$$\tilde{c}(x) = (e^{\frac{ih}{2}\sigma(D_x;D_y)}a(x)b(y))|_{y=x} = (e^{\frac{i}{2h}\sigma(hD_x;hD_y)}a(x)b(y))|_{y=x},$$

and where we shorten the notation by letting x, y denote points in $V \times V'$. More explicitly,

$$\tilde{c}(x) = \frac{1}{(2\pi h)^{4n}} \iint e^{\frac{i}{h}(x^* \cdot x + y^* \cdot x + \frac{1}{2}\sigma(x^*;y^*))}\widehat{a}(x^*)\widehat{b}(y^*)dx^*dy^*,$$

so

$$\widehat{\tilde{c}}(r) =$$
$$\frac{1}{(2\pi h)^{2n}} \iint (\frac{1}{(2\pi h)^{2n}} \int e^{\frac{i}{h}(x^* \cdot x + y^* \cdot x - r \cdot x)}dx)e^{\frac{i}{2h}\sigma(x^*;y^*)}\widehat{a}(x^*)\widehat{b}(y^*)dx^*dy^*.$$

Here the parenthesis is equal to $\delta(x^* + y^* - r)$, and $\sigma(x^*; y^*) = \{x^*, y^*\}$ when x^*, y^* are identified with linear forms. We then see that $\widehat{\tilde{c}} = \widehat{c}$, so that $\tilde{c} = c$. Going back to the original notation, we have shown:

Theorem 7.3. *If $a, b \in \mathcal{S}(V \times V')$, then $a^w(x, hD)b^w(x, hD) = c^w(x, hD)$, where*

$$c(x, \xi) = e^{\frac{ih}{2}\sigma(D_x, D_\xi; D_y, D_\eta)}(a(x, \xi)b(y, \eta))|_{y=x, \eta=\xi}. \tag{7.14}$$

This result can also be proved by a more direct method (exercise) but we have chosen a method that emphasizes the advantages of the Weyl quantization. By the same more direct method one can also show that for the classical quantization $(t = 1)$, we have

$$a(x, hD)b(x, hD) = c(x, hD),$$

where

$$c(x, \xi) = e^{ihD_\xi \cdot D_y}a(x, \xi)b(y, \eta)|_{y=x, \eta=\xi}. \tag{7.15}$$

Notice that the formula (7.14), contrary to the formula (7.15), is naturally invariant under composition of the symbols by an affine canonical transformation.

We next derive a formula which connects the symbols for the different quantizations. If $\ell = x^* \cdot x + \xi^* \cdot \xi$ is a real linear form, we have

$$\mathrm{Op}_{h,t}(e^{i\ell(x,\xi)/h}) = e^{\frac{it}{h}x^* \cdot x}\tau_{-\xi^*}e^{\frac{i(1-t)}{h}x^* \cdot x} = e^{\frac{i(s-t)}{h}\xi^* \cdot x^*}\mathrm{Op}_{h,s}(e^{i\ell/h}).$$

In particular,

$$\mathrm{Op}_{h,t}(e^{i\ell(x,\xi)/h}) = e^{\frac{i}{h}(\frac{1}{2}-t)\xi^* \cdot x^*}e^{i\ell(x,hD_x)/h}.$$

Using (7.13), we get

$$a^w(x, hD)$$
$$= \frac{1}{(2\pi h)^{2n}} \iint \widehat{a}(x^*, \xi^*) e^{\frac{i}{h}(t-\frac{1}{2})x^*\cdot\xi^*} \mathrm{Op}_{h,t}(e^{i(x^*\cdot x + \xi^*\cdot\xi)}) dx^* d\xi^* = \mathrm{Op}_{h,t}(a_t),$$

where

$$a_t = e^{\frac{i}{h}(t-\frac{1}{2})hD_x\cdot hD_\xi}a = e^{ih(t-\frac{1}{2})D_x\cdot D_\xi}a.$$

From this we easily get the more general relation:

$$a_t(x, \xi) = e^{i(t-s)hD_x\cdot D_\xi}a_s(x, \xi), \tag{7.16}$$

when $\mathrm{Op}_{h,t}(a_t) = \mathrm{Op}_{h,s}(a_s)$. It is now easy to obtain the composition formula (7.15) from (7.16) and Theorem 7.3. Denote the Weyl symbols by a_w, b_w, c_w (still of class $\mathcal{S}$) and let a, b, c be the classical symbols of the same operators: $a_w^w(x, hD) = a(x, hD)$, $b_w^w(x, hD) = b(x, hD)$, $c_w^w(x, hD) = c(x, hD)$, $c_w^w(x, hD) = a_w^w(x, hD) \circ b_w^w(x, hD)$, $c(x, hD) = a(x, hD) \circ b(x, hD)$. Then

$$c = e^{\frac{ih}{2}D_x\cdot D_\xi}c_w = e^{\frac{ih}{2}D_x\cdot D_\xi}(e^{\frac{ih}{2}\sigma(D_x, D_\xi; D_y, D_\eta)}a_w(x, \xi)b_w(y, \eta))|_{y=x, \eta=\xi}$$
$$= (e^{\frac{ih}{2}((D_x+D_y)\cdot(D_\xi+D_\eta)+D_\xi\cdot D_y - D_x\cdot D_\eta - D_x\cdot D_\xi - D_y\cdot D_\eta)}a(x, \xi)b(y, \eta))|_{y=x, \eta=\xi}$$
$$= (e^{ihD_y\cdot D_\xi}a(x, \xi)b(y, \eta))|_{y=x, \eta=\xi}.$$

$$\#$$

In order to prepare the symbolic calculus of pseudors, we consider in general $e^{ihA(D)}u$, for $u \in \mathcal{S}(\mathbf{R}^n)$, where $A(\xi) = \frac{1}{2}\langle Q\xi, \xi\rangle$ is a real quadratic form on $\mathbf{R}^n$. By Fourier's inversion formula, we have

$$\left| e^{ihA(D)}u(x) - \sum_{j\leq N-1} \frac{(ihA(D))^j}{j!}u(x) \right| \leq h^N C(s, N) \sum_{|\alpha|\leq s} \|D^\alpha A(D)^N u\|_{L^2},$$

$$\tag{7.17}$$

if $s > n/2$ is an integer. (Here we also use the fact that $|e^{it} - \sum_0^{N-1}\frac{(it)^j}{j!}| \leq \frac{|t|^N}{N!}$ for $t \in \mathbf{R}$).

We also want to estimate $e^{ihA(D)}u$ away from the support of u. Assume that A is non-degenerate and let $A^{-1}(x) = \frac{1}{2}\langle Q^{-1}x, x\rangle$ be the dual quadratic form on $\mathbf{R}^n$. Then

$$e^{ihA(D)}u = \left(\frac{1}{2\pi h}\right)^{\frac{n}{2}} \frac{e^{\frac{i\pi}{4}\mathrm{sgn}\,Q}}{|\det Q|^{\frac{1}{2}}}(e^{-iA^{-1}(x)/h} * u) = K_h * u,$$

where $f * g(x) = \int_{\mathbf{R}_y^n} f(x-y)g(y)dy$.

Let $\phi(x) = -A^{-1}(x)$, so that $|\phi'(x)| \sim |x|$. Let $\chi \in C_0^\infty(\mathbf{R}^n)$ be equal to 1 near 0, and make repeated integrations by parts, using

$$^tL = (\phi'(x))^{-2}\phi'(x) \cdot hD_x. \tag{7.18}$$

Then we get

$$|((1-\chi)K_h) * u(x)| \le C_N h^{N-\frac{n}{2}} \sum_{|\alpha|\le N} \|\langle x-y\rangle^{-N}D^\alpha u(y)\|_{L_y^1}. \tag{7.19}$$

Here we write $\langle x \rangle = \sqrt{1+x^2}$.

Definition 7.4. We say that $m : \mathbf{R}^n \to [0,\infty[$ is an order function if there are constants $C_0 > 0$, $N_0 > 0$, such that $m(x) \le C_0\langle x-y\rangle^{N_0}m(y)$.

Definition 7.5. Let m be an order function on $\mathbf{R}^n$. We let $S(\mathbf{R}^n, m) = S(m)$ be the set of $a \in C^\infty(\mathbf{R}^n)$ such that for every $\alpha \in \mathbf{N}^n$, there exists $C_\alpha > 0$, such that $|\partial^\alpha a(x)| \le C_\alpha m(x)$.

Notice that the product m_1m_2 of two order functions m_1, m_2 is an order function, that $S(m)$ is a Frechet space and that the map $S(m_1) \times S(m_2) \ni (a_1,a_2) \mapsto a_1a_2 \in S(m_1m_2)$ is continuous. Also $\mathcal{S}(\mathbf{R}^n)$ is dense in $S(m)$ for the topology of $S(m\langle x\rangle^\epsilon)$ for every $\epsilon > 0$. In fact, if $a \in S(m)$, consider the sequence $a_j(x) = \chi(\frac{x}{j})a(x)$, where $\chi \in \mathcal{S}(\mathbf{R}^n)$ and $\chi(0) = 1$.

If $a = a(x;h)$ depends on $h \in]0,1]$, we say that $a \in S(m)$ if $a(\cdot;h)$ is uniformly bounded in $S(m)$ when h varies in $]0,1]$. For $k \in \mathbf{R}$, we put $S^k(m) = h^{-k}S(m)$ (in the sense that the elements of $S^k(m)$ are functions of the form $h^{-k}a(x;h)$ for $a \in S(m)$). If $\delta \in [0,1]$, we let $S_\delta^k(m)$ be the space of functions $a(x;h)$ on $\mathbf{R}^n\times]0,1]$ which belong to $S(m)$ for every fixed h and satisfy

$$|\partial^\alpha a(x)| \le C_\alpha m(x)h^{-\delta|\alpha|-k}. \tag{7.20}$$

If $a_j \in S_\delta^{k_j}(m)$, $k_j \searrow -\infty$, we say that $a \sim \sum_{j=0}^\infty a_j$, if $a - \sum_{j=0}^N a_j \in S_\delta^{k_{N+1}}(m)$ for every $N \in \mathbf{N}$. For a given sequence a_j as above, we can always find such an asymptotic sum a (by the Borel argument, explained in Chapter 2) and a is unique up to an element in $S^{-\infty}(m) := \cap_k S_\delta^k(m)$.

Proposition 7.6. *Let A be a non-degenerate quadratic form on $(\mathbf{R}^n)^*$. Let $0 \le \delta \le \frac{1}{2}$ and let m be an order function. Then $e^{ihA(D)} : \mathcal{S}' \to \mathcal{S}'$*

is continuous: $S_\delta^0(m) \to S_\delta^0(m)$. *Moreover, if* $\delta < \frac{1}{2}$, *then* $e^{ihA(D)}u \sim \sum_0^\infty \frac{(ihA(D))^k}{k!} u$ *in* $S_\delta^0(m)$, *for every* $u \in S_\delta^0(m)$.

Proof. The proof is fairly straightforward in the case $\delta < \frac{1}{2}$, using (7.17) (stationary phase) and (7.19) (integration by parts). We here only treat the limiting case $\delta = \frac{1}{2}$, and we then have to review the integration by parts argument. As already noticed,

$$e^{ihA(D)}u(x) = C_n h^{-\frac{n}{2}} \int e^{i\phi(y)/h} u(x - y)dy.$$

Here, we split the integral into two parts by using the cutoff functions $\chi(\frac{y}{\sqrt{h}})$ and $1 - \chi(\frac{y}{\sqrt{h}})$, where $\chi \in C_0^\infty(\mathbf{R}^n)$ is equal to 1 near 0. The first integral is easy to estimate, and in the second integral, we integrate by parts, using ${}^t L$ in (7.18) with x replaced by y. The second integral becomes

$$\mathcal{O}(1) h^{-\frac{n}{2}} \int |(hD_y \cdot \frac{\phi'(y)}{|\phi'(y)|^2})^N ((1 - \chi(\frac{y}{\sqrt{h}}))u(x - y))| dy.$$

Putting $y = \sqrt{h}\tilde{y}$, we get

$$\mathcal{O}(1) \int |(D_{\tilde{y}} \cdot \frac{\phi'(\tilde{y})}{|\phi'(\tilde{y})|^2})^N ((1 - \chi(\tilde{y}))u(x - \sqrt{h}\tilde{y}))| d\tilde{y},$$

and here the integrand is

$$\mathcal{O}(1) \frac{m(x - \sqrt{h}\tilde{y})}{\langle \tilde{y} \rangle^N} \leq \mathcal{O}(1) \frac{m(x)\langle \sqrt{h}\tilde{y} \rangle^{N_0}}{\langle \tilde{y} \rangle^N},$$

which is integrable with integral $\mathcal{O}(m(x))$. The derivatives can be estimated similarly, since $\partial_x^\alpha e^{ihA(D)}u = e^{ihA(D)}\partial_x^\alpha u$.   #

Replacing n by $4n$, we obtain:

Proposition 7.7. *The map*

$$\mathcal{S} \times \mathcal{S} \ni (a_1, a_2) \mapsto a_1 \natural_h a_2 := (e^{\frac{ih}{2}\sigma(D_x, D_\xi; D_y, D_\eta)} a_1(x, \xi) a_2(y, \eta))|_{y=x, \eta=\xi}$$
(7.21)
has a bilinear continuous extension: $S_\delta^0(m_1) \times S_\delta^0(m_2) \to S_\delta^0(m_1 m_2)$ *for all* $\delta \in [0, \frac{1}{2}]$ *and all order functions* m_1, m_2. *When* $\delta < \frac{1}{2}$, *we have*

$$a_1 \natural_h a_2 \sim \sum_{k=0}^\infty \frac{1}{k!} ((\frac{ih}{2}\sigma(D_x, D_\xi; D_y, D_\eta))^k a_1(x, \xi) a_2(y, \eta))|_{y=x, \eta=\xi}$$

in $S_\delta^0(m_1 m_2)$ for all $a_j \in S_\delta^0(m_j)$, $j = 1, 2$.

One can also show that the extension in Proposition 7.7 is unique for a suitable topology.

Lemma 7.8. *Let $h = 1$ and let m be an order function. For $a \in S(m)$, Op (a) is continuous $\mathcal{S} \to \mathcal{S}$, and $\mathcal{S}' \to \mathcal{S}'$, and is a continuous function of $a \in S(m)$ with values in the space of continuous operators $\mathcal{S} \to \mathcal{S}$ and in the space of continous operators $\mathcal{S}' \to \mathcal{S}'$.*

Proof. We only give an outline of the proof. It is enough to consider the continuity in $\mathcal{S}$ since the one in $\mathcal{S}'$, will follow by duality. We first consider the case when $a \in \mathcal{S}$. Using repeated integrations by parts with the help of the operators $\langle x - y \rangle^{-2}(1 + (x - y) \cdot D_\eta)$ and $\langle \eta \rangle^{-2}(1 - \eta \cdot D_y)$, we see that Op (a) is uniformly continuous $\mathcal{S} \to L^\infty$, when a varies in a bounded set in $S(m)$. By a density argument, it follows that Op (a) can be defined as a continuous operator $\mathcal{S} \to L^\infty$, when $a \in S(m)$. This result, which holds not only for the Weyl-quantization, but also for the other t-quantizations with $0 \le t \le 1$, can then easily be extended to operators of the form $x^\alpha D^\beta \mathrm{Op}\,(a)$, and the lemma follows.   #

Theorem 7.9. *Let m_1, m_2 be order functions and let $0 \le \delta \le 1/2$. For $a_j \in S_\delta(m_j)$, we have $\mathrm{Op}_h(a_1)\mathrm{Op}_h(a_2) = \mathrm{Op}_h(a_1 \sharp_h a_2)$.*

Proof. Here it is enough to treat the case of a fixed value of h, say 1, and it is then clear how to combine Proposition 7.7, Lemma 7.8 and a simple density argument.   #

Exercise. Let m be an order function, let $\delta \in [0, \frac{1}{2}]$ and $0 \le t, s \le 1$. Show that if $a_s \in S_\delta^0(m)$ and a_t is given by (7.16), then $a_t \in S_\delta^0(m)$. Use this to extend the composition result in the preceding theorem to the case of operators of the form $\mathrm{Op}_{h,t}(a)$.

We next discuss L^2-continuity. For that we shall use the Cotlar–Stein

Lemma 7.10. *Let A_1, A_2, $\ldots \in \mathcal{L}(E, F)$, where E, F are Hilbert spaces, and assume that for some $M > 0$, we have:*

$$\sup_j \sum_{k=1}^\infty \|A_j^* A_k\|^{\frac{1}{2}} \le M$$

$$\sup_j \sum_{k=1}^\infty \|A_j A_k^*\|^{\frac{1}{2}} \le M.$$

Then $A = \sum_1^\infty A_j$ converges strongly and A is a bounded operator with $\|A\| \leq M$.

Proof. We first assume that only finitely many of the A_j are different from zero: $A_j = 0$ for $j \geq N + 1$. We have $\|A\|^2 = \|A^*A\|$, and more generally from the spectral theorem for bounded selfadjoint operators, we get $\|A\|^{2m} = \|(A^*A)^m\|$. Here,

$$(A^*A)^m = \sum_{j_1,\dots,j_{2m}} A_{j_1}^* A_{j_2} A_{j_3}^* \dots A_{j_{2m}}. \qquad (7.22)$$

The norm of $A_{j_1}^* A_{j_2} A_{j_3}^* \dots A_{j_{2m}}$ is bounded by $\|A_{j_1}^* A_{j_2}\| \cdot \|A_{j_3}^* A_{j_4}\| \dots \|A_{j_{2m-1}}^* A_{j_{2m}}\|$ and also by $\|A_{j_1}^*\| \cdot \|A_{j_2} A_{j_3}^*\| \dots \|A_{j_{2m}}\|$. Taking the geometric mean of the two bounds, we get:

$$\|A_{j_1}^* A_{j_2} \dots A_{j_{2m-1}}^* A_{j_{2m}}\| \leq M \cdot \|A_{j_1}^* A_{j_2}\|^{\frac{1}{2}} \|A_{j_2} A_{j_3}^*\|^{\frac{1}{2}} \dots \|A_{j_{2m-1}}^* A_{j_{2m}}\|^{\frac{1}{2}},$$

since $\|A_j\| \leq M$. Then by (7.22):

$$\|A\|^{2m} \leq M \sum_1^N \dots \sum_1^N \|A_{j_1}^* A_{j_2}\|^{\frac{1}{2}} \|A_{j_2}^* A_{j_3}\|^{\frac{1}{2}} \dots \|A_{j_{2m-1}}^* A_{j_{2m}}\|^{\frac{1}{2}}. \qquad (7.23)$$

Here we first estimate the sum over j_{2m} and obtain

$$\|A\|^{2m} \leq M^2 \sum_{j_1,\dots,j_{2m-1}} \|A_{j_1}^* A_{j_2}\|^{\frac{1}{2}} \|A_{j_2}^* A_{j_3}\|^{\frac{1}{2}} \dots \|A_{j_{2m-2}} A_{j_{2m-1}}^*\|^{\frac{1}{2}}.$$

We then estimate the j_{2m-1}-sum and so on, and we obtain:

$$\|A\|^{2m} \leq M^{2m-1} \sum_{j_1} M = N \cdot M^{2m}.$$

Hence $\|A\| \leq N^{\frac{1}{2m}} M$, and letting m tend to infinity, we get $\|A\| \leq M$.

We then consider the general case. If $u = A_k^* v$ for some k, then

$$\sum_1^\infty \|A_j u\| = \sum_{j=1}^\infty \|A_j A_k^* v\|$$

$$\leq \sum_{j=1}^\infty \|A_j A_k^*\|^{\frac{1}{2}} M \|v\| = (\sum_{j=1}^\infty \|A_k A_j^*\|^{\frac{1}{2}}) M \|v\| < \infty,$$

so $\sum A_j u$ converges when u is in the sum Σ of the ranges of the A_k^*. Since all partial sums $\sum_1^N A_j$ are uniformly bounded, we conclude that $\sum_1^\infty A_j u$ converges, when $u \in \overline{\Sigma}$.

If $u \in \Sigma^\perp$, then u is in the intersection of the kernels of all the A_j, so we trivially have convergence in that case too. #

We can now prove:

Theorem 7.11. *Let* $a \in S_\delta^0(1)$, $0 \leq \delta \leq \frac{1}{2}$. *Then* $\mathrm{Op}_h(a)$ *is bounded:* $L^2(\mathbf{R}^n) \to L^2(\mathbf{R}^n)$ *and there is a constant* C *independent of* h *such that* $\|\mathrm{Op}_h(a)\|_{\mathcal{L}(L^2(\mathbf{R}^n), L^2(\mathbf{R}^n))} \leq C$ *for* $0 < h \leq 1$.

Proof. We first reduce the proof to the case when $h = 1$ by observing that the change of variables $x = \sqrt{h}\tilde{x}$ can be used to define a unitary equivalence between $a^w(x, hD)$ and $\mathrm{Op}(a_h)$, where $a_h(x, \xi; h) = a(\sqrt{h}x, \sqrt{h}\xi; h)$ satisfies: $|\partial_x^\alpha \partial_\xi^\beta a_h(x, \xi)| \leq C_{\alpha, \beta}$, with constants $C_{\alpha, \beta}$ which are independent of h.

It remains then to show that if $a \in S(1)$ is independent of h, then $\mathrm{Op}(a)$ is bounded as an operator in L^2 and has a norm which can be bounded by some semi-norm of a in $S(1)$. Let $\chi \in C_0^\infty(\mathbf{R}^{2n})$ have the property that $\sum_{\alpha \in \mathbf{Z}^{2n}} \chi_\alpha(x, \xi) = 1$, where $\chi_\alpha(x, \xi) = \chi((x, \xi) - \alpha)$. Put $a_\alpha = \chi_\alpha a \in S$, $A_\alpha = \mathrm{Op}(a_\alpha)$. Then, as we saw in the beginning of this chapter, $\|A_\alpha\| \leq C_n \|\hat{a}_\alpha\|_{L^1} < \infty$, where $\hat{a}_\alpha$ denotes the standard Fourier transform of a_α, and the constant C_n depends on the dimension only.

Put $b_{\alpha, \beta} = \bar{a}_\alpha \natural a_\beta$. From the estimates (7.17), (7.19), it is easy to see that for every $N \in \mathbf{N}$ and every $\gamma \in \mathbf{N}^{2n}$:

$$|\partial_{(x, \xi)}^\gamma b_{\alpha, \beta}(x, \xi)| \leq p_{N, \gamma}(a)^2 \langle \alpha - \beta \rangle^{-N} \langle (x, \xi) - \frac{\alpha + \beta}{2} \rangle^{-N},$$

where $p_{N, \gamma}$ is a semi-norm on $S(1)$. It follows that

$$\|A_\alpha^* A_\beta\| \leq C_n \|\hat{b}_{\alpha, \beta}\|_{L^1} \leq p_N(a)^2 \langle \alpha - \beta \rangle^{-N},$$

where p_N is a semi-norm on $S(1)$. From this we immediately obtain that

$$\sup_\alpha \sum_\beta \|A_\alpha^* A_\beta\|^{\frac{1}{2}} \leq p(a), \quad \sup_\alpha \sum_\beta \|A_\alpha A_\beta^*\|^{\frac{1}{2}} \leq p(a),$$

where p is some semi-norm on $S(1)$. It then suffices to apply the Cotlar–Stein Lemma. #

Theorem 7.12. *(Semi-classical sharp Gårding inequality). Let* $P(x, \xi) \in S^0(\mathbf{R}^{2n})$ *satisfy:*

$$P(x, \xi) \geq 0, \quad \forall (x, \xi) \in \mathbf{R}^{2n}. \tag{7.24}$$

Then there exist $C, h_0 > 0$ such that:

$$(P^w(x, hD_x)u|u)_{L^2(\mathbf{R}^n)} \geq -Ch \, \|u\|^2_{L^2(\mathbf{R}^n)}, \qquad (7.25)$$

for every $u \in L^2(\mathbf{R}^n)$ and all $h \in \,]0, h_0]$.

Proof. Let $Y \in \mathbf{R}^{2n}$. By Taylor's formula we have ;

$$\begin{aligned} 0 \leq P((x, \xi) + Y) \\ = P(x, \xi) + \langle \nabla_{x,\xi} P(x, \xi), Y \rangle \\ + \int_0^1 (1 - s) \langle \partial^2_{x,\xi} P((x, \xi) + sY)Y, Y \rangle ds. \end{aligned} \qquad (7.26)$$

We take $Y = -\epsilon \nabla_{x,\xi} P(x, \xi)$ with $\epsilon > 0$ sufficiently small, using (7.26) and the fact that $\partial^2_{x,\xi} P$ is uniformly bounded, and get:

$$0 \leq P(x, \xi) - \epsilon |\nabla_{x,\xi} P|^2 + C\epsilon^2 |\nabla_{x,\xi} P|^2,$$

so for $C\epsilon < \frac{1}{2}$,

$$|\nabla_{x,\xi} P|^2 \leq \frac{2}{\epsilon} P(x, \xi).$$

Consequently, we get the standard estimate for non-negative smooth functions:

$$|\partial^\alpha_{x,\xi} P(x, \xi)| \leq C \, P(x, \xi)^{\frac{1}{2}}, \; \forall \, |\alpha| = 1.$$

Hence for $0 < \lambda < 1$ and $|\alpha| = 1$ we get:

$$|\partial^\alpha_{x,\xi} P(x, \xi)| \, (P(x, \xi) + \lambda)^{-1} \leq C_\alpha \lambda^{-\frac{1}{2}}. \qquad (7.27)$$

For $\alpha, \beta \in \mathbf{N}^n$, we have:

$$\partial^\alpha_x \partial^\beta_\xi (P(x, \xi) + \lambda)^{-1} = (P(x, \xi) + \lambda)^{-1}$$

$$\times \sum_{\substack{k \geq 1, \\ |\alpha| = |\alpha^1| + \cdots + |\alpha^k|, \\ |\beta| = |\beta^1| + \cdots + |\beta^k|, \\ |\alpha^j| + |\beta^j| \geq 1}} C_{\alpha^1, \beta^1 \cdots \alpha^k, \beta^k} \prod_{j=1}^k (P(x, \xi) + \lambda)^{-1} \partial^{\alpha^j}_x \partial^{\beta^j}_\xi P(x, \xi). (7.28)$$

Using (7.27), (7.28) and the fact that

$$(P(x, \xi) + \lambda)^{-1} |\partial^\alpha_x \partial^\beta_\xi P(x, \xi)| \leq C_{\alpha, \beta} \, \lambda^{-1}$$

for $|\alpha| + |\beta| \geq 2$, we get:

$$\partial^\alpha_x \partial^\beta_\xi (P(x, \xi) + \lambda)^{-1} \leq C_{\alpha, \beta} (P + \lambda)^{-1} \lambda^{-\frac{|\alpha| + |\beta|}{2}}, \qquad (7.29)$$

for all $\alpha, \beta \in \mathbf{N}^n$. Let $C \geq 1$ be a fixed constant and put

$$Q(x, \xi; h) = (P(x, \xi) + Ch)^{-1},$$

(7.29) shows that $hQ(x, \xi; h) \in S^0_{\frac{1}{2}}(\frac{1}{C})$ uniformly for $C \geq 1$. By Taylor's formula we have

$$(P + Ch) \natural_h Q(x, \xi; h) - 1 + \frac{ih}{2} \{P + Ch, Q\}$$

$$= \int_0^1 (1 - t) e^{\frac{iht}{2} \sigma(D_x, D_\xi; D_y, D_\eta)} \left(\frac{ih}{2} \sigma(D_x, D_\xi; D_y, D_\eta)\right)^2$$

$$(P(x, \xi) + Ch) Q(y, \eta; h) dt_{|x=y, \xi=\eta}.$$

Using the fact that for every $|\alpha| \leq 2$, $h^2 \partial^\alpha_{x, \xi} Q \in S^0_{\frac{1}{2}}(\frac{1}{C})$ uniformly for $C \geq 1$, as well as the fact that $\{P + Ch, Q\} = \{P, Q\} = 0$, we get from Proposition 7.7:

$$(P + Ch) \natural_h Q(x, \xi; h) - 1 \in S^0_{\frac{1}{2}}(\frac{1}{C}),$$

uniformly for $C \geq 1$. Now from Theorem 7.11 we get:

$$(P^w(x, hD_x) + Ch) \circ Q^w(x, hD_x; h) = I + R^w(x, hD_x; h),$$

with $\|R^w(x, hD_x; h)\|_{\mathcal{L}(L^2, L^2)} = \mathcal{O}(\frac{1}{C})$. Consequently for C large enough $(P^w(x, hD_x) + Ch) \circ Q^w(x, hD_x; h) \circ (I + R^w)^{-1} = I$, then $\sigma(P^w(x, hD_x)) \subset [-Ch, +\infty[$. From this we get (7.25).   #

Appendix: Metaplectic operators

This theory is an easy model theory to the theory of Fourier integral operators (sometimes fouriors from now on) and contains all the essential ideas. Let $\Phi(x, \theta)$ be a real quadratic form on $\mathbf{R}^n \times \mathbf{R}^N$ and assume

$$d\partial_{\theta_1} \Phi, \ldots, d\partial_{\theta_N} \Phi \text{ are linearly independent.} \qquad (A.1)$$

Then $C_\Phi := \{(x, \theta); \Phi'_\theta(x, \theta) = 0\}$ is an n-dimensional space, and it is easy to see that the linear map

$$C_\Phi \ni (x, \theta) \mapsto (x, \Phi'_x(x, \theta)) \in \mathbf{R}^n \times \mathbf{R}^n = T^* \mathbf{R}^n \qquad (A.2)$$

is injective and that the image Λ_Φ is a (linear) Lagrangian space. Indeed, the injectivity is a direct consequence of (A.1), and if we identify Λ_Φ with C_Φ by means of (A.2), we get

$$\xi \cdot dx_{|\Lambda_\Phi} \simeq \Phi'_x \cdot dx_{|C_\Phi} = (\Phi'_x \cdot dx + \Phi'_\theta \cdot d\theta)_{|C_\Phi} = d(\Phi_{|C_\Phi}),$$

so $\xi \cdot dx_{|\Lambda_\Phi}$ is closed and hence $\sigma_{|\Lambda_\Phi} = d(\xi \cdot dx)_{|\Lambda_\Phi} = 0$.

It also follows from (A.1), that $d\Phi(x,\theta) \neq 0$ for $x = 0$, $\theta \neq 0$, and making integrations by parts by means of the operators

$$(1 + (\Phi_x')^2 + (\Phi_\theta')^2)^{-1}(1 + \Phi_x' \cdot hD_x + \Phi_\theta' \cdot hD_\theta)$$

in a region $|x| \leq \mathcal{O}(1)|\theta|$, we see that

$$\langle I_\Phi, \phi \rangle = h^{-\frac{N}{2}-\frac{n}{4}} \iint e^{\frac{i}{\hbar}\Phi(x,\theta)} \phi(x) dx d\theta, \tag{A.3}$$

is well defined for $\phi \in \mathcal{S}(\mathbf{R}^n)$ (for instance as the limit when $\epsilon \to 0$ of the corresponding expression with the additional convergence factor $\chi(\epsilon(x,\theta))$, where $\chi \in \mathcal{S}(\mathbf{R}^{n+N})$, $\chi(0,0) = 1$), and defines a distribution $I_\Phi \in \mathcal{S}'(\mathbf{R}^n)$. Formally,

$$I_\Phi(x) = h^{-\frac{N}{2}-\frac{n}{4}} \int e^{\frac{i}{\hbar}\Phi(x,\theta)} d\theta. \tag{A.4}$$

Lemma A.1. *For almost all* [*] *real quadratic forms $q(x)$ on $\mathbf{R}^n$, $\Phi(x,\theta)-q(x)$ is a non-degenerate quadratic form on $\mathbf{R}^{n+N}$.*

Proof. $\Phi(x,\theta) - q(x)$ is degenerate precisely when $\det(\Phi'' - \begin{pmatrix} q_{xx}'' & 0 \\ 0 & 0 \end{pmatrix}) = 0$, which is an algebraic condition on q, so it is enough to find one quadratic form q which does not fulfill this condition, and for that, it is enough to find a complex q. Choose q with $\mathrm{Im}\, q < 0$ (negative definite). If $t = (t_x, t_\theta) \in \mathbf{C}^{n+N}$ is in the kernel of the Hessian of $\Phi - q$, we have

$$0 = \mathrm{Im}\,\langle(\Phi'' - \begin{pmatrix} q_{xx}'' & 0 \\ 0 & 0 \end{pmatrix})t, \overline{t}\rangle = -\mathrm{Im}\,\langle q''t_x, \overline{t_x}\rangle \geq \frac{1}{C}\|t_x\|^2,$$

so $t_x = 0$. Then we get $\Phi_{x\theta}'' t_\theta = 0$, $\Phi_{\theta\theta}'' t_\theta = 0$, which implies that $t_\theta = 0$, by (A.1).   #

Fix q as in the lemma. Then for every $\eta \in \mathbf{R}^n$, the function

$$(x,\theta) \mapsto \Phi(x,\theta) - q(x) - x \cdot \eta \tag{A.5}$$

has a unique critical point $(x(\eta), \theta(\eta))$. Equivalently, for every $\eta \in \mathbf{R}^n$, the (affine) Lagrangian space $\Lambda_\eta := \{(x,\xi); \xi = \eta + q'(x)\}$ intersects Λ_Φ at a unique point $(x(\eta), \xi(\eta))$ (the image of $(x(\eta), \theta(\eta)) \in C_\Phi$, under (A.2)).

---

[*] The precise sense of this is given in the proof

Moreover, the intersection is transversal. Let $H(\eta)$ be the critical value of (A.5). Then,

$$H'(\eta) = \left(\frac{\partial}{\partial \eta}(\Phi(x,\theta) - q(x) - x \cdot \eta)\right)_{(x,\theta)=(x(\eta),\theta(\eta))} = -x(\eta), \qquad (A.6)$$

and since $H(\eta)$ is a quadratic form, we see that $H(\eta)$ only depends on $\Lambda = \Lambda_\Phi$ and on q, but not on the choice of Φ for which $\Lambda = \Lambda_\Phi$.

Using the method of stationary phase (in its exact quadratic version) we see that

$$\mathcal{F}_h(e^{-iq(x)/h}I_\Phi)(\eta) = C_{\Phi,q}h^{n/4}e^{iH(\eta)/h}, \qquad (A.7)$$

where $C_{\Phi,q} \neq 0$ is independent of h. We conclude that if $\widetilde{\Phi}$ *is a second phase satisfying (A.1), such that* $\Lambda_{\widetilde{\Phi}} = \Lambda_\Phi$, *then there exists a constant* $C \neq 0$, *independent of* h, *such that* $I_\Phi = CI_{\widetilde{\Phi}}$. In a more fancy way, we can say that with Λ_Φ, we have associated a one-dimensional space of functions of h with values in $\mathcal{S}'$: $\{h \mapsto aI_\Phi; a \in \mathbf{C}\}$, which depends on Λ_Φ, but not on Φ.

To complete the picture, let us verify that if Λ is a linear Lagrangian space, then $\Lambda = \Lambda_\Phi$ for some Φ satisfying (A.1).

If $q(x)$ is a quadratic form, then the non-transversal intersection between Λ and $\Lambda_q = \{(x,\xi); \xi = q'(x)\}$ can be expressed as an algebraic condition on q. On the other hand, if we take q complex with $\operatorname{Im} q > 0$, then $\frac{1}{i}\langle\sigma, t \wedge \bar{t}\rangle > 0$ for all $t \in \Lambda_q$ (now viewed as a complex Lagrangian space), while it is easy to see that $\frac{1}{i}\langle\sigma, t \wedge \bar{t}\rangle = 0$ for all t in the complexification of Λ. Hence the complexifications of Λ and Λ_q intersect transversally. As in the proof of Lemma A.1, we conclude that Λ_q and Λ intersect transversally for most real quadratic forms q. We fix such a form q.

Define Λ_η as before (parallel to Λ_q). Then Λ and Λ_η intersect at a unique point $(x(\eta), \xi(\eta))$, and we can parameterize Λ by

$$\mathbf{R}^n \ni \eta \mapsto (x(\eta), \xi(\eta)) \in \Lambda. \qquad (A.8)$$

Using the fact that $\xi(\eta) = \eta + q'(x(\eta))$, we get

$$d(-x(\eta) \cdot d\eta) = \sum d\eta_j \wedge dx_j(\eta)$$
$$= \sum d(\xi_j(\eta)) \wedge d(x_j(\eta)) - \sum d(\partial_{x_j}q)(x(\eta)) \wedge d(x_j(\eta))$$
$$\simeq \sigma_{|\Lambda} - d(\sum(\partial_{x_j}q)(x(\eta))d(x_j(\eta))) = 0 - d^2(q(x(\eta))) = 0.$$

Let $H(\eta)$ be the real quadratic form with $dH(\eta) = -x(\eta) \cdot d\eta$, and put $\Phi(x,\eta) = x \cdot \eta + H(\eta) + q(x)$. Clearly Φ satisfies (A.1), and the corresponding critical space C_Φ is given by $x = -H'(\eta) = x(\eta)$, and Λ_Φ is the space of all $(x(\eta), \eta + q'(x(\eta))) = (x(\eta), \xi(\eta))$, so $\Lambda_\Phi = \Lambda$.

Let $L(x, \xi) = \sum x_j^* x_j + \sum \xi_j^* \xi_j = x^* \cdot x + \xi^* \cdot \xi$ be a linear form which vanishes on the linear Lagrangian space $\Lambda = \Lambda_\Phi$. Then

$$L(x, hD)I_\Phi = h^{-\frac{N}{2}-\frac{n}{4}} \int e^{i\Phi(x,\theta)/h}(x^* \cdot x + \xi^* \cdot \partial_x \Phi(x, \theta))d\theta.$$

Here the parenthesis in the integral is a linear form which vanishes on C_Φ and is therefore of the form $\sum_1^N a_j \frac{\partial \Phi}{\partial \theta_j}$ for some constants a_j. Hence,

$$L(x, hD)I_\Phi = \sum_1^N a_j h^{-\frac{N}{2}-\frac{n}{4}} \int e^{i\Phi(x,\theta)/h} \frac{\partial \Phi}{\partial \theta_j} d\theta = 0,$$

where the last identity follows by integrations by parts. (All the integrals are here given a sense as for the one in (A.3).)

As a special case, we consider operators. Replace n by $2n$ and $\mathbf{R}^n$ by $\mathbf{R}^{2n} = \mathbf{R}_x^n \times \mathbf{R}_y^n$. Let $\Phi = \Phi(x, y; \theta)$ satisfy (A.1). If

$$\Lambda'_\Phi = \{(x, \xi; y, -\eta); \ (x, \xi; y, \eta) \in \Lambda_\Phi\}$$

is the graph of a linear map and hence of a linear canonical transformation κ, and we let $J_\Phi : S \to S'$ be the operator with distribution kernel I_Φ, then we say that J_Φ is associated with κ. It is easy to see that in this case $J_\Phi : S \to S$. Conversely, if $\kappa : T^*\mathbf{R}^n \to T^*\mathbf{R}^n$ is a linear canonical transformation, then (graph κ)' is a linear Lagrangian space. Thus to every linear canonical transformation we have have a corresponding one-dimensional linear space of mappings $h \mapsto aJ_\Phi$, $a \in \mathbf{C}$, where $J_\Phi : S \to S$. We have:

– $\kappa = $ id corresponds the space of multiples of the identity operator.

– If J_Φ corresponds to κ, then the complex adjoint J_Φ^* corresponds to κ^{-1}.

– If J_{Φ_j} corresponds to κ_j, $j = 1, 2$, then $J_{\Phi_1} \circ J_{\Phi_2} = J_\Phi$ corresponds to $\kappa_1 \circ \kappa_2$. In fact, we get $\Phi(x, y; z, \theta_1, \theta_2) = \Phi_1(x, z, \theta_1) + \Phi_2(z, y, \theta_2)$, and we check that this function satisfies (A.1) with $\theta = (z, \theta_1, \theta_2)$ etc.

Combining these facts, we see that $J_\Phi^* J_\Phi = J_\Phi J_\Phi^* = C_\Phi I$, where $C_\Phi > 0$. Hence J_Φ extends to an operator which is bounded in L^2, and if $0 \neq a \in \mathbf{C}$ satisfies $|a|^2 C_\Phi = 1$, aJ_Φ is unitary. We say that aJ_Φ *is a metaplectic operator.*

From the earlier discussion, we see that if $L(x, \xi)$, $M(x, \xi)$ are linear forms with $L \circ \kappa = M$, and J_Φ is associated with κ, then $LJ_\Phi = J_\Phi M$, for if

$I_\Phi(x, y)$ is the kernel of J_Φ, then the kernel of $LJ_\Phi - J_\Phi M$ is $(L(x, hD_x) - M(y, -hD_y))I_\Phi$, and $L(x, \xi) - M(y, -\eta)$ vanishes on $C_\Phi = (\text{graph } \kappa)'$. Under the same assumptions, it follows that $e^{itL}J_\Phi = J_\Phi e^{itM}$, and since Weyl quantizations of symbols are superpositions of operators of the form $e^{itL(x, hD)}$ and the Weyl symbol of such an operator is $e^{itL(x, \xi)}$, we get

Theorem A.2. *Let* $a \in S(m)$, *where* m *is an order function on* $T^*\mathbf{R}^n$ *and let* $\kappa : T^*\mathbf{R}^n \to T^*\mathbf{R}^n$ *be a linear canonical transformation. Let* J_Φ *be an associated operator as above. Then* $\mathrm{Op}_h(a)J_\Phi = J_\Phi\mathrm{Op}_h(b)$, *where* $b = a \circ \kappa \in S(m \circ \kappa)$.

Notes

Among the books devoted to the theory of pseudodifferential operators and related topics, we can mention: Hörmander [Hö4], Treves [Tr], Taylor [Ta], Alinhac and Gérard [AlGé], Robert [Ro1], Ivrii [I1], Grigis–Sjöstrand [GrSj]. Many symbol classes have been introduced since the work by Kohn-Nirenberg [Ko-Ni]: Hörmander [Hö5], Beals–Fefferman [BeFe], [Be], [Sj2]. For the Gårding inequality and its various extensions, see [Gå], [Hö6], [LaNi], [CoFe], [Ta], [FePh], [Sj1]. The particular approach to Weyl-quantization of this chapter is inspired by [BGH], though the ideas are classical and many other references doubtless could be added. The same ideas appear in the study of operators with magnetic fields, see [HeSj4] and further references given there.

8. Functional calculus for pseudodifferential operators

If $f(\lambda)$ is a bounded continuous function on $\mathbf{R}$ and H is a selfadjoint operator, then by the spectral theorem (see Chapter 4) $f(H)$ is a well defined bounded operator. It is useful to obtain a more precise description of $f(H)$. Under appropriate conditions Strichartz [Str] showed that a function of an elliptic pseudor on a compact manifold is also a pseudor (see also Taylor [Ta]). The purpose of this chapter is to show that a smooth function of an h-pseudor H is also an h-pseudor. This result was obtained by Helffer–Robert via Mellin transformation. Our method is based on a standard Cauchy formula (see Theorem 8.1 below). One of its main advantages is that it allows us to pass easily from resolvent estimates to estimates of other functions of H. To know the properties of the resolvent as a pseudodifferential operator we shall use a characterization of pseudodifferential operators due to Beals [Be], and adapted to the h-pseudodifferential setting by Helffer–Sjöstrand [HeSj5], see also Robert [Ro1], Bony–Chemin [BonCh]. Our method works in the case of functions of several variables and we discuss functional calculus of several commuting selfadjoint operators. For further results in this direction, see Charbonnel [Char1], Colin de Verdière [CdV3] and the recent paper of Andersson [An].

If $f \in C_0^\infty(\mathbf{R})$ we can find an almost analytic extension $\tilde{f} \in C_0^\infty(\mathbf{C})$ with the properties

$$|\bar{\partial}\tilde{f}| \leq C_N |\text{Im } z|^N, \, \forall N \geq 0, \tag{8.1}$$

$$\tilde{f}_{|\mathbf{R}} = f. \tag{8.2}$$

Here $\bar{\partial} = \frac{\partial}{\partial \bar{z}} = \frac{1}{2}(\frac{\partial}{\partial x} + i\frac{\partial}{\partial y})$.

This idea was introduced by Hörmander [Hö2] and has subsequently been used by many people: Nirenberg [Ni], Melin–Sjöstrand [MeSj], Maslov [Ma2], Kucherenko [Ku], Dyn'kin [Dy], [Dy2], etc. The original approach by Hörmander was to adapt the Borel construction and put $\tilde{f}(x + iy) = \sum \frac{f^{(k)}(x)}{k!}(iy)^k \chi(\lambda_k y)$ with $\chi \in C_0^\infty(\mathbf{R})$ equal to 1 near 0 and with λ_k tending to $+\infty$ sufficiently fast when $k \to \infty$. A second construction was introduced by Mather [Mat] and more recently by Jensen and Nakamura [JeNa]. If $\psi(x) \in C_0^\infty$ is equal to 1 in a neighborhood of supp (f), and if χ is a standard cutoff function as above, then we can put

$$\tilde{f}(x + iy) = \frac{\psi(x)}{2\pi} \int e^{i(x+iy)\xi} \chi(y\xi) \hat{f}(\xi) d\xi,$$

where $\hat{f}$ is the Fourier transform of f. We check that the last formula produces an almost analytic extension (i.e. a function which verifies (8.1), (8.2) and which can be further truncated in y). First, we see that (8.2) follows from the Fourier inversion formula. Moreover,

$$\overline{\partial}\widetilde{f} = \frac{i}{2}\frac{\psi(x)}{2\pi}\int e^{i(x+iy)\xi}\chi'(y\xi)\xi\widehat{f}(\xi)d\xi + \frac{1}{2}\frac{\psi'(x)}{2\pi}\int e^{i(x+iy)\xi}\chi(y\xi)\widehat{f}(\xi)d\xi$$

$$= y^N\frac{i}{2}\frac{\psi(x)}{2\pi}\int e^{i(x+iy)\xi}\chi_N(y\xi)\xi^{N+1}\widehat{f}(\xi)d\xi$$

$$+ \frac{1}{2}\frac{\psi'(x)}{2\pi}\iint e^{i(x+iy-\widetilde{x})\xi}\chi(y\xi)f(\widetilde{x})d\widetilde{x}d\xi$$

$$= \text{I} + \text{II},$$

where $\chi_N(t) = t^{-N}\chi'(t) \in C_0^\infty(\mathbf{R})$. Here $|\text{I}| \le C_N|y|^N\|\xi^{N+1}\widehat{f}(\xi)\|_{L^1}$. To estimate II, we use the fact that $x - \widetilde{x} \neq 0$, on the support of $\psi'(x)f(\widetilde{x})$, and get by integrations by parts:

$$\text{II} = \frac{1}{4\pi}\psi'(x)\iint D_\xi(e^{i(x-\widetilde{x}+iy)\xi})\frac{\chi(y\xi)}{x-\widetilde{x}+iy}f(\widetilde{x})d\widetilde{x}d\xi$$

$$= \frac{i\psi'(x)}{4\pi}\iint e^{i(x-\widetilde{x}+iy)\xi}\frac{\chi'(y\xi)y}{x-\widetilde{x}+iy}f(\widetilde{x})d\widetilde{x}d\xi$$

$$= \frac{i\psi'(x)}{4\pi}y^N\iint e^{i(x-\widetilde{x}+iy)\xi}\xi^N(\xi+i)^2\frac{\chi_N(y\xi)y}{(x-\widetilde{x}+iy)(\xi+i)^2}f(\widetilde{x})d\widetilde{x}d\xi$$

$$= \frac{i\psi'(x)}{4\pi}y^N\iint (i-D_{\widetilde{x}})^2(-D_{\widetilde{x}})^N(e^{i(x-\widetilde{x}+iy)\xi})\frac{\chi_N(y\xi)y}{x-\widetilde{x}+iy}f(\widetilde{x})\frac{1}{(\xi+i)^2}d\widetilde{x}d\xi$$

$$= \frac{i\psi'(x)y^N}{4\pi}\iint e^{i(x-\widetilde{x}+iy)\xi}\frac{\chi_N(y\xi)y}{(\xi+i)^2}(i+D_{\widetilde{x}})^2 D_{\widetilde{x}}^N(\frac{f(\widetilde{x})}{x-\widetilde{x}+iy})d\widetilde{x}d\xi$$

$$= \mathcal{O}(|y|^N).$$

In both constructions we notice that we can take $\widetilde{f}$ with support in an arbitrarily small neighborhood of the support of f. We also notice that if g is the difference of two almost analytic extensions of the same function, so that $g|_{\mathbf{R}} = 0$, $\overline{\partial}g = \mathcal{O}_N(|\text{Im }x|^N)$, $\forall N$, then $g = \mathcal{O}_N(|\text{Im }x|^N)$, $\forall N$.

Theorem 8.1. *Let P be a selfadjoint operator on a Hilbert space $\mathcal{H}$. Let $f \in C_0^2(\mathbf{R})$ and let $\widetilde{f} \in C_0^1(\mathbf{C})$ be an extension of f with $\overline{\partial}\widetilde{f} = \mathcal{O}(|\text{Im }z|)$. Then*

$$f(P) = -\frac{1}{\pi}\int \overline{\partial}\widetilde{f}(z)(z-P)^{-1}L(dz). \qquad (8.3)$$

Here $L(dz) = dxdy$ is the Lebesgue measure on $\mathbf{C}$, and we notice that the integral in (8.3) converges as a Riemann integral for functions with values in $\mathcal{L}(\mathcal{H}, \mathcal{H})$.

Notice that if we replace P by a complex number, then (8.3) simply reflects that $\frac{1}{\pi z}$ is a fundamental solution of $\overline{\partial}$.

Proof. (Dimassi [Di1]) Let $Q \in \mathcal{L}(\mathcal{H}, \mathcal{H})$ be the RHS of (8.3). Let $u, v \in \mathcal{H}$ and write

$$((z - P)^{-1}u|v) = \int (z - t)^{-1}(dE_t u|v),$$

where $E_t = 1_{]-\infty, t]}(P)$ is the family of spectral projections associated with P. (Here, we use the fact that $(z - P)^{-1} = \int (z - t)^{-1} dE_t$.) Consequently,

$$(Qu|v) = -\frac{1}{\pi} \int \overline{\partial}\tilde{f}(z) \int (z - t)^{-1}(dE_t u|v)L(dz),$$

and we can apply Fubini's theorem to get

$$(Qu|v) = \int (-\frac{1}{\pi} \int \overline{\partial}\tilde{f}(z)(z - t)^{-1}L(dz))(dE_t|v).$$

The inner integral is equal to $\int \tilde{f}(z)\overline{\partial}_z(\frac{1}{\pi}(z - t)^{-1})L(dz) = f(t)$, since $\frac{1}{\pi z}$ is a fundamental solution of $\overline{\partial}$ (which is proved by means of Green–Riemann's formula). Consequently,

$$(Qu|v) = \int f(t)(dE_t u|v)$$

and hence $Q = f(P)$.   #

An even shorter proof can be given, using the representation of P as a multiplication operator (Theorem 4.13). This result can be generalized to classes of non selfadjoint operators acting in Banach spaces. Then we may use (8.3) as a definition. In such a case, we should write $\tilde{f}(P)$ to the left, and require that $\overline{\partial}_z \tilde{f}$ vanishes to a sufficiently high order on the spectrum of P (which now is some closed subset of the complex plane). See Dyn'kin [Dy], Droste [Dr], Davies [Da1,2], Jensen–Nakamura [JeNa].

We now develop R. Beals' characterization of pseudodifferential operators in its semi-classical variant. Assume that $a \in \mathcal{S}'(\mathbf{R}^{2n})$ and that $a(x, D)$ (in the standard ($t = 1$) quantization) is L^2-bounded. Let $\phi, \psi \in \mathcal{S}$ be fixed and consider

$$C\|\phi\|\|\psi\| \geq |(a(x, D)\psi|\phi)| = |\frac{1}{(2\pi)^n} \iint a(x, \xi)e^{ix\xi}\overline{\phi}(x)\widehat{\psi}(\xi)dxd\xi|.$$

Here we can replace $\overline{\phi}$ by $\overline{\phi}e^{-ix \cdot x^*}$ and $\widehat{\psi}$ by $\widehat{\psi}e^{-i\xi \cdot \xi^*}$ (corresponding to new functions ϕ, ψ with the same L^2 norms), and we get

$$|\mathcal{F}(\overline{\phi}\widehat{\psi}ae^{ix \cdot \xi})(x^*, \xi^*)| \leq C\|\phi\|\|\psi\|.$$

If $\chi(x,\xi) \in C_0^\infty(\mathbf{R}^{2n})$, choose ϕ, ψ with $\overline{\phi}(x)\widehat{\psi}(\xi) \neq 0$ in a neighborhood of supp χ so that $\chi = \Phi(x,\xi)\overline{\phi}(x)\widehat{\psi}(\xi)e^{ix\xi}$ for some $\Phi \in C_0^\infty$. It follows that

$$|\mathcal{F}(\chi a)(x^*,\xi^*)| \leq C_\chi \|a\|_{\mathcal{L}(L^2,L^2)}. \tag{8.4}$$

Here the constant C_χ only depends on the support of χ and on $\sum_{|\alpha| \leq 2n+1} \|\partial^\alpha \chi\|_{L^\infty}$. We get the same constant C_χ if we replace χ by any translate of χ, or equivalently if we replace a by any translate of a. The reason for this is that if $(y,\eta) \in \mathbf{R}^{2n}$, then $a(y+x, \eta+D_x)$ is obtained from $a(x,D_x)$ by conjugation by the unitary operator $\tau_y e^{ix\cdot\eta}$.

If $(\partial_{x,\xi}^\alpha a)(x,D)$ is L^2-bounded for $|\alpha| \leq 2n+1$, we get from (8.4) and the subsequent remark that

$$|\mathcal{F}(\chi a)(x^*,\xi^*)| \leq C_\chi \langle (x^*,\xi^*) \rangle^{-(2n+1)} \sum_{|\alpha| \leq 2n+1} \|(\partial_{x,\xi}^\alpha a)(x,D)\|_{\mathcal{L}(L^2,L^2)},$$
$$\tag{8.5}$$

where C_χ is invariant under translations of χ. In particular,

$$\|a\|_{L^\infty} \leq C \sum_{|\alpha| \leq 2n+1} \|(\partial_{x,\xi}^\alpha a)(x,D)\|_{\mathcal{L}(L^2,L^2)}.$$

Let a_w be the Weyl symbol of $a(x,D)$ so that $a(x,D) = a_w^w(x,D)$ and $a_w = e^{iD_x \cdot D_\xi / 2} a$. We shall now prove that with a new 'translation invariant' constant C_χ, we have for every $\chi \in C_0^\infty(\mathbf{R}^{2n})$:

$$|\mathcal{F}(\chi a_w)(x^*,\xi^*)| \leq C_\chi \langle (x^*,\xi^*) \rangle^{-(2n+1)} \sum_{|\alpha| \leq 2n+1} \|(\partial_{x,\xi}^\alpha a_w)^w(x,D)\|_{\mathcal{L}(L^2,L^2)}.$$
$$\tag{8.6}$$

Here $(\partial_{x,\xi}^\alpha a)(x,D) = (\partial_{x,\xi}^\alpha a_w)^w(x,D)$, so the sums in (8.5) and (8.6) are equal.

Up to a constant factor, which we shall neglect, we have $a_w = e^{i\Phi} * a$, where Φ is a real non-degenerate quadratic form on $\mathbf{R}^{2n}$. Let $1 = \sum_{j \in \mathbf{Z}^{2n}} \chi_j$ be a partition of unity on $\mathbf{R}^{2n}$ with $\chi_j = \tau_j \chi_0$, $\chi_0 \in C_0^\infty$. Let $\widetilde{\chi}_0$ satisfy $\widetilde{\chi}_0 \chi_0 = \chi_0$ and put $\widetilde{\chi}_k = \tau_k \widetilde{\chi}_0$. Temporarily we simplify the notation by writing x instead of (x,ξ) and ξ instead of (x^*,ξ^*). Then for $j,k \in \mathbf{Z}^{2n}$, we have:

$$\mathcal{F}(\chi_j(e^{i\Phi} * (\chi_k a)))(\xi)$$

$$= \iiint e^{i(-x\cdot\xi + \Phi(x-y) + y\cdot\eta)} \chi_j(x)\widetilde{\chi}_k(y)\widehat{\chi_k a}(\eta)\, d\eta\, dy\, dx / (2\pi)^{2n}$$

$$= \frac{e^{iF(j,k)}}{(2\pi)^{2n}} \iiint e^{i(-x\cdot(\xi - \partial_x \Phi(j-k)) + y\cdot(\eta - \partial_x \Phi(j-k)))} \chi_{j,k}(x,y)\widehat{\chi_k a}(\eta)\, dy\, dx\, d\eta,$$

where F is real-valued and where

$$\chi_{j,k}(x,y) = \chi_j(x)\widetilde{\chi}_k(y)e^{i\Phi((x-j)-(y-k))}.$$

Here we also used the Taylor sum formula:

$$\Phi(x-y) = \Phi(j-k)+\partial_x\Phi(j-k)\cdot(x-j)-\partial_x\Phi(j-k)\cdot(y-k)+\Phi((x-y)-(j-k)).$$

Notice that the modulus of any derivative of $\chi_{j,k}(x,y)$ can be bounded by a constant which is independent of x,y,j,k.

We make $2N$ integrations by parts using the operators

$$\frac{1 - (\xi - \partial_x\Phi(j-k))\cdot D_x}{\langle\xi - \partial_x\Phi(j-k)\rangle^2}, \quad \frac{1 + (\eta - \partial_x\Phi(j-k))\cdot D_y}{\langle\eta - \partial_x\Phi(j-k)\rangle^2}.$$

After estimating the resulting x,y integrals in a straightforward way, we get:

$$\mathcal{F}(\chi_j(e^{i\Phi}*\chi_k a))(\xi)$$
$$= \mathcal{O}_N(1) \int \langle\xi - \partial_x\Phi(j-k)\rangle^{-N}\langle\eta - \partial_x\Phi(j-k)\rangle^{-N}|\widehat{\chi_k a}(\eta)|d\eta. \quad (8.7)$$

Since Φ is non-degenerate, $\langle\xi - \partial_x\Phi(j-k)\rangle$ is of the same order of magnitude as $\langle\Phi''^{-1}\xi - j + k\rangle$, and similarly for $\langle\eta - \partial_x\Phi(j-k)\rangle$. With $N > 2n$, we get:

$$\sum_{k\in\mathbf{Z}^{2n}} \langle\xi - \partial_x\Phi(j-k)\rangle^{-N}\langle\eta - \partial_x\Phi(j-k)\rangle^{-N} \leq C_N\langle\xi - \eta\rangle^{-N}.$$

Summing over k in (8.7), we get

$$|\mathcal{F}(\chi_j(e^{i\Phi}*a))(\xi)| \leq C_N \int \langle\xi - \eta\rangle^{-N} \sup_k |\widehat{\chi_k a}(\eta)|d\eta.$$

Here we use (8.5) with C_{χ_k} independent of k and conclude that (8.6) holds with $\chi = \chi_j$ and C_{χ_j} independent of j. To get (8.6) for arbitrary χ, we simply represent χ as a linear combination with smooth coefficients of a finite number of the χ_j.

As a special case, we have proved:

Proposition 8.2. *Let* $a \in \mathcal{S}'(\mathbf{R}^{2n})$ *and assume that* $\mathrm{Op}\,(\partial_{x,\xi}^\alpha a) \in \mathcal{L}(L^2, L^2)$ *for* $|\alpha| \leq 2n + 1$. *Then* $a \in L^\infty$ *and*

$$\|a\|_{L^\infty} \leq C_n \sum_{|\alpha|\leq 2n+1} \|\mathrm{Op}\,(\partial_{x,\xi}^\alpha a)\|_{\mathcal{L}(L^2,L^2)}.$$

The proof above gives the same result for any quantization $\text{Op}_{1,t}$ with $0 \leq t \leq 1$.

Remark. Let $B \subset \mathcal{D}'(\mathbf{R}^{2n})$ be a translation invariant Banach space with a translation invariant norm, and assume that the inclusion $B \to \mathcal{D}'$ is sequentially continuous. Then it can be proved, using the Banach–Steinhaus theorem, that for every $\chi \in C_0^\infty(\mathbf{R}^{2n})$, there are constants C_χ, $N_\chi \geq 0$, such that

$$|\mathcal{F}(\chi u)(\xi)| \leq C_\chi \|u\|_B \langle \xi \rangle^{N_\chi},$$

and it is easy to see that C_χ and N_χ can be chosen invariant under translations of χ. Bony [Bon] made essentially this observation and applied it with B equal to the space of Weyl symbols of L^2 bounded operators, which in our case leads to Proposition 8.2 with $2n + 1$ replaced by some finite unspecified number $N \geq 0$.

If $A = \text{Op}(a)$, then the commutators $[x_j, A]$ and $[D_{x_j}, A]$ have the symbols $\frac{1}{i}\{x_j, a\} = i\partial_{\xi_j} a$ and $\frac{1}{i}\{\xi_j, a\} = \frac{1}{i}\partial_{x_j} a$ respectively, so the assumption in the proposition can be reformulated as: $\text{ad}_{\ell_1(x,D)} \cdots \text{ad}_{\ell_k(x,D)} \text{Op}(a) \in \mathcal{L}(L^2, L^2)$ for all $k \leq 2n + 1$ and all linear forms $\ell_j(x, \xi)$. (Here we use the standard notation: $\text{ad}_A B = [A, B]$, and notice that by the Jacobi identity for commutators: $[A, [B, C]] + [B, [C, A]] + [C, [A, B]] = 0$, we have $\text{ad}_{[A,B]} = [\text{ad}_A, \text{ad}_B]$). We next turn to the h-dependent case:

Proposition 8.3. *Let $A = A_h : \mathcal{S}(\mathbf{R}^n) \to \mathcal{S}'(\mathbf{R}^n)$, $0 < h \leq 1$. The following two statements are equivalent:*

(1) $A = \text{Op}_h(a)$, *for some* $a = a(x, \xi; h) \in S^0(1)$.

(2) For every $N \in \mathbf{N}$ and for every sequence $\ell_1(x, \xi), \ldots, \ell_N(x, \xi)$ of linear forms on $\mathbf{R}^{2n}$, the operator $\text{ad}_{\ell_1(x,hD)} \circ \cdots \circ \text{ad}_{\ell_N(x,hD)} A_h$ belongs to $\mathcal{L}(L^2, L^2)$ and is of norm $\mathcal{O}(h^N)$ in that space.

Proof. That (1)$\Rightarrow$(2) follows from the calculus in Chapter 7. In the opposite direction, we assume (2), and notice that we have the general identity $U_h b^w(x, hD) U_h^{-1} = b^w(\sqrt{h}x, \sqrt{h}D)$, if U_h is the unitary operator given by: $U_h u(x) = h^{n/4} u(\sqrt{h}x)$. Then (2) can be reformulated as:

$$\text{ad}_{\ell_1(h^{1/2}x, h^{1/2}D)} \circ \cdots \circ \text{ad}_{\ell_N(h^{1/2}x, h^{1/2}D)} a_h^w(x, D) = \mathcal{O}(h^N) \text{ in } \mathcal{L}(L^2, L^2),$$

with $a_h(x, \xi) = a(h^{1/2}x, h^{1/2}\xi; h)$, or rather as:

$$\text{ad}_{\ell_1(x,D)} \circ \cdots \circ \text{ad}_{\ell_N(x,D)} a_h^w(x, D) = \mathcal{O}(h^{\frac{N}{2}}) \text{ in } \mathcal{L}(L^2, L^2).$$

Applying Proposition 8.2 to a_h and to its derivatives, we get

$$\|\partial_{x,\xi}^\alpha a_h\|_{L^\infty} \le C_\alpha h^{|\alpha|/2},$$

which implies that $a \in S^0(1)$. #

We will also need the following generalization of the implication (2)$\Rightarrow$(1) in the preceding proposition. Assume that $a = a(x, \xi, z; h)$ depends on the additional parameters z and that for some function $\delta = \delta(z)$ with values in $]0, 1]$ we have

$$\mathrm{ad}_{\ell_1(x,hD)} \circ \ldots \circ \mathrm{ad}_{\ell_N(x,hD)} \mathrm{Op}_h(a) = \mathcal{O}(\delta^{-N} h^N) \text{ in } \mathcal{L}(L^2, L^2) \qquad (*)$$

for all $N \ge 0$ and all linear forms $\ell_1, \ldots, \ell_N$ on $\mathbf{R}^{2n}$.

Proposition 8.4. *Under the assumption* (*), *we have*

$$|\partial_x^\alpha \partial_\xi^\beta a(x, \xi, z; h)| \le C_{\alpha,\beta} \max(1, \frac{\sqrt{h}}{\delta})^{2n+1} \delta^{-|\alpha|-|\beta|}. \qquad (8.8)$$

Proof. As in the proof of Proposition 8.3, we introduce $a_h(x, \xi) = a(\sqrt{h}x, \sqrt{h}\xi, z; h)$ and reformulate (*) as:

$$\mathrm{ad}_{\ell_1(x,D)} \circ \ldots \circ \mathrm{ad}_{\ell_N(x,D)} \mathrm{Op}(a_h) = \mathcal{O}(\delta^{-N} h^{N/2}). \qquad (8.9)$$

Proposition 8.2 then gives

$$\|\partial_x^\alpha \partial_\xi^\beta a_h\|_{L^\infty} \le C_n \sum_{|\alpha'+\beta'|\le 2n+1} \|\mathrm{Op}(\partial_x^{\alpha+\alpha'} \partial_\xi^{\beta+\beta'} a_h)\|_{\mathcal{L}(L^2,L^2)}$$

$$\le C_n \max(\delta^{-|\alpha+\beta|} h^{|\alpha+\beta|/2}, \delta^{-|\alpha+\beta|-(2n+1)} h^{(|\alpha+\beta|+2n+1)/2})$$

$$= C_n \max(1, \frac{\sqrt{h}}{\delta})^{2n+1} \delta^{-|\alpha+\beta|} h^{|\alpha+\beta|/2},$$

and since $\partial_x^\alpha \partial_\xi^\beta a_h = h^{|\alpha+\beta|/2} (\partial_x^\alpha \partial_\xi^\beta a)(\sqrt{h}x, \sqrt{h}\xi, z; h)$, we get (8.8). #

Let $p(x, \xi; h) \in S^0(1)$ be elliptic in the sense that there is a constant $C > 0$ such that $|p| \ge 1/C$. Then $\tilde{q} = 1/p$ belongs to $S^0(1)$ and $p^w(x, hD) \circ \tilde{q}^w(x, hD) = 1 - hr^w(x, hD)$, where $r \in S^0(1)$, so for $h \le h_0 \ge 0$ small enough, $\|hr^w(x, hD)\|_{\mathcal{L}(L^2,L^2)} < 1/2$ and $(I - hr^w(x, hD))^{-1}$ exists in $\mathcal{L}(L^2, L^2)$ and has norm ≤ 2. Put $Q_1 = \tilde{q}^w(x, hD)(I - hr^w(x, hD))^{-1}$, so that $\|Q_1\|_{\mathcal{L}(L^2,L^2)} \le C$ and $p^w \circ Q_1 = I$ (where we drop '(x, hD)' in order to shorten the notation). Similarly, we can construct $Q_2 \in \mathcal{L}(L^2, L^2)$ of norm $\le C$ such that $Q_2 \circ p^w = I$. Then,

$$Q_2 = Q_2(p^w Q_1) = (Q_2 p^w)Q_1 = Q_1 =: Q.$$

We claim that

$$Q = q^w(x, hD), \quad q \in S^0(1). \tag{8.10}$$

Proof. Let ℓ_1, ℓ_2, ... be linear forms on $\mathbf{R}^{2n}$ and put $L_j = \ell_j(x, hD)$. If L is one of the L_j, we first recall that $e^{itL} \circ p^w \circ e^{-itL} = p_t^w$, where $p_t = p \circ \exp(tH_\ell)$. It follows that this operator is a C^1 function of t with values in the space $\mathcal{L}(L^2, L^2)$. If $A : \mathcal{S} \to \mathcal{S}'$ is any continuous operator, we notice that $(D_t)_{t=0}(e^{itL} A e^{-itL}) = \mathrm{ad}_L(A)$. It follows that

$$\begin{aligned}
\mathrm{ad}_L(Q) &= (D_t)_{t=0}(e^{itL} p^w e^{-itL})^{-1} \\
&= -Q((D_t)_{t=0}(e^{itL} p^w e^{-itL}))Q = -Q \, \mathrm{ad}_L(p^w) Q.
\end{aligned}$$

By iteration we then see that $\mathrm{ad}_{L_k} \circ \ldots \circ \mathrm{ad}_{L_1} Q$ belongs to $\mathcal{L}(L^2, L^2)$ and is equal to a finite linear combination of terms of the form:

$$Q((\mathrm{ad}_L)^{J_1} P)Q((\mathrm{ad}_L)^{J_2} P)Q \ldots Q((\mathrm{ad}_L)^{J_\ell} P)Q,$$

where $J_1, \ldots, J_\ell$ is a partition of $\{1, 2, \ldots, k\}$ and where we write $(\mathrm{ad}_L)^J = \prod_{j \in J} \mathrm{ad}_{L_j}$. This notation is justified by the fact that the ad_{L_j} commute, which can be seen from the Jacobi identity for commutators, and the fact that $\mathrm{ad}_{[L_j, L_k]} = \mathrm{ad}_{\mathrm{const.}} = 0$. Then $\mathrm{ad}_{L_k} \circ \ldots \circ \mathrm{ad}_{L_1} Q = \mathcal{O}(h^k)$ in $\mathcal{L}(L^2, L^2)$ and we can use Proposition 8.3 to conclude.   #

We can also obtain an asymptotic expansion for the symbol of the inverse operator. Write $P\widetilde{Q} = I - hR$, where $P = p^w(x, hD)$, $\widetilde{Q} = (\frac{1}{p})^w(x, hD)$, $R = r^w(x, hD)$. Put $Q_N = \widetilde{Q}(I + hR + h^2 R^2 + \ldots + h^N R^N)$. Then $PQ_N = I - h^{N+1} R^{N+1}$, so $Q_N = QPQ_N = Q - Qh^{N+1} R^{N+1} \equiv Q$ $\mathrm{mod}\,(S^{-(N+1)}(1))$. If $Q = \mathrm{Op}_h(q)$, we obtain

$$q \sim \frac{1}{p} + h(\frac{1}{p} \sharp_h r) + h^2(\frac{1}{p} \sharp_h r \sharp_h r) + \ldots \tag{8.11}$$

More generally, let m be an order function, and let $p \in S^0(m)$ be elliptic in the sense that $|p(x, \xi; h)| \geq \frac{1}{C} m(x, \xi)$ for some $C > 0$. Define $\widetilde{q} = \frac{1}{p} \in S^0(\frac{1}{m})$ and observe as above that $p^w(x, hD)\widetilde{q}^w(x, hD) = I - hr^w(x, hD)$, $r \in S^0(1)$. Again $I - hr^w$ is invertible as an L^2 bounded operator when $h \leq h_0$ for some $h_0 > 0$ small enough, and we now apply Beals' lemma directly to $I - hr^w(x, hD)$, to see that $(I - hr^w)^{-1} \in \mathrm{Op}_h(S^0(1))$. Then $\widetilde{q}^w \circ (I - hr^w)^{-1} \in \mathrm{Op}_h(S^0(m^{-1}))$ is a right inverse. Similarly we get a left inverse, and as before we get $Q \in \mathrm{Op}_h(S^0(m^{-1}))$ with $PQ = QP = I$, writing $P = p^w(x, hD)$.

We now apply the preceding results to the functional calculus of pseudodifferential operators. Let $m \geq 1$ be an order function on $\mathbf{R}^{2n}$ and let

$P = p^w(x, hD; h)$, where $p \in S^0(m)$ is real-valued. When m is unbounded, we also assume that $p + i$ is elliptic, and in that case, the discussion below will be valid only for $0 < h \leq h_0$ for some sufficiently small $h_0 > 0$. These assumptions will be valid throughout the discussion of the functional calculus for one operator below. We will also frequently write p^w for $p^w(x, hD; h)$

We know that $(p^w \pm i)^{-1}$ exists and belongs to $\mathrm{Op}_h(\frac{1}{m})$ (provided that h is sufficiently small in the case when m is not bounded). It is easy to see that $(p^w \pm i)^{-1}(L^2(\mathbf{R}^n)) =: \mathcal{D}_P$ is a space independent of the choice of the sign in front of the 'i' and that $u \in \mathcal{D}_P$ can be equipped with any of the equivalent norms $\|(p^w + i)u\|$ or $\|(p^w - i)u\|$. (Of course, $\mathcal{D}_P = L^2$ in the case when m is bounded.) We may view $P = p^w$ as a symmetric operator with domain $\mathcal{S}(\mathbf{R}^n)$.

Proposition 8.5. *P is essentially selfadjoint and the unique selfadjoint extension is given by p^w equipped with the domain $\mathcal{D}_P$.*

Proof. Let $\overline{P}$ denote the closure of P. Then $\overline{P}u = v$ means that $u_j \to u$, $v_j \to v$ in L^2, where $Pu_j = v_j$, with $u_j, v_j \in \mathcal{S}$. Then $p^w u = v$, $(p^w + i)u = v + iu$, so $u = (p^w + i)^{-1}(v + iu) \in \mathcal{D}_P$. Conversely, if $u \in \mathcal{D}_P$, and $v = p^w u$, we let $f_j \in \mathcal{S}$ with $f_j \to (p^w + i)u$ in L^2 and put $u_j = (p^w + i)^{-1}f_j$, $v_j = p^w(p^w + i)^{-1}f_j$, so that $u_j \to u$, $v_j \to v$ in L^2. We have then shown that $\overline{P}$ is given by $p^w(x, hD; h)$ with domain $\mathcal{D}_P$.

Let $u \in \mathcal{D}(P^*)$, and $P^*u = v$. It follows that $(p^w + i)u = v + iu$, so $u = (p^w + i)^{-1}(v + iu) \in \mathcal{D}_P$. Hence $\mathcal{D}(P^*) = \mathcal{D}_P$ and $P^* = \overline{P}$. #

From now on, we let P denote also the selfadjoint extension. We shall consider the resolvent $(z - P)^{-1}$ for $\mathrm{Im}\, z \neq 0$.

Proposition 8.6. *For $|z| \leq$ const., $\mathrm{Im}\, z \neq 0$, we have $(z - P)^{-1} = r^w(x, hD, z; h)$, where*

$$|\partial_x^\alpha \partial_\xi^\beta r(x, \xi, z; h)| \leq C_{\alpha,\beta} \max(1, \frac{h^{\frac{1}{2}}}{|\mathrm{Im}\, z|})^{2n+1} |\mathrm{Im}\, z|^{-(|\alpha|+|\beta|)-1}. \tag{8.12}$$

The same result holds for $(P + i)(z - P)^{-1}$.

Proof. We treat $(P + i)(z - P)^{-1}$. Let $\ell_1, \ell_2, \ldots$ be linear forms on $\mathbf{R}^{2n}$ and put $L_j = \ell_j(x, hD)$. Then $\mathrm{ad}_{L_k} \circ \ldots \circ \mathrm{ad}_{L_1}(P + i)(z - P)^{-1}$ is a finite linear combination of terms of the form,

$$(\mathrm{ad}_L^{J_0}(P + i))(z - P)^{-1}(\mathrm{ad}_L^{J_1}P)(z - P)^{-1} \ldots (\mathrm{ad}_L^{J_\ell}P)(z - P)^{-1},$$

where $J_0, \ldots, J_\ell$ is a partition of $\{1, \ldots, k\}$, $J_j \neq \emptyset$ for $j \neq 0$ and we allow J_0 to be empty and use the convention that $\mathrm{ad}_L^0(P + i) = P + i$. This is clear, since we know that $(z - P)^{-1} = \mathrm{Op}_h(q_{h,z})$, with $q_{h,z} \in S(m^{-1})$ for $\mathrm{Im}\, z \neq 0$ and for every fixed h. We also see that $(\mathrm{ad}_L^J P)(z - P)^{-1}$ is $\mathcal{O}(\frac{h^{\sharp J}}{|\mathrm{Im}\, z|})$ in $\mathcal{L}(L^2, L^2)$ and we obtain that

$$\mathrm{ad}_{L_k} \circ \ldots \circ \mathrm{ad}_{L_1}(P + i)(z - P)^{-1} = \mathcal{O}(\frac{h^k}{|\mathrm{Im}\, z|^{k+1}})$$

in $\mathcal{L}(L^2, L^2)$. The estimates (8.12) for the symbol of $(P + i)(z - P)^{-1}$ then follow from Proposition 8.4.   #

Theorem 8.7. *Let $f \in C_0^\infty(\mathbf{R})$. Then $f(P) \in \mathrm{Op}_h(S^0(m^{-k}))$, for every $k \in \mathbf{N}$.*

Proof. We apply (8.3) with $\widetilde{f}$ satisfying (8.1). Let $r(x, \xi, z; h)$ be the symbol in (8.12). Then $f(P) = \mathrm{Op}_h(a)$, where

$$a(x, \xi; h) = -\frac{1}{\pi} \int \overline{\partial} \widetilde{f}(z) r(x, \xi, z; h) L(dz), \qquad (8.13)$$

and from (8.1) and (8.12) we easily obtain that $a \in S^0(1)$ so $f(P) \in \mathrm{Op}_h(S^0(1))$. Writing $f(t) = (t + i)^{-k} f_k(t)$, we see that $f(P) = (P + i)^{-k} f_k(P) = \mathrm{Op}_h(S^0(m^{-k}))$.   #

We finally show how to get an asymptotic expansion in powers in h of the symbol of $f(P)$, when $P = \mathrm{Op}_h(p)$, with $p \sim p_0(x, \xi) + h p_1(x, \xi) + h^2 p_2(x, \xi) + \ldots$ in $S^0(m)$ and $p + i$ is elliptic in the case when m is not bounded. We notice that if $\delta > 0$ and if we restrict the integral in (8.13) to the domain $|\mathrm{Im}\, z| \leq h^\delta$ then we get an element in the symbol class $S^{-\infty}(1) = \cap_k S^k(1)$. On the other hand, if $\delta < \frac{1}{2}$, and we restrict our attention to the domain $|\mathrm{Im}\, z| \geq h^\delta$, then by Proposition 8.6, we have $r \in S_\delta^\delta(1)$ and we want to find the asymptotic expansion of r in this space. Clearly we can find a formal asymptotic expansion:

$$\widetilde{r}(x, \xi, z; h) \sim \frac{1}{z - p_0(x, \xi)} + h \frac{q_1(x, \xi, z)}{(z - p_0(x, \xi))^3} + h^2 \frac{q_2(x, \xi, z)}{(z - p_0(x, \xi))^5} + \cdots$$
$$(8.14)$$

with $q_j(x, \xi, z)$ a polynomial in z with smooth coefficients, so that in the sense of formal asymptotic expansions in powers of h, we have

$$(z - p) \sharp_h \widetilde{r} = 1, \quad \widetilde{r} \sharp_h (z - p) = 1,$$

where we put

$$a\widetilde{\sharp}_h b \sim \sum \frac{1}{k!} ((\frac{ih}{2}\sigma(D_x, D_\xi; D_y, D_\eta))^k (a(x,\xi)b(y,\eta)))_{\big|_{x=y,\xi=\eta}}.$$

(When $p_1 = 0$, we also have $q_1 = 0$.) Letting $|z|$ tend to infinity, we see that q_j is a polynomial of degree $\leq 2j$. If we let z take different values of the order of magnitude $m(x,\xi)$, we see that

$$q_j(x,\xi,z) = \sum_{k=0}^{2j} q_{j,k}(x,\xi)z^k, \quad q_{j,k} \in S(m^{2j-k}).$$

If we restrict our attention to $|z| \leq \text{const.}$, $|\text{Im } z| \geq h^\delta$, we see that we can give a meaning to (8.14) in $S_\delta^\delta(\frac{1}{m})$ and that $(z-p)\sharp_h \widetilde{r} = 1 - k$, $k \in S^{-\infty}(1)$. Then (by the Beals lemma) $(1 - k^w)^{-1} = (1 - \widetilde{k}^w)$, $\widetilde{k} \in S^{-\infty}(1)$ and consequently $r = \widetilde{r}\sharp_h(1 - \widetilde{k})$ belongs to $S_\delta^\delta(\frac{1}{m})$ and also has the asymptotic expansion (8.14).

It follows that $f(P) = \text{Op}_h(a)$, $a \in S^0(\frac{1}{m})$, $a \sim \widetilde{a}_0 + h\widetilde{a}_1 + h^2\widetilde{a}_2 + \dots$, $\widetilde{a}_j \in S(\frac{1}{m})$,

$$\widetilde{a}_j = -\frac{1}{\pi} \int_{|\text{Im } z| \geq h^\delta} \overline{\partial}\widetilde{f}(z) \frac{q_j(x,\xi,z)}{(z - p_0(x,\xi))^{2j+1}} L(dz). \tag{8.15}$$

Modulo $S^{-\infty}(\frac{1}{m})$, we can replace $\widetilde{a}_j$ by

$$\begin{aligned}
a_j &= -\frac{1}{\pi} \int \overline{\partial}\widetilde{f} \frac{q_j}{(z - p_0)^{2j+1}} L(dz) = -\frac{1}{(2j)!} \frac{1}{\pi} \int \overline{\partial}\widetilde{f} q_j(-\partial_z)^{2j} \frac{1}{z - p_0} L(dz) \\
&= \frac{1}{(2j)!} \int (\partial_z)^{2j} (q_j(x,\xi,z)\widetilde{f}(z)) \overline{\partial}(\frac{1}{\pi} \frac{1}{z - p_0}) L(dz) \\
&= \frac{1}{(2j)!} \partial_t^{2j} (q_j(x,\xi,t)f(t))_{t=p_0(x,\xi)}.
\end{aligned} \tag{8.16}$$

In particular, $a_0 = f(p_0(x,\xi))$, $a_1 = p_1(x,\xi)f'(p_0(x,\xi))$.

We end this chapter by discussing functional calculus for several commuting (formally) selfadjoint operators. In order to avoid some abstract difficulties, we consider right away the case of pseudodifferential operators. Let $M(x,\xi) \geq 1$ be an order function on $\mathbf{R}^{2n}$ and let $p_1, \dots, p_m \in S^0(M)$ be real-valued. Put $P_j = \text{Op}_h(p_j)$ and define

$$Q := mI + \sum_1^m P_j^2 \in \text{Op}_h(S^0(M^2)). \tag{8.17}$$

When M is unbounded, we assume that Q is elliptic and restrict our attention to a region $h > 0$ small enough. If $f \in C_0^\infty(\mathbf{R}^m)$ or more generally if $f \in C^\infty(\mathbf{R}^m)$ is constant near infinity, we want to define and study $f(P_1, \ldots, P_m)$. We assume that P_j commute: $[P_j, P_k] = 0$ for all j, k.

To start with we need some formalism from the theory of several complex variables. If $u(z)$ is a distribution defined on some open set in $\mathbf{C}^m$, we define the $(1, 0)$ and $(0, 1)$ forms

$$\partial_z u = \sum_1^m (\partial_{z_j} u) dz_j, \quad \overline{\partial}_z u = \sum_1^m (\overline{\partial}_{z_j} u) \overline{dz_j},$$

where $\partial_{z_j} = \frac{1}{2}(\partial_{x_j} + \frac{1}{i}\partial_{y_j})$, $\overline{\partial}_{z_j} = \frac{1}{2}(\partial_{x_j} - \frac{1}{i}\partial_{y_j})$, when writing $z_j = x_j + iy_j$. The formal complex adjoint of ∂_z is given by

$$\partial_z^* = \sum -\overline{\partial}_{z_j} dz_j^\rfloor = {}^t\overline{\partial}_z,$$

when $dz_1, \ldots, dz_m$ is considered to be an orthonormal basis for the $(1, 0)$ forms at each point. More explicitly:

$$\partial_z^* v = \sum -\overline{\partial}_{z_j} v_j, \quad \text{when } v = \sum v_j(z) dz_j.$$

Then $\partial_z^* \partial_z = -\frac{1}{4}\Delta$, where Δ is the standard Laplacian on $\mathbf{R}^{2m} \approx \mathbf{C}^m$. Let $E_0(z)$ be the standard fundamental solution of Δ, so that $E_0(z) = C_1 \log(|z|^2)$ when $m = 1$, and $E_0(z) = -C_m |z|^{2(1-m)}$ when $m \geq 2$. Here $C_m > 0$. Put $F_0(z) = -4\partial_z E_0$ so that $\partial_z^* F_0 = \delta_0$. Notice that

$$F_0 = \sum \frac{4C_m(1 - m)}{|z|^{2m}} \overline{z}_j dz_j = -\widetilde{C}_m \sum \frac{\overline{z}_j dz_j}{|z|^{2m}}, \quad \text{when } m \geq 2,$$

$$F_0 = -\frac{4C_1 \overline{z} dz}{|z|^2} = -\widetilde{C}_1 \frac{\overline{z} dz}{|z|^2}, \quad \text{when } m = 1.$$

As in the case of one complex variable, if $f \in C_0^\infty(\mathbf{R}^m)$, it is easy to construct $\widetilde{f} \in C_0^\infty(\mathbf{C}^m)$ with $\widetilde{f}|_{\mathbf{R}^m} = f$, $\overline{\partial}\widetilde{f} = \mathcal{O}(|\text{Im } z|^N)$, for every $N \in \mathbf{N}$. We now put,

$$f(P_1, \ldots, P_m) = -\widetilde{C}_m \int \langle \overline{\partial}\widetilde{f}, \frac{(z - P)^* \cdot dz}{(\sum_1^m (z_\nu - P_\nu)^*(z_\nu - P_\nu))^m} \rangle L(dz)$$

$$= -\widetilde{C}_m \sum_{j=1}^m \int \overline{\partial}_{z_j}\widetilde{f}(z)(z_j - P_j)^* (\sum_1^m (z_\nu - P_\nu)^*(z_\nu - P_\nu))^{-m} L(dz)$$

$$= -\widetilde{C}_m \int ((z - P)^* \cdot \overline{\partial}\widetilde{f})(\sum_1^m (z_\nu - P_\nu)^*(z_\nu - P_\nu))^{-m} L(dz). \qquad (8.18)$$

Notice that if we replace P by $(p_1, p_2, \ldots, p_m) \in \mathbf{R}^m$, then (8.18) holds.

For the understanding of (8.18) we make two comments:

(1) Assuming as before that $h > 0$ is sufficiently small when M is not bounded, we see that $|i - P|^2 = \sum (i - P_j)^*(i - P_j)$ is essentially selfadjoint and invertible, and that the inverse is a pseudor. We next look at $|z - P|^2 = Q(z) = \sum (z_j - P_j)^*(z_j - P_j)$. For $u \in \mathcal{S}$, we have

$$(|z - P|^2 u | u) = \sum \|(z_j - P_j)u\|^2 \geq |\operatorname{Im} z|^2 \|u\|^2.$$

Also

$$\sum \|(z_j - P_j)u\|^2 = (|z - P|^2 u | u) \leq \||z - P|^2 u\| \, \|u\| \leq \frac{\epsilon}{2}\||z - P|^2 u\|^2 + \frac{1}{2\epsilon}\|u\|^2,$$

for every $\epsilon > 0$, so we get

$$\left(\sum \|(z_j - P_j)u\|^2\right)^{\frac{1}{2}} \leq \epsilon \||z - P|^2 u\| + C_\epsilon \|u\|.$$

We can then compare two different operators $|z - P|^2$ and $|w - P|^2$. Since

$$(|z-P|^2 - |w-P|^2) = \sum(|z_j - w_j|^2 + \overline{(z_j - w_j)}(w_j - P_j) + (z_j - w_j)(w_j - P_j)^*),$$

we get for every $u \in \mathcal{S}$, when z, w belong to some bounded set:

$$\|(|z - P|^2 - |w - P|^2)u\| \leq \epsilon \||z - P|^2 u\| + C_\epsilon \|u\|.$$

From this estimate it is clear that the closures of $|z - P|^2$ and $|w - P|^2$ have the same domain, namely the closure of $\mathcal{S}$ for the norm $\||i - P|^2 u\|$. Let Q denote the unique selfadjoint extension of $|i - P|^2$ and let $\widetilde{Q}$ be the closure of $|z - P|^2$ with the same domain. Then for $0 < R < +\infty$:

$$(Ri + \widetilde{Q})(Ri + Q)^{-1} = I + (\widetilde{Q} - Q)(Ri + Q)^{-1},$$

and

$$\|(\widetilde{Q} - Q)(Ri + Q)^{-1}u\| \leq \epsilon \|Q(Ri + Q)^{-1}u\| + C(\epsilon)\|(Ri + Q)^{-1}u\|$$
$$\leq (\epsilon + \frac{C(\epsilon)}{R})\|u\|.$$

Here $\epsilon + \frac{C(\epsilon)}{R} < 1$ if we first choose $\epsilon = \frac{1}{2}$ and then take $R > 0$ sufficiently large. It follows that $Ri + \widetilde{Q}$ is surjective if R is sufficiently large. Similarly $-Ri + \widetilde{Q}$ is surjective. We conclude that the operators $Q(z)$ with the same domain as the unique selfadjoint extension of $Q(i)$ are selfadjoint. From now on we let $|z - P|^2$ denote these selfadjoint operators, and we also let $|z - P|$ denote the corresponding non-negative square-roots. It also follows that $|z - P|^2$ is invertible for $\operatorname{Im} z \neq 0$ and that the norm of the inverse is $\leq |\operatorname{Im} z|^{-2}$. We conclude that the integrals in (8.18) are well-defined and give rise to a bounded operator.

(2) For $|\operatorname{Im} z| \neq 0$, we notice that

$${}^t\overline{\partial}_z((z - P)^*(\sum_1^m (z_\nu - P_\nu)^*(z_\nu - P_\nu))^{-m} \cdot dz) = 0,$$

or equivalently that

$$\sum_j \overline{\partial}_{z_j}((z_j - P_j)^*(\sum_1^m (z_\nu - P_\nu)^*(z_\nu - P_\nu))^{-m}) = 0,$$

or still:

$$d_z(\sum_{j=1}^m \widetilde{f}(z)(z_j - P_j)^*(\sum_1^m (z_\nu - P_\nu)^*(z_\nu - P_\nu))^{-m} d\overline{z}_j^{\rfloor}(dz \wedge d\overline{z}))$$

$$= \sum_{j=1}^m \overline{\partial}_{z_j}\widetilde{f}(z)(z_j - P_j)^*(\sum_1^m (z_\nu - P_\nu)^*(z_\nu - P_\nu))^{-m} dz \wedge d\overline{z},$$

where
$$dz \wedge d\overline{z} = dz_1 \wedge dz_2 \ldots \wedge dz_m \wedge d\overline{z}_1 \wedge \ldots \wedge d\overline{z}_m,$$

and

$$d\overline{z}_j^{\rfloor}(dz \wedge d\overline{z}) = (-1)^{m+j-1} dz_1 \wedge \ldots \wedge dz_m \wedge d\overline{z}_1 \wedge \ldots \widehat{d\overline{z}_j} \ldots \wedge d\overline{z}_m,$$

the hat indicating an absent factor. Writing the integrals as limits of the corresponding integrals over $|\operatorname{Im} z| \geq \epsilon$, when $\epsilon \to 0$, we then obtain from Stokes' formula:

$$f(P_1, \ldots, P_m)$$
$$= \widehat{C}_m \lim_{\epsilon \to 0} \int_{|\operatorname{Im} z| = \epsilon} \sum_{j=1}^m \widetilde{f}(z)(z_j - P_j)^*((z - P)^*(z - P))^{-m} d\overline{z}_j^{\rfloor}(dz \wedge d\overline{z}).$$

We conclude that if $f = 0$, then since $|\widetilde{f}| \leq C_N |\operatorname{Im} z|^N$, $\forall N \in \mathbf{N}$, we have $f(P_1, \ldots, P_m) = 0$. In other words, for arbitrary $f \in C_0^\infty(\mathbf{R}^m)$, our definition of $f(P_1, \ldots, P_m)$ does not depend on the choice of the almost analytic extension $\widetilde{f}$.

We next show that $f(P_1, \ldots, P_m) \in \mathrm{Op}_h(S^0(1))$ with the help of Beals' lemma. Let $L_j = x_j$, $1 \le j \le n$, $L_j = hD_{x_{j-n}}$, $n+1 \le j \le 2n$. Let

$$\mathrm{ad}_L^\alpha = \mathrm{ad}_{L_1}^{\alpha_1} \ldots \mathrm{ad}_{L_{2n}}^{\alpha_{2n}},$$

$\alpha \in \mathbf{N}^{2n}$. Then

$$\mathrm{ad}_L^\alpha((z_j - P_j)^* |z - P|^{-2m})$$

is a finite linear combination of terms of the form

$$(\mathrm{ad}_L^{\alpha^0}(z_j - P_j)^*)|z - P|^{-2k_1} \mathrm{ad}_L^{\alpha^1}(|z - P|^2)|z - P|^{-2k_2} \ldots$$
$$|z - P|^{-2k_\ell} \mathrm{ad}_L^{\alpha^\ell}(|z - P|^2)|z - P|^{-2k_{\ell+1}}, \quad (8.19)$$

with $k_1 + \ldots + k_{\ell+1} = m + \ell$, $k_1, \ldots, k_{\ell+1} \ne 0$, $\alpha = \alpha^0 + \alpha^1 + \ldots + \alpha^\ell$, $\alpha^1, \ldots, \alpha^\ell \ne 0$. Here α^0 may vanish and we then put $\mathrm{ad}_L^0((z_j - P_j)^*) = (z_j - P_j)^*$. We notice that $\mathrm{ad}_L^{\alpha^j}(|z - P|^2)$ is a linear combination of terms of the form $(\mathrm{ad}_L^{\beta^j}(z_\nu - P_\nu)^*)(\mathrm{ad}_L^{\gamma^j}(z_\nu - P_\nu))$ with $\alpha^j = \beta^j + \gamma^j$. Recall that $|z - P|$ denotes the positive square root of $|z - P|^2$ and let $|z - P|^{-1}$ be the inverse of $|z - P|$, and hence the square root of $|z - P|^{-2}$, the inverse of $|z - P|^2$. If $R \in \mathrm{Op}_h(S^0(M))$, we let $A_j = R|i - P|^{-2}(i - P)_j^* \in \mathrm{Op}_h(S^0(1))$ so that $R = \sum A_j(i - P_j)$. Then, $R = \sum A_j(z_j - P_j) + B$, with $B = \sum A_j(i - z_j) \in \mathrm{Op}_h(S^0(1))$, so

$$\|Ru\| \le C(\sum \|(z_j - P_j)u\| + \|u\|) \le \tilde{C}(\||z - P|u\| + \|u\|)$$

since $\||z - P|u\|^2 = \sum \|(z_j - P_j)u\|^2$. Since $\|u\| \le \frac{1}{|\mathrm{Im}\, z|}\||z - P|u\|$, we get (with z in some compact set in $\mathbf{C}$): $\|Ru\| \le \frac{C}{|\mathrm{Im}\, z|}\||z - P|u\|$, so $\|R|z - P|^{-1}\| \le \frac{C}{|\mathrm{Im}\, z|}$.

Using this we can now estimate the norm of the expression (8.19) by

$$\mathcal{O}(1)h^{|\alpha|}|\mathrm{Im}\, z|^{-2m}\frac{|\mathrm{Im}\, z|}{|\mathrm{Im}\, z|^{|\alpha^0| + \ldots + |\alpha^\ell|}} = \mathcal{O}(1)\frac{1}{|\mathrm{Im}\, z|^{2m-1}}\left(\frac{h}{|\mathrm{Im}\, z|}\right)^{|\alpha|}.$$

As in Proposition 8.4, we then get

$$(z_j - P_j)|z - P|^{-2m} = \mathrm{Op}_h(a_j),$$

with

$$|\partial_x^\alpha \partial_\xi^\beta a_j| \le C_{\alpha,\beta}(\max(1, \frac{\sqrt{h}}{|\mathrm{Im}\, z|}))^{2n+1}|\mathrm{Im}\, z|^{-|\alpha|-|\beta|-(2m-1)}. \quad (8.20)$$

We have then proved the case $N = 0$ in:

Theorem 8.8. *Let* $f \in C_0^\infty(\mathbf{R}^m)$. *Then* $f(P_1, \ldots, P_m) \in \mathrm{Op}_h(S^0(M^{-N}))$ *for every* $N \in \mathbf{N}$.

Proof. For larger N, we write

$$f(\lambda_1, \ldots, \lambda_m) = (m^2 + \lambda_1^2 + \ldots + \lambda_m^2)^{-k} f_k(\lambda_1, \ldots, \lambda_m),$$

where we also notice that $m^2 + \lambda_1^2 + \ldots + \lambda_m^2 = \sum_1^m |i - \lambda_j|^2$ when λ_j are real. We will clearly get the theorem for $N = 2k$, if we manage to prove:

$$f(P_1, \ldots, P_m) = |i - P|^{-2k} f_k(P_1, \ldots, P_m),$$

or equivalently (when acting on functions in $\mathcal{S}$):

$$|i - P|^{2k} f(P_1, \ldots, P_m) = f_k(P_1, \ldots, P_m).$$

It then suffices to show that

$$P_j f(P_1, \ldots, P_m) = g_j(P_1, \ldots, P_m),$$

with $g_j(\lambda) = \lambda_j f(\lambda)$. Here,

$$g_j(P) - P_j f(P) = -\tilde{C}_m \int ((z - P)^* \cdot \overline{\partial}_z \tilde{f}(z))(z_j - P_j)|z - P|^{-2m} L(dz),$$

and

$$(z_j - P_j)|z - P|^{-2m} = \frac{1}{1 - m} \overline{\partial}_{z_j} |z - P|^{2-2m},$$

so we get a constant times

$$\int ((z - P)^* \cdot \overline{\partial}_z \tilde{f}(z)) \overline{\partial}_{z_j} (|z - P|^{2-2m}) L(dz)$$

$$= -\int (\overline{\partial}_{z_j} \tilde{f}(z) + (z - P)^* \cdot \overline{\partial}_z \overline{\partial}_{z_j} \tilde{f}(z))|z - P|^{2-2m} L(dz)$$

$$= -\int (\overline{\partial}_{z_j} \tilde{f})|z - P|^{2-2m} L(dz) + \int \overline{\partial}_{z_j} \tilde{f}(z) \overline{\partial}_z \circ (z - P)^* |z - P|^{2-2m} L(dz).$$

Here

$$\overline{\partial}_z \circ (z - P)^* |z - P|^{2-2m}$$
$$= m|z - P|^{2-2m} + (z - P)^* \cdot (z - P)(1 - m)|z - P|^{-2m} = |z - P|^{2-2m},$$

so we get $g_j(P) - P_j f(P) = 0$.   #

When $p_j \sim p_{j,0} + h p_{j,1} + \ldots$ in $S^0(M)$, we also get, as in the case of functions of one operator, that

$$f(P_1, \ldots, P_m) = \mathrm{Op}_h(a),$$

with

$$a \sim a_0 + h a_1 + h^2 a_2 + \ldots \text{ in } S^0(M^{-N}), \forall N,$$

where

$$a_0(x, \xi) = f(p_{1,0}, \ldots, p_{m,0}).$$

Assume that $f(P)$ has only a discrete spectrum in some open interval I and let $\mu \in I$ be an eigenvalue. Let $F_\mu \subset L^2$ be the corresponding finite dimensional eigenspace. Since P_j and $f(P)$ commute, we see that $P_j : F_\mu \to F_\mu$. It is then easy to see that F_μ has a basis $e_1, \ldots, e_N$ such that $P_j e_k = \lambda_{j,k} e_k$. From this we get

$$f(P_1, \ldots, P_m) e_k = \mu e_k = f(\lambda_{1,k}, \ldots, \lambda_{m,k}) e_k. \tag{8.21}$$

In fact, let us prove the last equality in the slightly more general case when $e \in L^2$, $P_j e = \lambda_j e$ for all j. Then e is in the domain of any power of $|z - P|$, and

$$f(P)e = -\widetilde{C}_m \int \overline{\partial} \widetilde{f} \cdot (z - P)^* |z - P|^{-2m} e L(dz)$$

$$= -\widetilde{C}_m \int \overline{\partial} \widetilde{f}(z) \cdot \overline{(z - \lambda)} |z - \lambda|^{-2m} L(dz) e$$

$$= f(\lambda)e.$$

Notes

Functional calculus occurs frequently in spectral theory and will be useful in the next chapters. There are several different approaches to stating a functional calculus for a pseudodifferential operator. In the case of pseudodifferential operators with discrete spectrum one can use the method of Strichartz [Str], which was adapted by Robert [Ro2] to the case of h-pseudors with compact resolvents. Functions of elliptic selfadjoint operators are studied by Taylor [Ta], using the formula: $f(A) = (2\pi)^{-1} \int_{-\infty}^{+\infty} e^{itA} \hat{f}(t) dt$, and considering e^{itA} as a Fourier integral operator. The Mellin transform was used by Helffer–Robert [HeRo1] to prove Theorem 8.7. Using this method Charbonnel [Char2] developed a functional calculus for functions of several commuting h-pseudors.

A functional calculus for bounded operators on a Banach space, based on the Cauchy formula (8.3), was constructed by Dyn'kin [Dy1] and in the case of several commuting operators by Droste [Dr]. Any formula which represents a function as a superposition of simpler functions can (at least in principle) be taken as a starting point for a functional calculus (by replacing the argument by an operator). The considerable usefulness of the Cauchy formula (8.3) in practical applications was exploited in [HeSj6] (there combined with so-called Grushin reductions) and in many subsequent works. See for instance [HeSj4], [Da2,3], [JeNa], [Gé], [Sk], [Sj4], [SjZw] and [Di1].

9. Trace class operators and applications
of the functional calculus

We start by reviewing some standard facts about Hilbert–Schmidt and trace class operators, and refer for instance to [GoKr] for more details. Let E, F be separable Hilbert spaces and let $A \in \mathcal{L}(E, F)$ be compact. Then A^*A and AA^* are compact selfadjoint operators ≥ 0, and they have the same non-vanishing eigenvalues, $s_1(A)^2$, $s_2(A)^2$, ..., with $s_j(A) \searrow 0$, $j \to \infty$ in the case when there are infinitely many such values. The $s_j(A)$ are called the characteristic or singular values of A.

Definition 9.1. A is Hilbert-Schmidt if $\|A\|_{\mathrm{HS}} := (\sum_1^\infty s_j(A)^2)^{\frac{1}{2}} < \infty$, and A is of trace class if $\|A\|_{\mathrm{tr}} = \sum_1^\infty s_j(A) < \infty$.

The space of Hilbert–Schmidt (HS) and trace class operators are complete normed (i.e. Banach) spaces, and a bounded operator A is HS respective of trace class iff A^* is HS respective of trace class. If (e_j), (f_j) are O.N. bases in E and F respectively, and $(a_{j,k})$ is the corresponding matrix of A, then

$$\|A\|_{\mathrm{HS}}^2 = \sum\sum |a_{j,k}|^2 = \sum \|Ae_j\|^2 = \sum \|A^*f_j\|^2.$$

If $B \in \mathcal{L}(F, H)$ and A is HS, then BA is HS, and $\|BA\|_{\mathrm{HS}} \leq \|B\|\|A\|_{\mathrm{HS}}$. Similarly, if $C \in \mathcal{L}(D, E)$ and A is HS, then AC is HS and $\|AC\|_{\mathrm{HS}} \leq \|A\|_{\mathrm{HS}}\|C\|$.

If $E = L^2(Y, \nu)$, $F = L^2(X, \mu)$, where (Y, ν) and (X, μ) are some measure spaces, then $A : E \to F$ is Hilbert–Schmidt iff A is an integral operator of the form: $Au(x) = \int K(x, y)u(y)\nu(dy)$ with a kernel K in $L^2(X \times Y; \mu \times \nu)$. Moreover, we have $\|K\|_{L^2} = \|A\|_{\mathrm{HS}}$, when A is such an operator.

For trace class operators, we have

$$\|A\|_{\mathrm{tr}} = \sup_{(e_j),(f_j)} \sum |(Ae_j|f_j)|,$$

where (e_j) and (f_j) are O.N. bases in E, F. (We only consider the non-trivial case, when both E and F are of infinite (countable) dimension.)

If B is a bounded operator and A is of trace class, then BA is of trace class and $\|BA\|_{\mathrm{tr}} \leq \|B\|\|A\|_{\mathrm{tr}}$, and we have the similar result for compositions of the type AB, still with B bounded and A of trace class.

If $A : E \to E$ is of trace class, then $\operatorname{tr} A := \sum (Ae_j|e_j)$ is independent of the choice of the O.N. basis (e_j) and $|\operatorname{tr} A| \leq \|A\|_{\mathrm{tr}}$. If $A : E \to F$ is of trace class and $B : F \to E$ is bounded, then $\operatorname{tr} AB = \operatorname{tr} BA$.

If $A : E \to F$ and $B : F \to G$ are HS, then BA is of trace class and $\|BA\|_{\mathrm{tr}} \le \|B\|_{\mathrm{HS}}\|A\|_{\mathrm{HS}}$. In the case when $G = E$, we also have $\operatorname{tr} BA = \operatorname{tr} AB$.

The HS operators $E \to F$ form a Hilbert space with scalar product: $(A|B) = \operatorname{tr} B^*A$.

If $A = \mathrm{Op}_t(a)$, then the distribution kernel of A is $K(x,y) = \breve{a}(tx + (1-t)y, x-y)$, where $\breve{a}$ denotes the inverse Fourier transform with respect to the last variable. The change of variables, $\widetilde{x} = tx + (1-t)y$, $\widetilde{y} = x - y$ has a Jacobian of absolute value 1, so we get:

Proposition 9.2. *Let $a \in \mathcal{S}'(\mathbf{R}^{2n})$. Then $\mathrm{Op}_t(a)$ is HS iff $a \in L^2(\mathbf{R}^{2n})$, we also have*

$$\|\mathrm{Op}_t(a)\|_{\mathrm{HS}}^2 = \frac{1}{(2\pi)^n} \iint |a(x,\theta)|^2 dx d\theta.$$

The study of pseudors of trace class is slightly more delicate, and we cannot give a complete characterization as in the case of HS operators. We shall only give a sufficient condition for a pseudor to be of trace class. We start by looking at a class of integral operators. Let $Au(x) = \int K(x,y)u(y)dy$ be an integral operator $L^2(\mathbf{R}^n) \to L^2(\mathbf{R}^n)$, say with $K \in \mathcal{S}(\mathbf{R}^{2n})$, to start with. We shall estimate the trace class norm of A. Let $1 = \sum_{(j,k)\in\mathbf{Z}^{2n}} \chi_{j,k}$ be a partition of unity with $\chi_{j,k} = \tau_{(j,k)}\chi_{0,0}$, $\chi_{0,0} \in C_0^\infty(\mathbf{R}^{2n})$. Then $K(x,y) = \sum K_{j,k}(x,y)$, where $K_{j,k}(x,y) = \chi_{j,k}(x,y)K(x,y)$. We estimate the trace norm of the operator corresponding to $K_{j,k}$, and without loss of generality, we may assume that $(j,k) = (0,0)$. Choose $\psi \in C_0^\infty(\mathbf{R}^n)$ such that $\psi(x)\psi(y) = 1$ near $\mathrm{supp}\,(K_{0,0})$. Then

$$K_{0,0}(x,y) = \frac{1}{(2\pi)^{2n}} \iint \widehat{K}_{0,0}(\xi,\eta)\psi(x)e^{ix\cdot\xi}\psi(y)e^{iy\cdot\eta}d\xi d\eta$$

represents $K_{0,0}$ as a superposition of rank 1 kernels. In general, if $B : u \mapsto (u|e)f$ is a rank 1 operator with f,e in some Hilbert space $\mathcal{H}$, then $\|B\|_{\mathrm{tr}} = \|e\|\|f\|$. Thus if $A_{0,0}$ is the operator corresponding to $K_{0,0}$, we get

$$\|A_{0,0}\|_{\mathrm{tr}} \le \frac{1}{(2\pi)^{2n}} \iint |\widehat{K}_{0,0}(\xi,\eta)|d\xi d\eta \|\psi\|^2.$$

We get the same estimates for $A_{j,k}$ given by $K_{j,k}$ and we conclude that

$$\|A\|_{\mathrm{tr}} \le C \sum_{(j,k)\in\mathbf{Z}^{2n}} \|\widehat{\chi_{j,k}K}\|_{L^1}, \tag{9.1}$$

where C depends on the partition of unity. From this we get

$$\|A\|_{\mathrm{tr}} \leq C \sum_{|\alpha| \leq 2n+1} \|\partial_{x,y}^{\alpha} K\|_{L^1}. \tag{9.2}$$

By density, we conclude that if $K \in \mathcal{S}'(\mathbf{R}^{2n})$ and if the right hand side of (9.1) is finite, then A is of trace class and we have (9.1) and (9.2) respectively. Also notice that when the RHS of (9.1) (or (9.2)) is finite, then K is a bounded continuous function so we can define $K(x,x)$ which is integrable. When $K \in \mathcal{S}(\mathbf{R}^{2n})$, we see (for instance by approximating K by finite rank kernels) that

$$\mathrm{tr}\,(A) = \int K(x,x)dx, \tag{9.3}$$

and this extends to general Ks for which the RHS of (9.1) is finite.

We next turn to pseudors, and we start with the case of the classical quantization ($t = 1$). If $A = \mathrm{Op}_1(a)$, we can write

$$Au(x) = \frac{1}{(2\pi)^n} \int e^{ix\cdot\xi} a(x,\xi) \mathcal{F}u(\xi)d\xi,$$

and since $\mathcal{F}$ is a bounded operator, it is enough to estimate the trace class norm of the operator with kernel $K(x,\xi) = e^{ix\cdot\xi}a(x,\xi)$. Writing $x\cdot\xi = -j\cdot k + j\cdot\xi + x\cdot k + (x-j)\cdot(\xi-k)$, we get

$$\mathcal{F}(\chi_{j,k}(x,\xi)e^{ix\cdot\xi}a(x,\xi))(x^*,\xi^*) = e^{-ij\cdot k}\mathcal{F}(e^{i(x-j)\cdot(\xi-k)}\chi_{j,k}a)(x^*-k,\xi^*-j).$$

Let $\widetilde{\chi}_{0,0} \in C_0^{\infty}$ be equal to 1 near the support of $\chi_{0,0}$ and put $\widetilde{\chi}_{j,k} = \tau_{j,k}\widetilde{\chi}_{0,0}$. Then the expression above becomes

$$e^{-ij\cdot k}\mathcal{F}(\psi_{j,k}\chi_{j,k}a)(x^*-k,\xi^*-j) = e^{-ij\cdot k}\frac{1}{(2\pi)^{2n}}(\widehat{\psi}_{j,k} * \widehat{\chi_{j,k}a})(x^*-k,\xi^*-j),$$

where $\psi_{j,k} = \tau_{j,k}\psi_{0,0}$, $\psi_{0,0} = e^{ix\cdot\xi}\widetilde{\chi}_{0,0}$. Since $\widehat{\psi}_{j,k}$ and $\widehat{\psi}_{0,0}$ have the same L^1 norm, we get

$$\|\mathcal{F}(\chi_{j,k}e^{ix\cdot\xi}a)\|_{L^1} \leq C\|\widehat{\chi_{j,k}a}\|_{L^1}.$$

We conclude that A is of trace class if

$$\sum_{(j,k)\in\mathbf{Z}^{2n}} \|\widehat{\chi_{j,k}a}\|_{L^1} < \infty, \tag{9.4}$$

or if we make the stronger assumption

$$\sum_{|\alpha| \leq 2n+1} \|\partial_{x,\theta}^{\alpha}a\|_{L^1} < \infty. \tag{9.5}$$

By a density argument, we also see that

$$\operatorname{tr} \operatorname{Op}_1(a) = \frac{1}{(2\pi)^n} \iint a(x,\xi) dx d\xi, \qquad (9.6)$$

when (9.4) holds. The corresponding formula for the h-quantization is:

$$\operatorname{tr} \operatorname{Op}_{1,h}(a) = \frac{1}{(2\pi h)^n} \iint a(x,\xi) dx d\xi. \qquad (9.7)$$

We next look at the Weyl quantization.

Lemma 9.3. *Let $\Phi(x)$ be a non-degenerate quadratic form on $\mathbf{R}^m$ and let $1 = \sum_{j \in \mathbf{Z}^m} \chi_j$, $\chi_j = \tau_j \chi_0$, $\chi_0 \in C_0^\infty$. Then there is a constant $C > 0$ such that*

$$\sum_j \|\mathcal{F}(\chi_j(e^{i\Phi} * u))\|_{L^1} \le C \sum_j \|\widehat{\chi_j u}\|_{L^1}, \qquad (9.8)$$

for all $u \in \mathcal{S}$.

Proof. As in (8.7) we have

$$\mathcal{F}(\chi_j(e^{i\Phi} * (\chi_k u)))(\xi)$$
$$= \mathcal{O}(1) \frac{1}{\langle \partial_x \Phi(j-k) - \xi \rangle^N} \int \frac{1}{\langle \eta - \partial_x \Phi(j-k) \rangle^N} |\widehat{\chi_k u}(\eta)| d\eta.$$

Consequently, if we fix $N \le m + 1$:

$$\|\mathcal{F}(\chi_j(e^{i\Phi} * (\chi_k u)))\|_{L^1} = \mathcal{O}(1) \int \frac{1}{\langle \eta - \partial_x \Phi(j-k) \rangle^N} |\widehat{\chi_k u}(\eta)| d\eta$$

and if we sum over j and use the fact that Φ is non-degenerate, we get

$$\sum_j \|\mathcal{F}(\chi_j(e^{i\Phi} * (\chi_k u)))\|_{L^1} = \mathcal{O}(1) \|\widehat{\chi_k u}\|_{L^1}.$$

Summing over k, we get (9.8). #

If $\operatorname{Op}(a_w) = \operatorname{Op}_1(a)$, then $a = C e^{i\Phi} * a_w$, where Φ is a non-degenerate quadratic form, so if (9.4) holds for a_w, then it also holds for a, and

$$\sum_{j,k} \|\widehat{\chi_{j,k} a}\|_{L^1} \le C \sum_{j,k} \|\widehat{\chi_{j,k} a_w}\|_{L^1}.$$

We then get

Theorem 9.4. *Let $a \in S'(\mathbf{R}^{2n})$ satisfy (9.4) (which is a weaker assumption than (9.5)). Then $\mathrm{Op}(a)$ is of trace class and*

$$\|\mathrm{Op}(a)\|_{\mathrm{tr}} \leq C_n \sum_{j,k} \|\widehat{\chi_{j,k}a}\|_{L^1}, \tag{9.9}$$

$$\mathrm{tr}\,\mathrm{Op}(a) = \frac{1}{(2\pi)^n} \iint a(x,\xi)dxd\xi. \tag{9.10}$$

Let $m \geq 1$ be an order function and let $p \in S^0(m)$ be real-valued. For simplicity, we assume p to be independent of h. If m is unbounded, we also assume that $p+i$ is elliptic. Let $I \subset\subset \mathbf{R}$ be an open interval and assume that $\underline{\lim}_{|(x,\xi)|\to\infty}\mathrm{dist}\,(p(x,\xi),I) > 0$. We can assume, without loss of generality, that $\underline{\lim}\,p(x,\xi) - \sup I > 0$. Let $\widetilde{p}(x,\xi) \in S(m)$ be real-valued coinciding with p for large (x,ξ), and have the property that $\widetilde{p} > \sup I$ everywhere. For $0 < h \leq h_0 > 0$ sufficiently small, $p^w = p^w(x,hD)$ and $\widetilde{p}^w = \widetilde{p}^w(x,hD)$ are essentially selfadjoint, with the same domain, and $(z-\widetilde{p})^{-1}$ is well defined and holomorphic for z in some fixed complex neighborhood of $\overline{I}$. If $f \in C_0^\infty(I)$ and if we choose the almost analytic extension $\widetilde{f}$ with support sufficiently close to I, we see that $f(\widetilde{p}^w) = 0$, either by abstract theory, since $\widetilde{p}^w$ has no spectrum in I, or by an integration by parts in the formula analogous to (8.3). Continuing to follow a trick of Dimassi [Di1], we write the resolvent identity for $\mathrm{Im}\,z \neq 0$:

$$(z - p^w)^{-1} = (z - \widetilde{p}^w)^{-1} + (z - p^w)^{-1}(p^w - \widetilde{p}^w)(z - \widetilde{p}^w)^{-1},$$

substitute into (8.3) and integrate. We get:

$$f(p^w) = -\frac{1}{\pi} \int \overline{\partial}\widetilde{f}(z)(z - p^w)^{-1}(p^w - \widetilde{p}^w)(z - \widetilde{p}^w)^{-1}L(dz). \tag{9.11}$$

Here $p^w - \widetilde{p}^w$ is of trace class, with trace class norm $= \mathcal{O}(h^{-n})$, so we conclude that $f(p^w)$ is of trace class and of trace class norm $= \mathcal{O}(h^{-n})$. It follows that $\mathrm{tr}\,f(p^w) = \sum f(\lambda_j) = \mathcal{O}(h^{-n})$. In this argument, we may replace I by a slightly larger interval, and choose $f \geq 0$ of class C_0^∞ with support in the slightly larger interval and equal to 1 on I. It follows that *the number of eigenvalues of p^w in I is $\mathcal{O}(h^{-n})$.*

In order to get further, we observe the following simple result:

Proposition 9.5. *Let $a_j \in S^0(m_j)$, $j = 1,2$ and assume that $\mathrm{supp}\,(a_1) \subset K_1 \subset\subset \mathbf{R}^{2n}$, where K_1 is independent of h. Then for every $N \in \mathbf{N}$ and $C > 0$:*

$$|\partial^\alpha(a_1 \natural_h a_2)| \leq C_N(h^N \mathrm{dist}\,((x,\xi),K_1)^{-N}),$$

for dist $((x, \xi), K_1) \geq \frac{1}{C}$.

This can be proved either directly by integrations by parts, or by using the pseudor calculus of Chapter 7, with suitable new choices of order functions.

Let $\chi \in C_0^\infty(\mathbf{R}^{2n})$ be equal to 1 in a neighborhood of $\operatorname{supp}(p - p^w)$ and consider

$$f(p^w)(1 - \chi^w) = -\frac{1}{\pi} \int \bar{\partial} \tilde{f}(z)(z - p^w)^{-1}(p^w - \tilde{p}^w)(z - \tilde{p}^w)^{-1}(1 - \chi^w)L(dz).$$

Using the preceding proposition, we see that all derivatives of the symbol of $(p - \tilde{p})^w(z - \tilde{p}^w)^{-1}(1 - \chi^w)$ are $\mathcal{O}(h^N \langle (x, \xi) \rangle^{-N})$ for every $N \in \mathbf{N}$. The trace class norm of this expression is therefore $\mathcal{O}(h^\infty)$, and consequently

$$\|f(p^w)(1 - \chi^w)\|_{\mathrm{tr}} = \mathcal{O}(h^\infty),$$

in particular,

$$\operatorname{tr} f(p^w) = \operatorname{tr}(f(p^w)\chi^w) + \mathcal{O}(h^\infty). \tag{9.12}$$

Now recall that

$$f(p^w) = \operatorname{Op}_h(a),$$

where $a \in S^0(1)$, $a \sim a_0 + ha_1 + h^2 a_2 + \ldots$ with $a_0 = f(p(x, \xi))$ and with $a_1 = 0$ (assuming that f is independent of h). Moreover, $a_j(x, \xi)$ vanishes outside $\operatorname{supp}(p - \tilde{p})$. Consequently, we obtain

Theorem 9.6. *We have*

$$\operatorname{tr} f(p^w) \sim \frac{1}{(2\pi h)^n} \sum_{j=0}^\infty h^j \int a_j(x, \xi) dx d\xi.$$

Here a_j have the properties recalled above. In particular $a_0 = f(p)$ and $a_1 = 0$.

Corollary 9.7. *Let $[\alpha, \beta] \subset\subset I$ be an h-independent sub-interval, and let $N([\alpha, \beta]; h)$ denote the number of eigenvalues of p^w in $[\alpha, \beta]$. Then when $h \to 0$:*

$$\frac{1}{(2\pi h)^n}(\underline{V}([\alpha, \beta]) + o(1)) \leq N([\alpha, \beta]; h) \leq \frac{1}{(2\pi h)^n}(\overline{V}([\alpha, \beta]) + o(1)),$$

where

$$\overline{V}([\alpha, \beta]) = \lim_{\epsilon \searrow 0} \int\!\!\int_{p(x, \xi) \in [\alpha - \epsilon, \beta + \epsilon]} dx d\xi,$$

$$\underline{V}([\alpha, \beta]) = \lim_{\epsilon \searrow 0} \int\!\!\int_{p(x, \xi) \in [\alpha + \epsilon, \beta - \epsilon]} dx d\xi.$$

Proof. Assume for simplicity that $\alpha < \beta$. For every small $\epsilon > 0$, choose $\overline{f}, \underline{f} \in C_0^\infty(I; [0,1])$ with

$$1_{[\alpha+\epsilon,\beta-\epsilon]} \leq \underline{f} \leq 1_{[\alpha,\beta]} \leq \overline{f} \leq 1_{[\alpha-\epsilon,\beta+\epsilon]}.$$

It then suffices to observe that

$$\operatorname{tr} \underline{f}(p^w) \leq N([\alpha,\beta]; h) \leq \operatorname{tr} \overline{f}(p^w),$$

and to apply the preceding theorem.   #

Notes

Theorem 9.4 and its proof are close to the arguments of [Sj1,2]. See also Boulkhemair [Bo]. Theorem 9.6 is due to Helffer–Robert [HeRo] using a functional calculus based on the Mellin transformation. Here we used the Cauchy formula (8.3) together with an idea from [Di1].

10. More precise spectral asymptotics for non-critical Hamiltonians

Under an additional assumption, we shall improve Corollary 9.7. The idea is to study the unitary group $e^{-itP/h}$ and more precisely, we shall study the trace of $e^{-itP/h}f(P)$, where $f \in C_0^\infty(\mathbf{R})$ is independent of h. We assume that P, f, I satisfy the assumptions given after Theorem 9.4. As we could have done already in the preceding chapter, we allow the symbol p of P to depend on h and assume that it has an asymptotic expansion $p \sim p_0 + hp_1 + \dots$ in $S^0(m)$, where $m \geq 1$ is an order function. As noticed there, we assume without loss of generality that $\varliminf_{|(x,\xi)|\to\infty} p_0(x,\xi) - \sup I > 0$. Then the full symbol is bounded from below, and it is easy to prove that for $h > 0$ sufficiently small, the operator P is bounded from below: $P \geq -C_0$ for some constant $C_0 > 0$. Increasing C_0 if necessary, we may also assume that $I \subset]-C_0+1, C_0-1[$. We can find a a smooth real-valued function $g(\lambda)$ such that $g(\lambda) = \lambda$ in a neighborhood of $\overline{I}$, g is increasing on $[-C_0, \infty[$ and g is equal to a constant plus a C_0^∞ function. Then $g(P)$ and P will have the same eigenvalues in I. Moreover $\{\rho \in \mathbf{R}^{2n}; p_0(\rho) \in I\} = \{\rho \in \mathbf{R}^{2n}; g(p_0(\rho)) \in I\}$ and $p_0 = g(p_0)$ on this set. Also, $g(P) = a_g^w$, where $a_g \in S^0(1)$ has an asymptotic expansion in powers of h, and this asymptotic expansion coincides with that of p in a neighborhood of the (compact) set above. This means that as long as we are only interested in the eigenvalues of P in I, we may replace P by $g(P)$, and m by 1. So from now on we may assume that $m = 1$.

Let $f(P) = a_f^w(x, hD_x; h)$ and let $\chi \in C_0^\infty(\mathbf{R}^{2n})$ be equal to 1 near $\overline{p_0^{-1}(I)}$. From the results of Chapter 9, it is easy to show that

$$f(P)u(x) = \frac{1}{(2\pi h)^n} \iint e^{i(x-y)\cdot\theta/h}\chi(x,\theta)a_f(\frac{x+y}{2}, \theta; h)\chi(y,\theta)u(y)dyd\theta$$
$$+K(h)u, \qquad (10.1)$$

where $\|K(h)\|_{\mathrm{tr}} = \mathcal{O}(h^\infty)$.

We want to approximate $V_f(t) := e^{-itP/h}f(P)$ for $|t|$ small by a Fourier integral operator of the form

$$U_f(t)u(x) = \frac{1}{(2\pi h)^n} \iint e^{i(\phi(t,x,\theta)-y\cdot\theta)/h}b(t,x,y,\theta;h)u(y)dyd\theta, \qquad (10.2)$$

where $b \in C^\infty(]-\frac{1}{C}, \frac{1}{C}[; S^0(1))$ should have uniformly compact support in (x, y, θ), and $\phi(t, x, \theta)$ should be real, smooth and well-defined near the support of b. Trying to solve the equation $(hD_t + P)U_f = 0$, we are led to the problem:

$$e^{-i\phi/h}(hD_t + P)(e^{i\phi/h}b) \in C^\infty(]-\frac{1}{C}, \frac{1}{C}[; S^{-\infty}(1)), \qquad (10.3)$$

$$e^{i\phi/h}b|_{t=0} = e^{ix\cdot\theta/h}\chi(x,\theta)a_f(\frac{x+y}{2},\theta;h)\chi(y,\theta), \qquad (10.4)$$

where P acts in the x variables. We choose ϕ to solve

$$\partial_t\phi + p_0(x,\partial_x\phi) = 0, \quad \phi|_{t=0} = x\cdot\theta, \qquad (10.5)$$

for $|t|$ small enough and (x,θ) in some sufficiently large but fixed compact set. (Actually, the problem can be solved globally in (x,θ), with $\phi(t,x,\theta) - x\cdot\theta \in C^\infty(]-\frac{1}{C},\frac{1}{C}[;S(1))$.) By integrations by parts and the method of stationary phase, we see that if $b \in C^\infty(]-\frac{1}{C},\frac{1}{C}[;S^0(1))$ has uniformly compact support in (x,y,θ), and $b \sim b_0 + hb_1 + \ldots$, then

$$P(e^{i\phi/h}b) = e^{i\phi/h}d,$$

where $d \in C^\infty(]-\frac{1}{C},\frac{1}{C}[;S^0(1))$ has an asymptotic expansion

$$d \sim d_0(t,x,y,\theta) + d_1(t,x,y,\theta)h + \ldots,$$

with

$$d_0(t,x,y,\theta) = p_0(x,\partial_x\phi)b_0(t,x,y,\theta),$$

$$d_1(t,x,y,\theta) = \frac{1}{i}\partial_\xi p_0(x,\partial_x\phi)\cdot\partial_x c_0 + q_0(x,\theta)c_0$$

plus terms depending on p_1 and c_1.

Here q_0 is a smooth function which can be expressed in terms of p_0, p_1, ϕ. Moreover, c_j vanish for (x,y,θ) outside some fixed compact set, and c and its derivatives are $\mathcal{O}(h^N\langle(x,y,\theta)\rangle^{-N})$ away from that set. Here we have assumed that ϕ has been suitably defined far away, either by solving the eikonal equation globally or by making ϕ equal to a constant far away.

Then $(hD_t + P)(e^{i\phi/h}b) = e^{i\phi/h}hc$, $c \in C^\infty(]-\frac{1}{C},\frac{1}{C}[;S^0(1))$, and $c \sim c_0 + hc_1 + \ldots$, where

$$c_0 = \frac{1}{i}\partial_t b_0 + (\frac{1}{i}\partial_\xi(p_0)(x,\partial_x\phi)\cdot\partial_x + q_0(t,x,\theta))b_0. \qquad (10.6)$$

By solving the usual transport equations, we get $b \in C^\infty(]-\frac{1}{C},\frac{1}{C}[;S^0(1))$ with uniformly compact support in (x,y,θ), such that $(hD_t + P)(e^{i\phi/h}b) = e^{i\phi/h}hc$, with $c \in C^\infty(]-\frac{1}{C},\frac{1}{C}[;S^{-N}(\langle(x,y,\theta)\rangle^{-N}))$ for every $N \in \mathbf{N}$. Then

$$(hD_t + P)U_f(t) = \widetilde{K}(t;h), \quad U_f(0) = f(P) - K(h), \qquad (10.7)$$

with $\widetilde{K}$, $K = \mathcal{O}(h^\infty)$ in trace norm.

We want to compare $U_f(t)$ with $V_f(t) := e^{-itP/h}f(P)$. Put $W_f(t) = U_f(t) - V_f(t)$. Then $(hD_t + P)W_f(t) = \mathcal{O}(h^\infty)$, $W_f(0) = \mathcal{O}(h^\infty)$ in trace norm. Consider $e^{itP/h}W_f(t)$:

$$hD_t(e^{itP/h}W_f(t)) = e^{itP/h}(hD_t + P)W_f(t) = \mathcal{O}(h^\infty) \text{ in trace norm.}$$

Hence, $e^{itP/h}W_f(t) = \mathcal{O}(h^\infty)$ and consequently $W_f(t) = \mathcal{O}(h^\infty)$ in trace norm, so we obtain:

$$e^{-itP/h}f(P) = U_f(t) + K(h), \text{ where } K(h) = \mathcal{O}(h^\infty) \text{ in trace norm.} \quad (10.8)$$

We assume, (possibly after shrinking I) that

$$p_0(x,\xi) \in \overline{I} \Rightarrow dp_0(x,\xi) \neq 0, \quad (10.9)$$

and we want to study

$$\text{tr}\,\frac{1}{2\pi h}\int e^{it(\lambda-P)/h}\psi(t)f(P)dt = \text{tr}\,(\mathcal{F}_h^{-1}(\psi)(\lambda - P))f(P), \quad (10.10)$$

when $\psi \in C_0^\infty(]-\frac{1}{C},\frac{1}{C}[)$, $\lambda \in I$. In view of (10.8) we see that the expression (10.10) is equal to $\mathcal{O}(h^\infty)$ plus

$$\frac{1}{(2\pi h)}\frac{1}{(2\pi h)^n}\iiint e^{i(\lambda t+\phi(t,x,\theta)-x\cdot\theta)/h}\psi(t)b(t,x,x,\theta;h)dxd\theta dt. \quad (10.11)$$

We look for critical points of the phase

$$\lambda t + \phi(t,x,\theta) - x\cdot\theta = t(\lambda - p_0(x,\theta) + \mathcal{O}(t)) \quad (10.12)$$

with t small. Putting the t derivative equal to 0, we see that $p_0(x,\theta) = \lambda + \mathcal{O}(t)$, and since $dp_0 \neq 0$, when $p_0 = \lambda$, the phase is critical for $|t|$ small precisely when $t = 0$, $p_0(x,\theta) = \lambda$. Near any such critical point we choose local coordinates $t, p_0, \omega_1, \ldots, \omega_{2n-1}$ and consider the Hessian of (10.12) with respect to t, p_0 at the critical point:

$$\begin{pmatrix} * & -1 \\ -1 & 0 \end{pmatrix}.$$

This is a non-degenerate matrix of determinant -1 and of signature 0. After introducing some partition of unity on the energy surface $p_0(x,\theta) = \lambda$ and applying the method of stationary phase with respect to the t, p_0-integration, we see that the expression (10.11) has an asymptotic expansion in powers of h:

$$\sim t_0(\lambda)h^{-n} + t_1(\lambda)h^{1-n} + \dots \qquad (10.13)$$

Here we use the fact that $\chi = 1$, $b = f(p_0) + \mathcal{O}(h)$ in the interesting region, and get

$$t_0(\lambda) = \psi(0)f(\lambda)\frac{1}{(2\pi)^n}\int_{p_0=\lambda} \left|\det \frac{d(x,\theta)}{d(p_0,\omega_1,\dots,\omega_{2n-1})}\right|d\omega_1\dots d\omega_{2n-1}, \qquad (10.14)$$

where in order to simplify the notation we neglected to write down the intermediate step with the partition of unity on the energy surface.

Recall that the Liouville form on the hypersurface $p_0 = \lambda$ is the $2n-1$ form $L_\lambda = g_\lambda(\omega)d\omega_1 \wedge \dots \wedge d\omega_{2n-1}$ such that $L_\lambda \wedge dp_0 = dx \wedge d\theta$, and we shall view L_λ as a density. The definition of the Liouville form does not depend on the choice of local coordinates, since $dx \wedge d\theta$ is the symplectic volume element and since we have $(M \wedge dp_0)|_{p_0=\lambda} = 0$ at the points where $p_0 = \lambda$, if M is any $2n-1$ form whose restriction to $p_0 = \lambda$ is zero. (10.14) becomes

$$t_0(\lambda) = \frac{\psi(0)f(\lambda)}{(2\pi)^n}\int_{p_0=\lambda}L_\lambda(d\omega). \qquad (10.15)$$

The right hand side of (10.10) can be written as

$$(\mathcal{F}_h^{-1}\psi) * \mu_f(\lambda), \qquad (10.16)$$

where $\mathcal{F}_h^{-1}\psi(\lambda) = \frac{1}{h}\tilde\psi(\frac{\lambda}{h}) = \tilde\psi_h(\lambda)$, $\tilde\psi = \mathcal{F}_1^{-1}\psi$, $\mu_f = \sum f(\lambda_j)\delta(\lambda - \lambda_j)$, and λ_j denote the eigenvalues of P in I. If ψ is of the form $\frac{1}{2\pi}g * \check g$, with $g \in C_0^\infty(]-\frac{1}{2C},\frac{1}{2C}[)$ real, $\check g(t) = g(-t)$, then $\tilde\psi = |\hat g|^2 \geq 0$. We may also arrange that $\psi(0) = 1$ or, equivalently, that $\int \tilde\psi d\lambda = 1$.

A consequence of (10.16,13,14) is that if $N(J;h)$ denotes the number of eigenvalues of P in a subinterval J of I and if the length $|J|$ of J is h, then $N(J;h) = \mathcal{O}(h^{1-n})$.

Put $\ell(\lambda) = \int_{p_0=\lambda}L_\lambda(d\omega)$, so that

$$(\mathcal{F}_h^{-1}\psi * \mu_f)(\lambda) = \frac{1}{(2\pi h)^n}(f(\lambda)\ell(\lambda) + \mathcal{O}(h)), \quad \lambda \in I. \qquad (10.17)$$

This relation extends to all $\lambda \in \mathbf{R}$, and we can there replace $\mathcal{O}(h)$ by $\mathcal{O}(\frac{h}{\langle\lambda\rangle^n})$. We integrate this from $-\infty$ to $\lambda \in I$:

$$\int\left(\int_{-\infty}^{\lambda}\tilde\psi_h(\lambda' - y)d\lambda'\right)\mu_f(dy)$$
$$= \frac{1}{(2\pi h)^n}\left(\iint_{p_0(x,\xi)\leq\lambda}f(p_0(x,\xi))dxd\xi + \mathcal{O}(h)\right), \qquad (10.18)$$

and here

$$\int_{-\infty}^{\lambda} \tilde{\psi}_h(\lambda' - y)d\lambda' = \int_{-\infty}^{0} \tilde{\psi}(\frac{\lambda - y}{h} + t)dt = 1_{]-\infty,\lambda]}(y) + \mathcal{O}(\langle\frac{\lambda - y}{h}\rangle^{-\infty}).$$

According to the observation prior to (10.17), we have

$$\int \langle\frac{\lambda - y}{h}\rangle^{-\infty}\mu_f(dy) = \mathcal{O}(1)(\sum_{k=0}^{\infty}\langle k\rangle^{-\infty})h^{1-n} = \mathcal{O}(h^{1-n}),$$

so (10.18) gives

$$\int_{-\infty}^{\lambda} \mu_f(dy) = \frac{1}{(2\pi h)^n}(\iint_{p_0(x,\xi)\leq\lambda} f(p_0(x,\xi))dxd\xi + \mathcal{O}(h)). \qquad (10.19)$$

We now drop the assumption that all values in $\overline{I}$ are non-critical for p_0.

Theorem 10.1. *Let* $[a,b] \subset I$, $a < b$ *and assume that* a, b *are not critical values of* p_0. *Then the number of eigenvalues of* P *in* $[a,b]$ *is*

$$N([a,b];h) = \frac{1}{(2\pi h)^n}(\iint_{p_0(x,\xi)\in[a,b]} dxd\xi + \mathcal{O}(h)).$$

Proof. Choose $f_j \in C_0^\infty(\mathbf{R})$ such that f_2 has its support in $]a,b[$, such that f_1 and f_3 have their supports in small neighborhoods of a and b respectively and such that $f_1 + f_2 + f_3 = 1$ on $[a,b]$. Then we can use (10.19) and its obvious analogue to study

$$\sum_{\lambda_j\leq b} f_3(\lambda_j) \text{ and } \sum_{\lambda_j\geq a} f_1(\lambda_j),$$

while $\sum f_2(\lambda_j)$ has already been studied in Chapter 9.   #

Notes

Theorem 10.1 was proved in the case of Schrödinger operators with compact resolvent by Chazarain [Ch], who constructed parametrices for small times for the associated evolution equation. In the general case the theorem is due to Helffer–Robert [HeRo2], Ivrii [I1]. The leading term in the expansion, 'the Weyl term', can be obtained by many other methods. In the case of a selfadjoint elliptic operator of arbitrary positive order on a compact manifold, a Weyl asymptotic with small remainder is due to Hörmander [Hö3].

11. Improvement when the periodic trajectories form a set of measure 0

In this chapter we shall estimate the large time behaviour of the unitary group and see that, under an additional assumption, we can get a two term asymptotic result for the number of eigenvalues in an interval.

As a preparation, we discuss (a special case of) the Egorov theorem for conjugation of a pseudor with a Fourier integral operator (from now on fourior for short). For simplicity we work in trace classes, and only prove what we shall need later.

Let $p \sim p_0 + hp_1 + \ldots \in S^0(1)$ be real-valued. Let $q \sim q_0 + hq_1 + \ldots \in S^0(m)$, where the order function m is bounded and integrable. We shall study $Q_t = e^{itP/h} Q e^{-itP/h}$ modulo $\mathcal{O}(h^\infty)$ in trace norm, where $P = p^w$, $Q = q^w$. To do this we construct an approximation for Q_t. Notice that

$$hD_t(Q_t) = \mathrm{ad}_P Q_t, \tag{11.1}$$

or if we write $Q_t = q_t^w$:

$$\partial_t q_t = \frac{i}{h}(p \sharp_h q_t - q_t \sharp_h p), \quad q_{|t=0} = q. \tag{11.2}$$

We look for an approximate solution $\widetilde{q} \sim \widetilde{q}_0(t, x, \xi) + h\widetilde{q}_1(t, x, \xi) + \ldots \in S^0(m)$ of (11.2) with an error in $S^{-\infty}(m)$. We first get

$$\partial_t \widetilde{q}_0 = H_{p_0} \widetilde{q}_0, \quad \widetilde{q}_0|_{t=0} = q_0, \tag{11.3}$$

which has the unique solution

$$\widetilde{q}_0(t, x, \xi) = q_0(\exp(tH_{p_0})(x, \xi)) \in S(m),$$

and the higher order symbols are obtained by solving transport equations of similar type. We get $\mathrm{supp}\,(\widetilde{q}_j(t, \cdot)) \subset \exp(-tH_{p_0})(K)$, if K is the union of the supports of the q_j. Choosing a suitable asymptotic sum for $\widetilde{q}$, we obtain a smooth family of operators $\widetilde{Q}_t$ with $(hD_t - \mathrm{ad}_P)(\widetilde{Q}_t) = R_t$ with $R_t \in \mathrm{Op}_h(S^{-\infty}(m))$ uniformly for t in any compact interval, and with $\widetilde{Q}_0 = Q$. In particular, $\|R_t\|_{\mathrm{tr}} = \mathcal{O}(h^\infty)$ uniformly for t in any bounded interval. We write

$$hD_t(e^{-itP/h}\widetilde{Q}_t e^{itP/h}) = e^{-itP/h} R_t e^{itP/h}$$

and integrate in t, and obtain:

$$\|e^{-itP/h}\widetilde{Q}_t e^{itP/h} - Q\|_{\mathrm{tr}} = \mathcal{O}(h^\infty)$$

uniformly on any bounded interval. Consequently,

$$e^{itP/h}Qe^{-itP/h} = \widetilde{Q}_t + \mathcal{O}(h^\infty) \text{ in trace norm,} \qquad (11.4)$$

uniformly for t in any bounded interval. (The more precise and general forms of Egorov's theorem describe $e^{itP/h}Qe^{-itP/h}$ more completely as a pseudor, even with $e^{-itP/h}$ replaced by a more general fourior. We give such a result in the appendix to this chapter.)

Let $P = p^w$ satisfy essentially the same assumptions as in Chapter 10: $p \sim p_0 + hp_1 + h^2 p_2 + \ldots$ in $S^0(m)$, $m \geq 1$, p real, $p+i$ elliptic and h sufficiently small when m is unbounded. We also let $I \subset\subset \mathbf{R}$ be an open interval and assume that $\underline{\lim}_{|(x,\xi)|\to\infty}\text{dist}\,(p_0(x,\xi),I) > 0$, $0 \leq f \in C_0^\infty(I;\mathbf{R})$. As before, we may assume that $\underline{\lim}\,p_0 - \sup I > 0$, and replacing P by $g(P)$ for a suitable function g, we may also assume that $m = 1$. Assume that every value in $\overline{I}$ is non-critical for p_0 (an assumption that will be relaxed in the final result), and take $0 \leq f \in C_0^\infty(I;\mathbf{R})$.

We shall study the trace norm of $(\mathcal{F}_h^{-1}\psi)(\lambda - P)f(P)\chi^w$ for $\lambda \in I$, $\chi \in C_0^\infty(\mathbf{R}^{2n};[0,1])$, $\psi \in C_0^\infty(\mathbf{R})$ also when $\text{supp}\,(\psi)$ is large. We start, however, with the case when $\psi \in C_0^\infty(]-\frac{1}{C},\frac{1}{C}[)$ for $C > 0$ large enough so that the analysis of Chapter 10 can be applied. We can find $\phi \in C_0^\infty(]-\frac{1}{C},\frac{1}{C}[)$ of the form $g * \check{g}$ with $0 \leq g \in C_0^\infty(]-\frac{1}{2C},\frac{1}{2C}[)$ such that $\phi > 0$ on the support of ψ, and write $\psi = k\phi$ for $k \in C_0^\infty(]-\frac{1}{C},\frac{1}{C}[)$. Then $\mathcal{F}_h^{-1}\psi = \mathcal{F}_h^{-1}k * \mathcal{F}_h^{-1}\phi$ and

$$\|(\mathcal{F}_h^{-1}\psi)(\lambda - P)f(P)\chi^w\|_{\text{tr}}$$
$$\leq \int |\mathcal{F}_h^{-1}k(\lambda')|\|\mathcal{F}_h^{-1}\phi(\lambda - \lambda' - P)f(P)\chi^w\|_{\text{tr}}d\lambda'. \qquad (11.5)$$

Here the integral over $|\lambda'| \geq \sqrt{h}$ is $\mathcal{O}(h^\infty)$, and for $|\lambda'| \leq \sqrt{h}$, we write λ instead of $\lambda - \lambda'$ and estimate:

$$\|(\mathcal{F}_h^{-1}\phi)(\lambda - P)f(P)\chi^w\|_{\text{tr}} = \|\frac{1}{2\pi h}\int e^{it(\lambda-P)/h}\phi(t)f(P)\chi^w dt\|_{\text{tr}}.$$

Let $\widetilde{\chi} \in C_0^\infty(\mathbf{R}^{2n};\mathbf{R})$ be equal to 1 in a neighborhood of $\{\exp(tH_{p_0})(\rho); \rho \in \text{supp}\,\chi, |t| \leq \frac{1}{C}\}$. Then by Egorov's theorem above:

$$(1 - \widetilde{\chi}^w)e^{it(\lambda-P)/h}f(P)\chi^w = \mathcal{O}(h^\infty) \text{ in trace norm, when } |t| \leq \frac{1}{C},$$

and consequently,

$$\|(\mathcal{F}_h^{-1}\phi)(\lambda - P)f(P)\chi^w\|_{\text{tr}} = \|\widetilde{\chi}^w(\mathcal{F}_h^{-1}\phi)(\lambda - P)f(P)\chi^w\|_{\text{tr}} + \mathcal{O}(h^\infty)$$
$$= \|\widetilde{\chi}^w(\mathcal{F}_h^{-1}\phi)(\lambda - P)f(P)\widetilde{\chi}^w\chi^w\|_{\text{tr}} + \mathcal{O}(h^\infty)$$
$$\leq \|\chi^w\|\|\widetilde{\chi}^w(\mathcal{F}_h^{-1}\phi)(\lambda - P)f(P)\widetilde{\chi}^w\|_{\text{tr}} + \mathcal{O}(h^\infty).$$

By construction $\mathcal{F}_h^{-1}\phi \geq 0$. The trace norm and the trace of a positive selfadjoint operator coincide, moreover, using the fact that $|\chi| \leq 1$ we see by the semi-classical sharp Gårding inequality (Theorem 7.12) that $\|\chi^w\| \leq 1 + \mathcal{O}(h)$, so we get

$$\|(\mathcal{F}_h^{-1}\phi)(\lambda - P)f(P)\chi^w\|_{\mathrm{tr}}$$
$$\leq (1 + \mathcal{O}(h))\mathrm{tr}\,(\widetilde{\chi}^w(\mathcal{F}_h^{-1}\phi)(\lambda - P)f(P)\widetilde{\chi}^w) + \mathcal{O}(h^\infty).$$

Here the trace can be evaluated as in Chapter 10, and we get

$$\|(\mathcal{F}_h^{-1}\phi)(\lambda - P)f(P)\chi^w\|_{\mathrm{tr}} \leq C(\phi)h^{-n}(\int_{p_0=\lambda} |\widetilde{\chi}|^2 L_\lambda(d\omega) + o(1)).$$

Notice that the integral to the right changes only by $o(1)$ if we replace λ by $\lambda + o(1)$. Returning to (11.5), we get

$$\|(\mathcal{F}_h^{-1}\psi)(\lambda - P)f(P)\chi^w\|_{\mathrm{tr}} \leq C(\psi)h^{-n}(\int_{p_0=\lambda} |\widetilde{\chi}|^2 L_\lambda(d\omega) + o(1)). \quad (11.6)$$

For $t_0 \in \mathbf{R}$, we have $\mathcal{F}_h^{-1}(\tau_{t_0}\psi)(\lambda) = e^{it_0\lambda/h}(\mathcal{F}_h^{-1}\psi)(\lambda)$ and since $e^{it_0(\lambda-P)}$ is unitary, we see that the LHS of (11.6) does not change, if we replace ψ by $\tau_{t_0}\psi$.

If $\psi \in C_0^\infty(\mathbf{R})$ is arbitrary, we can write ψ as a finite sum of translates of functions in $C_0^\infty(]-\frac{1}{C}, \frac{1}{C}[)$ and applying (11.6) to each of these functions, we see that *(11.6) is valid for general $\psi \in C_0^\infty(J)$ ($J \subset \mathbf{R}$ is an open bounded interval), with $\widetilde{\chi} \in C_0^\infty(\mathbf{R}^{2n})$ equal to 1 near* $\mathrm{supp}\,\chi$.

Looking now at a slightly different situation, we keep the assumptions on ψ and χ, and assume in addition that ψ vanishes near 0 and that

$$\exp(tH_p)(\mathrm{supp}\,\chi) \cap \mathrm{supp}\,\chi = \emptyset, \text{ for } t \in \mathrm{supp}\,\psi. \quad (11.7)$$

Then for $t \in \mathrm{supp}\,\psi$, if we let $\chi_t \in C_0^\infty$ be equal to 1 near $\exp(tH_{p_0})(\mathrm{supp}\,\chi)$, we have by Egorov's theorem:

$$\|(1 - \chi_t^w)e^{-itP/h}f(P)\chi^w\|_{\mathrm{tr}} = \mathcal{O}(h^\infty).$$

Hence

$$\mathrm{tr}\,(e^{-itP/h}f(P)\chi^w) = \mathrm{tr}\,(\chi_t^w e^{-itP/h}f(P)\chi^w) + \mathcal{O}(h^\infty)$$
$$= \mathrm{tr}\,(e^{-itP/h}f(P)\chi^w\chi_t^w) + \mathcal{O}(h^\infty).$$

We can choose χ_t with support disjoint from that of χ, and conclude that

$$\mathrm{tr}\,(e^{-itP/h}f(P)\chi^w) = \mathcal{O}(h^\infty), \text{ for } t \text{ in a neighborhood of } \mathrm{supp}\,\psi,$$

and consequently, under the assumption (11.7):

$$\operatorname{tr}(\mathcal{F}_h^{-1}\psi)(\lambda - P)f(P)\chi^w = \mathcal{O}(h^\infty). \tag{11.8}$$

Notice that we get the same conclusion if we replace the assumption (11.7) by the assumption that $\operatorname{supp}\chi$ is disjoint from the energy surface $p_0 = \lambda$ (in view of the results of Chapter 9).

We now fix an energy $\lambda_0 \in I$ and assume

The union of periodic H_{p_0} trajectories in $p_0 = \lambda_0$ is of Liouville measure 0. (11.9)

Fix $R > 0$. Then $\{\rho; p_0(\rho) = \lambda_0, \exp t H_{p_0}(\rho) = \rho, \text{ for some } 0 < |t| \leq R\}$ is a *closed* set of measure 0, and for every $\epsilon > 0$, we can find $\chi_1 = \chi_{R,\epsilon} \in C_0^\infty(\mathbf{R}^{2n}; [0,1])$ equal to 1 near that set, and a corresponding $\widetilde{\chi} \in C_0^\infty(\mathbf{R}^{2n})$, which is equal to 1 near $\operatorname{supp}\chi$, such that $\int_{p_0=\lambda_0} |\widetilde{\chi}|^2 L_{\lambda_0}(d\omega) \leq \epsilon$. We can then find $\chi_2, \ldots, \chi_N \in C_0^\infty$ such that

$$\chi := \sum_1^N \chi_j \text{ is equal to 1 near } p_0^{-1}(\operatorname{supp} f), \tag{11.10}$$

and for $j \geq 2$, we have either

$$\exp(t H_{p_0})(\operatorname{supp}\chi_j) \cap \operatorname{supp}\chi_j = \emptyset \text{ for } \frac{1}{2C} \leq |t| \leq R,$$

or

$$\operatorname{supp}\chi_j \cap \{p_0 = \lambda_0\} = \emptyset.$$

Let $\psi \in C_0^\infty(]-R, R[)$ and let $\psi_0 \in C_0^\infty(]-\frac{1}{C}, \frac{1}{C}[)$ be equal to ψ near $[-\frac{1}{2C}, \frac{1}{2C}]$. Then for $\lambda - \lambda_0 = o(1)$ (or more precisely for $|\lambda - \lambda_0| \leq \epsilon(h)$, where $\epsilon(h) > 0$ is some fixed function tending to 0, when $h \to 0$):

$$\operatorname{tr}((\mathcal{F}_h^{-1}(\psi - \psi_0))(\lambda - P)f(P))$$
$$= \operatorname{tr}((\mathcal{F}_h^{-1}(\psi - \psi_0))(\lambda - P)f(P)\chi^w) + \mathcal{O}(h^\infty)$$
$$= \sum_1^N \operatorname{tr}((\mathcal{F}_h^{-1}(\psi - \psi_0))(\lambda - P)f(P)\chi_j^w) + \mathcal{O}(h^\infty)$$
$$= \operatorname{tr}((\mathcal{F}_h^{-1}(\psi - \psi_0))(\lambda - P)f(P)\chi_1^w) + \mathcal{O}(h^\infty)$$
$$= \mathcal{O}_\psi(1)h^{-n}(\epsilon + o(1)), \quad h \to 0.$$

Here $\operatorname{tr}(\mathcal{F}_h^{-1}(\psi_0)(\lambda - P)f(P))$ has been studied in Chapter 10, and we get for general $\psi \in C_0^\infty(\mathbf{R})$:

$$\operatorname{tr}\mathcal{F}_h^{-1}\psi(\lambda - P)f(P) = h^{-n}\left(\frac{\psi(0)f(\lambda_0)}{(2\pi)^n}\int_{p_0=\lambda_0} L_{\lambda_0}(d\omega) + o(1)\right), \tag{11.11}$$

when $\lambda - \lambda_0 = o(1)$, $h \to 0$.

Let $\psi \in C_0^\infty(] - \frac{1}{C}, \frac{1}{C}[)$ be even with $\psi = 1$ near 0, with $\tilde{\psi} := \mathcal{F}_1^{-1}(\psi) \geq 0$, and put $\tilde{\psi}_h = \frac{1}{h}\tilde{\psi}(\frac{\cdot}{h}) = \mathcal{F}_h^{-1}\psi$. One of the results of the preceding chapter is then that:

$$\tilde{\psi}_h * \mu_f(\lambda) = t_0(\lambda)h^{-n} + t_1(\lambda)h^{1-n} + \mathcal{O}(h^{2-n}), \qquad (11.12)$$

where μ_f is defined in Chapter 10 and where $t_0(\lambda)$ is given in (10.15), $t_1(\lambda)$ has similar properties and depends on the subprincipal symbol of P. See Remark 11.2 for a computation. As with (10.17), this relation extends to all real λ and we can replace $\mathcal{O}(h^{2-n})$ by $\mathcal{O}(h^{2-n}\langle\lambda\rangle^{-N})$ for every N. Since $\mathcal{F}_h^{-1}\psi(\epsilon\cdot) = \tilde{\psi}_{\epsilon h}$, we can also apply (11.11) and get

$$\tilde{\psi}_{\epsilon h} * \mu_f(\lambda) = (t_0(\lambda) + o(1))h^{-n}, \quad h \to 0, \ \lambda - \lambda_0 = o(1) \qquad (11.13)$$

for every fixed $\epsilon > 0$. This implies that if we restrict our attention to some interval around λ_0 which shrinks to $\{\lambda_0\}$ when $h \to 0$, then the number of eigenvalues in every subinterval of length ϵh is $\mathcal{O}(1)\epsilon h^{1-n}$, provided that h is small enough depending on ϵ.

Let $H = 1_{[0,+\infty[}$ be the Heaviside function and define ϕ_h by $H*\tilde{\psi}_h - H = h\phi_h$. Using the fact that $\int \tilde{\psi}dy = 1$, we get $\phi_h(\lambda) = \frac{1}{h}\phi(\frac{\lambda}{h})$, where ϕ is an odd function with a jump discontinuity at 0 and which is of class S on each half axis. We have $\phi(\lambda) = \int_{-\infty}^{\lambda} \tilde{\psi}(y)dy$ when $\lambda < 0$. We want to study

$$(H * \mu_f)(\lambda) = \sum_{\lambda_j \leq \lambda} f(\lambda_j),$$

and we know from (11.12) that

$$H * \tilde{\psi}_h * \mu_f(\lambda) = T_0(\lambda)h^{-n} + T_1(\lambda)h^{1-n} + \mathcal{O}(h^{2-n}), \qquad (11.14)$$

where $T_j = H * t_j$. We are therefore interested in

$$H * \tilde{\psi}_h * \mu_f - H * \mu_f = h\phi_h * \mu_f. \qquad (11.15)$$

The idea is to approximate $h\phi_h$ by an odd superposition of translations of the function $\tilde{\psi}_{\epsilon h}$ and to use (11.13) to get a cancellation. Let $\chi \in C_0^\infty(\mathbf{R}; [0, 1])$ be equal to 1 near 0 and even. Put $\phi^R = \chi(\frac{\cdot}{R})\phi$, $\phi_h^R = \frac{1}{h}\phi^R(\frac{\cdot}{h})$. We first look at

$$h(\phi_h - \phi_h^R) * \mu_f(\lambda) = \int (1 - \chi(\frac{\lambda - t}{Rh}))\phi(\frac{\lambda - t}{h})\mu_f(dt)$$

$$= \mathcal{O}(1) \sum_{|k|\geq R} \langle k\rangle^{-N}h^{1-n} = \mathcal{O}(\frac{1}{R})h^{1-n}. \qquad (11.16)$$

We next want to replace ϕ_h^R by $\phi_h^R * \widetilde{\psi}_{\epsilon h} = (\phi^R * \widetilde{\psi}_\epsilon)_h$. Here

$$\phi^R * \widetilde{\psi}_\epsilon(x) - \phi^R(x) = \int (\phi^R(x - \epsilon y) - \phi^R(x))\widetilde{\psi}(y)dy.$$

Assume for instance that $x \geq 0$. Then the last integral is

$$\mathcal{O}_N(1)\int_{y \leq x/\epsilon} |\phi^R(x - \epsilon y) - \phi^R(x)|\langle y \rangle^{-N}dy + \mathcal{O}_N(1)\int_{y \geq x/\epsilon} \langle y \rangle^{-N}dy$$

$$= \mathcal{O}_{N,M}(1)\int_{-\infty}^{x/\epsilon} (\max_{0 \leq t \leq 1} \langle x - t\epsilon y \rangle^{-M})\epsilon |y|\langle y \rangle^{-N}dy + \mathcal{O}_N(1)\langle \frac{x}{\epsilon} \rangle^{1-N}.$$

Here $C\langle x - t\epsilon y \rangle\langle t\epsilon y \rangle \geq \langle x \rangle$, so

$$\langle x - t\epsilon y \rangle^{-M} \leq C\langle x \rangle^{-M}\langle t\epsilon y \rangle^M \leq C\langle x \rangle^{-M}\langle y \rangle^M$$

and choosing M, N suitably we get for any (new) given $N \in \mathbf{N}$:

$$|\phi^R * \widetilde{\psi}_\epsilon(x) - \phi^R(x)| = \mathcal{O}_N(1)(\epsilon\langle x \rangle^{-N} + \langle \frac{x}{\epsilon} \rangle^{-N}). \tag{11.17}$$

We then look at

$$\langle \frac{\cdot}{h\epsilon} \rangle^{-N} * \mu_f(\lambda_0)$$

$$= (1_{[-\widetilde{R},\widetilde{R}]}(\frac{\cdot}{\epsilon h})\langle \frac{\cdot}{h\epsilon} \rangle^{-N}) * \mu_f(\lambda_0) + ((1 - 1_{[-\widetilde{R},\widetilde{R}]})(\frac{\cdot}{\epsilon h})\langle \frac{\cdot}{h\epsilon} \rangle^{-N}) * \mu_f(\lambda_0).$$

In order to estimate the first term to the right, we notice that the support of the function $1_{[-\widetilde{R},\widetilde{R}]}(\frac{\cdot}{h\epsilon})$ can be covered by $\mathcal{O}(\widetilde{R})$ intervals of the length ϵh, and by arguments already used before and the observation after (11.13), we see that the first term to the right in the last equation is $\mathcal{O}(\widetilde{R}\epsilon h^{1-n})$. As for the second term, we observe that

$$(1 - 1_{[-\widetilde{R},\widetilde{R}]})(\frac{x}{\epsilon h})\langle \frac{x}{h\epsilon} \rangle^{-N} = \mathcal{O}(\frac{1}{\widetilde{R}})\langle \frac{x}{h} \rangle^{1-N},$$

and the corresponding convolution is then $\mathcal{O}(\widetilde{R}^{-1}h^{1-n})$. We conclude that for $h > 0$ sufficiently small depending on $\epsilon > 0$, we have

$$\langle \frac{\cdot}{\epsilon h} \rangle^{-N} * \mu_f(\lambda_0) = \mathcal{O}(\delta(\epsilon)h^{1-n})$$

for some function $\delta(\epsilon)$ which tends to 0 when ϵ tends to 0. We get the same estimate for $\epsilon\langle \frac{\cdot}{h} \rangle^{-N} * \mu_f(\lambda_0)$, so

$$h(\phi_h^R * \widetilde{\psi}_{\epsilon h} - \phi_h^R) * \mu_f(\lambda_0) = \mathcal{O}(\delta(R,\epsilon)h^{1-n}), \quad \delta(\epsilon,R) \to 0, \epsilon \to 0. \tag{11.18}$$

Now look at

$$h\phi_h^R * \widetilde{\psi}_{\epsilon h} * \mu_f(\lambda_0) = h \int \phi_h^R(y)(\widetilde{\psi}_{\epsilon h} * \mu_f)(\lambda_0 - y)dy.$$

On the support of $\phi_h^R(y)$ we have $y = \mathcal{O}_R(h)$, so we can use (11.13) and get

$$h\phi_h^R * \widetilde{\psi}_{\epsilon h} * \mu_f(\lambda_0) = h \int \phi_h^R(y)dy\, t_0(\lambda_0)h^{-n} + o_{R,\epsilon}(1)h^{1-n} = o_{R,\epsilon}(1)h^{1-n},$$

$$(11.19)$$

using also the fact that ϕ_h^R is odd. From (11.16,18,19), we get

$$h\phi_h * \mu_f(\lambda_0) = o(1)(h^{1-n}), \qquad (11.20)$$

and combining this with (11.15) and (11.14), we get

$$H * \mu_f(\lambda_0) = T_0(\lambda_0)h^{-n} + T_1(\lambda_0)h^{1-n} + o(h^{1-n}). \qquad (11.21)$$

Theorem 11.1. *Let $m \geq 1$ be an order function, $p \sim p_0 + hp_1 + h^2p_2 + \ldots \in S^0(m)$ real valued with $p + i$ elliptic when m is unbounded. Let $[a, b] \subset \mathbf{R}$ be an interval such that $\underline{\lim}_{|(x,\xi)| \to \infty} \text{dist}\,(p_0(x, \xi), [a, b]) > 0$ and assume that a, b are not critical values, and that the unions of periodic H_{p_0} trajectories in the energy surfaces $p_0 = a$ and $p_0 = b$ are of measure 0. Then for $h > 0$ small enough, the number $N([a, b]; h)$ of eigenvalues of p^w in $[a, b]$ satisfies:*

$$N([a, b]; h) = \frac{1}{(2\pi h)^n}\left(\iint_{p_0(x,\xi)\in[a,b]} dx d\xi + \alpha h + o(h)\right), \qquad (11.22)$$

where

$$\alpha = \int_{\{p_0=a\}} p_1(x, \xi)L_a(d(x, \xi)) - \int_{\{p_0=b\}} p_1(x, \xi)L_b(d(x, \xi)).$$

Proof. Let $\sigma > 0$ be small enough so that p_0 has no critical values in $[a - \sigma, a + \sigma] \cup [b - \sigma, b + \sigma]$. Let $f_1 \in C_0^\infty(]a - \sigma, a + \sigma[; [0, 1])$, $f_2 \in C_0^\infty(]a + \frac{\sigma}{2}, b - \frac{\sigma}{2}[; [0, 1])$, $f_3 \in C_0^\infty(]b - \sigma, b + \sigma[; [0, 1])$ satisfy $f_1 + f_2 + f_3 = 1$ on $[a - \frac{\sigma}{2}, b + \frac{\sigma}{2}]$. Let $\mu_0(h) \leq \mu_1(h) \leq \ldots \leq \mu_{N(h)}(h)$ be the eigenvalues of $p^w(x, hD_x; h)$ (counted with their multiplicities) in $[a - \sigma, b + \sigma]$. We have

$$N([a, b]; h) = \sum_{a \leq \mu_j(h) \leq b} (f_1 + f_2 + f_3)(\mu_j(h))$$

$$= \sum_{a \leq \mu_j(h)} f_1(\mu_j(h)) + \sum f_2(\mu_j(h)) + \sum_{\mu_j(h) \leq b} f_3(\mu_j(h))$$

$$= \sum_{a \leq \mu_j(h)} f_1(\mu_j(h)) + \text{tr}\, f_2(p^w(x, hD_x; h)) + \sum_{\mu_j(h) \leq b} f_3(\mu_j(h)). \quad (11.23)$$

According to Theorem 9.6 we have

$$\operatorname{tr} f_2(p^w(x, hD_x; h))$$
$$= \frac{1}{(2\pi h)^n} \left(\int f_2(p_0(x, \xi)) dx d\xi \right.$$
$$\left. + h \int p_1(x, \xi) f_2'(p_0(x, \xi)) dx d\xi + \mathcal{O}(h^2) \right). \qquad (11.24)$$

From (11.21) we have:

$$\sum_{\mu_j(h) \leq b} f_3(\mu_j(h)) = H * \mu_{f_3}(b) = T_0(b) h^{-n} + T_1(b) h^{1-n} + o(h^{1-n}), \qquad (11.25)$$

and

$$\sum_{a \leq \mu_j(h)} f_1(\mu_j(h)) = \operatorname{tr} f_1(p^w) - \sum_{\mu_j(h) < a} f_1(\mu_j(h))$$
$$= \left(\int f_1(p_0(x, \xi)) \frac{dx d\xi}{(2\pi)^n} - T_0(a) \right) h^{-n}$$
$$+ \left(\int p_1(x, \xi) f_1'(p_0(x, \xi)) \frac{dx d\xi}{(2\pi)^n} - T_1(a) \right) h^{1-n} + o(h^{1-n}). \qquad (11.26)$$

Combining (11.23–26), we get (11.22), where it remains to compute α.

Let $\varphi \in C_0^\infty(]b - \frac{\sigma}{2}, b + \frac{\sigma}{2}[)$, using (11.14) and the fact that $\mathcal{O}(h^{2-n})$ is uniform on $\lambda \in]b - \frac{\sigma}{2}, b + \frac{\sigma}{2}[$, we get

$$\int \left(\int \varphi(\tau + \lambda) \widetilde{\psi}_h(\tau) d\tau \right) (H * \mu_{f_3})(\lambda) d\lambda = \int \varphi(\lambda) H * \widetilde{\psi}_h * \mu_{f_3}(\lambda) d\lambda$$
$$= \int \varphi(\lambda) T_0(\lambda) d\lambda \, h^{-n} + \int \varphi(\lambda) T_1(\lambda) d\lambda \, h^{1-n} + \mathcal{O}(h^{2-n}). \qquad (11.27)$$

As $\int \tau \mathcal{F}^{-1} \psi(\tau) d\tau = 0$ (which follows from the fact that $\psi = 1$ near zero) and $\int \widetilde{\psi}(\tau) d\tau = 1$, it follows that

$$\int \varphi(\tau + \lambda) \widetilde{\psi}_h(\tau) d\tau = \int \varphi(h\tau + \lambda) \widetilde{\psi}(\tau) d\tau = \varphi(\lambda) + \mathcal{O}(h^2). \qquad (11.28)$$

Using (11.27), (11.28) and the fact that $H * \mu_{f_3}(\lambda) = \mathcal{O}(h^{-n})$, we get:

$$\int \varphi(\lambda)(H * \mu_{f_3})(\lambda) d\lambda$$
$$= \int \varphi(\lambda) T_0(\lambda) d\lambda \, h^{-n} + \int \varphi(\lambda) T_1(\lambda) d\lambda \, h^{1-n} + \mathcal{O}(h^{2-n}). \qquad (11.29)$$

If $\phi(\lambda) = \int_\lambda^{+\infty} \varphi(\tau)d\tau$, then the left member in the last equation is equal to tr $(f_3\phi)(P)$. From the functional calculus of Chapter 8 (formula (8.16)) we have:

$$\text{tr}\,(f_3\phi) = \frac{1}{(2\pi h)^n} \iint \phi(p_0(x,\xi))f_3(p_0(x,\xi))dxd\xi$$
$$+ \frac{h}{(2\pi h)^n} \iint p_1(x,\xi)[\phi(p_0(x,\xi))f_3'(p_0(x,\xi))$$
$$- \varphi(p_0(x,\xi))f_3(p_0(x,\xi))]dxd\xi + \mathcal{O}(h^{2-n}).\quad (11.30)$$

Using the fact that

$$\int p_1(x,\xi)(\varphi f_3)(p_0(x,\xi))dxd\xi = \int \varphi(\lambda)f_3(\lambda)(\int_{p_0=\lambda} p_1(x,\xi)L_\lambda(d\omega))d\lambda,$$

we deduce from (11.29), (11.30) that

$$T_1(\lambda) = \int_{\{p_0(x,\xi)\le\lambda\}} p_1(x,\xi)f_3'(p_0(x,\xi))dxd\xi - f_3(\lambda)\int_{\{p_0=\lambda\}} p_1(x,\xi)L_\lambda(d\omega).$$
$$(11.31)$$

Now (11.23-26), (11.31) and the fact that $f_1 + f_2 + f_3 = 1$ on $[a,b]$ give

$$\alpha = \int_{\{p_0=a\}} p_1(x,\xi)L_a(d\omega) - \int_{\{p_0=b\}} p_1(x,\xi)L_b(d\omega).$$

Remark 11.2. By definition $T_1(\lambda) = H * t_1(\lambda) = \int_{-\infty}^\lambda t_1(y)dy$. From (11.31) we deduce that

$$t_1(\lambda) = \frac{\partial}{\partial\lambda}T_1(\lambda) = -f_3(\lambda)\frac{\partial}{\partial\lambda}(\int_{p_0=\lambda} p_1(x,\xi)(L_\lambda d\omega)).\quad (11.32)$$

By the same method we can compute all the coefficients, $t_j(\lambda)$, of the expression (10.13), see [Ro1].

Appendix: Egorov's theorem

We proceed in the spirit of the general theory of fouriors, followed in the appendix about metaplectic operators in Chapter 7, but for simplicity we do not develop the global theory.

A smooth real valued function, defined in a neighborhood Ω of $(x_0,\theta_0) \in \mathbf{R}^n \times \mathbf{R}^N$, with $\phi_\theta'(x_0,\theta_0) = 0$ is called a non-degenerate phase function if

$$d\partial_{\theta_1}\phi,\ldots,d\partial_{\theta_N}\phi \text{ are linearly independent at every point of } C_\phi,\quad (A.1)$$

where
$$C_\phi := \{(x,\theta) \in \Omega; \phi'_\theta(x,\theta) = 0\}. \tag{A.2}$$

Then C_ϕ is a smooth n-dimensional manifold and the map

$$C_\phi \ni (x,\theta) \mapsto (x, \phi'_x(x,\theta)) \in T^*\mathbf{R}^n \tag{A.3}$$

has injective differential at every point. Possibly after shrinking Ω around (x_0, θ_0), we can assume that (A.3) is injective and that the image $\Lambda_\phi \subset T^*\mathbf{R}^n$ is an n-dimensional manifold. As in the appendix to Chapter 7, we can identify $\xi \cdot dx_{|\Lambda_\phi}$ with $d(\phi_{|C_\phi})$, so $\xi \cdot dx_{|\Lambda_\phi}$ is an exact form and $\sigma_{|\Lambda_\phi} = d(\xi \cdot dx_{|\Lambda_\phi}) = 0$. Consequently Λ_ϕ is a Lagrangian submanifold of $T^*\mathbf{R}^n$. Let $\xi_0 = \phi'_x(x_0, \theta_0)$, so that (x_0, ξ_0) is the image of (x_0, θ_0) under (A.3).

As in Lemma A.1 of Chapter 7, we see that for almost all real quadratic forms $q(x)$:
$$\det \text{Hess}\,((\phi(x,\theta) - q(x))_{x_0,\theta_0} \neq 0. \tag{A.4}$$

Let q be such a quadratic form. Then for η in a small neighborhood of ξ_0, the function
$$(x,\theta) \mapsto \phi(x,\theta) - q(x - x_0) - x \cdot \eta \tag{A.5}$$

has a unique critical point $(x(\eta), \theta(\eta))$ in some neighborhood of (x_0, θ_0), and $(x(\xi_0), \theta(\xi_0)) = (x_0, \theta_0)$. Moreover, this critical point is non-degenerate and (as in the appendix to Chapter 7) the critical value $H(\eta)$ depends only on Λ_ϕ, q and $\phi(x_0, \theta_0)$ (i.e. a normalization constant in ϕ).

Let $a(x, \theta; h) \in S^0(1)$ have its support in a small neighborhood of (x_0, θ_0), and put
$$I(a, \phi; h)(x) = h^{-\frac{N}{2} - \frac{n}{4}} \int e^{i\phi(x,\theta)/h} a(x, \theta; h)\,d\theta. \tag{A.6}$$

For η in a neighborhood of ξ_0, the method of stationary phase (Chapter 5) shows that
$$\mathcal{F}_h(e^{-iq(x-x_0)/h} I(a, \phi; h))(\eta; h) = e^{iH(\eta)/h} A(\eta; h), \tag{A.7}$$

where $A \in S^{-n/4}$ satisfies

$$A(\eta; h) \equiv (2\pi)^{\frac{N}{2} + \frac{n}{4}} \frac{h^{n/4} e^{i\frac{\pi}{4}\iota(\phi,q)}}{\det(\phi'' - q'')(x(\eta), \theta(\eta))} a(x(\eta), \theta(\eta); h) \mod S^{-1-n/4}. \tag{A.8}$$

Here $\iota(\phi, q)$ is the signature of the matrix in (A.4), i.e. the number of positive eigenvalues minus the number of negative eigenvalues. In (A.8) we have given only the first term in an asymptotic expansion, this term as well as the others all vanish when $(x(\eta), \theta(\eta))$ is outside the support of a and A is of class $S^{-\infty}$ in that region.

For η outside a neighborhood of ξ_0, we can estimate the Fourier transform in (A.7) by integration by parts, and we get in this region:

$$|\partial_\eta^\alpha \mathcal{F}_h(e^{-iq(\cdot - x_0)/h} I(a, \phi; h))| \le C_{N,\alpha} h^N \langle \eta \rangle^{-N}, \qquad (A.9)$$

for all α, N.

Let $\psi(x, w)$ be a second non-degenerate phase function, defined in a neighborhood of $(x_0, w_0) \in \mathbf{R}^n \times \mathbf{R}^M$ with $(x_0, w_0) \in C_\psi$ and assume that $\psi'_x(x_0, w_0) = \xi_0$, $\psi(x_0, w_0) = \phi(x_0, \theta_0)$ and that $\Lambda_\phi = \Lambda_\psi$ near (x_0, ξ_0). Then we have

Proposition A.1. *For every sufficiently small neighborhood Ω of (x_0, θ_0), there exists a neighborhood $\tilde{\Omega}$ of (x_0, w_0) such that for every $a \in S^0(\mathbf{R}^n \times \mathbf{R}^N, 1)$ with support in Ω, there exists $b \in S^0(\mathbf{R}^n \times \mathbf{R}^M, 1)$ with support in $\tilde{\Omega}$, such that*

$$I(a, \phi) = I(b, \psi) + r, \qquad (A.10)$$

with

$$|\partial_x^\alpha r(x; h)| \le C_{N,\alpha} h^N \langle x \rangle^{-N}, \quad N \ge 0, \ \alpha \in \mathbf{N}^n \qquad (A.11)$$

and conversely for every $b \ldots$, there exists $a \ldots$

Proof. Observe that

$$\mathcal{F}_h(e^{-iq(x - x_0)/h} I(b, \psi))(\eta; h) = e^{iH(\eta)/h} B(\eta; h), \qquad (A.12)$$

for η in a neighborhood of ξ_0, where $B \in S^{-n/4}$ is given by a formula analogous to (A.8). For a given a with support close enough to (x_0, θ_0), we can then construct b so that B coincides with A (in (A.7)) modulo $S^{-\infty}$ in a neighborhood of ξ_0. Outside such a neighborhood, we have (A.9) for $I(a, \phi)$ and similarly for $I(b, \psi)$. Then (A.10) follows from Fourier's inversion formula.   #

Note that the proposition remains valid for classical symbols $a \in S^0_{cl}(\mathbf{R}^n \times \mathbf{R}^N, 1)$, $b \in S^0_{cl}(\mathbf{R}^n \times \mathbf{R}^M)$, where in general we define $S^k_{cl}(\mathbf{R}^n, m)$ to be the space of symbols $a(x; h)$ in $S^k(\mathbf{R}^n, m)$ (with m some order function and $k \in \mathbf{R}$) with an asymptotic expansion in the latter space:

$$a(x; h) \sim h^{-k} \sum_{j=0}^\infty h^j a_j(x).$$

To complete the picture, we notice (by adapting the argument in the appendix of Chapter 7) that if $\Lambda \subset T^* \mathbf{R}^n$ is a Lagrangian manifold, then in a neighborhood of any given point $(x_0, \xi_0) \in \Lambda$, we have $\Lambda = \Lambda_\phi$ for some non-degenerate phase function.

We now discuss Fourier integral operators, and replace $\mathbf{R}_x^n$ by $\mathbf{R}_{x,y}^{2n} = \mathbf{R}_x^n \times \mathbf{R}_y^n$. Let $\phi(x, y, \theta)$ be a non-degenerate phase function defined near $(x_0, y_0, \theta_0) \in \mathbf{R}^{2n+N}$ with $\phi'_\theta(x_0, y_0, \theta_0) = 0$ and assume that

$$\Lambda'_\phi := \{(x, \xi; y, -\eta); (x, \xi; y, \eta) \in \Lambda_\phi\}$$

is the graph of a diffeomorphism κ_ϕ mapping a neighborhood of (y_0, η_0) onto a neighborhood of (x_0, ξ_0), where $\xi_0 = \phi'_x(x_0, y_0, \theta_0)$, $-\eta_0 = \phi'_y(x_0, y_0, \theta_0)$. Then κ is a canonical transformation, i.e. a local diffeomorphism with the property that the pull-back of the symplectic 2 form under κ is again the symplectic 2 form. Let $J(a, \phi)$ be the operator with distribution kernel $I(a, \phi)$, where a is a symbol of class S^0 with support close to (x_0, y_0, θ_0). We say that the fourior $J(a, \phi)$ is associated with $\kappa = \kappa_\phi$ and of order 0. Notice that the complex adjoint $J(a, \phi)^*$ is associated with κ^{-1}, and that h-pseudors can be viewed as fouriors associated with the identity transformation $\kappa = \mathrm{id}$.

Here and below, the following observation is useful (and standard in the theory of pseudors). Let m be an order function on $\mathbf{R}^{3n}$ and let $b \in S^0(\mathbf{R}^{3n}, m)$. Then the proof of Lemma 7.8 shows that

$$Au(x) := \frac{1}{(2\pi h)^n} \iint e^{i(x-y)\cdot\theta/h} b(x, y, \theta; h) u(y) \, dy \, d\theta$$

is a well-defined and continuous operator $\mathcal{S} \to \mathcal{S}$ and $\mathcal{S}' \to \mathcal{S}'$ which depends continuously on b.

Applying A under the sign of integration in Fourier's inversion formula for $u \in \mathcal{S}$, we see that

$$A = \mathrm{Op}_{h,1}(a),$$

where

$$a(x, \xi; h) = e^{-ix\cdot\xi/h} A(e^{i(\cdot)\cdot\xi/h}).$$

Proposition 7.6, applied with the parameter dependent order function $m_x(y, \theta) = m(x, y, \theta)$, shows that

$$a(x, \xi; h) = \frac{1}{(2\pi h)^n} \iint e^{\frac{i}{h}(x-y)\cdot(\theta-\xi)} b(x, y, \theta; h) \, dy \, d\theta$$

belongs to $S^0(\mathbf{R}^{2n}, \tilde{m})$, with $\tilde{m}(x, \xi) = m(x, x, \xi)$, and has the asymptotic expansion in that space:

$$a(x, \xi; h) \sim \sum_0^\infty \frac{h^k}{k!} (\partial_\theta \cdot D_y)^k (b(x, y, \theta; h))_{|(y,\theta)=(x,\xi)}$$

$$\sim \sum_{\alpha \in \mathbf{N}^n} \frac{h^{|\alpha|}}{\alpha!} (\partial_\xi^\alpha D_y^\alpha b(x, y, \xi))_{|y=x}.$$

Proposition A.2. *Let $J(a, \phi)$ be as above and let $J(b, \psi)$ be a second fourior of order 0 associated with a second canonical transformation $\tilde{\kappa}$, mapping a neighborhood of (x_0, ξ_0) onto a neighborhood of (z_0, ζ_0). Then $J(b, \psi) \circ J(a, \phi)$ is a fourior of order 0, associated with $\tilde{\kappa} \circ \kappa$.*

This result is an immediate consequence of the fact that $\psi(z, x, w) + \phi(x, y, \theta)$ is a non-degenerate phase function with (x, w, θ) as fibre variables.

Let $J(a, \phi)$ be as above, and assume in addition that a is elliptic near (x_0, y_0, θ_0) and (mainly to fix the ideas) that $\phi(x_0, y_0, \theta_0) = 0$. Then $J(a, \phi)^*$ $J(a, \phi)$ and $J(a, \phi)J(a, \phi)^*$ are h-pseudors, elliptic near (y_0, η_0) and (x_0, ξ_0) respectively, and we can compose them to the left and the right respectively to get h-pseudors with symbols in $S^0(1)$ which are equal to $1 \bmod S^{-\infty}$ in some neighborhoods of (y_0, η_0) and (x_0, ξ_0) respectively. If $\psi(x, y, \theta) = -\phi(y, x, \theta)$ is a phase associated with $J(a, \phi)^*$ (generating κ^{-1}), it follows that there exists $b \in S^0$ with support near (y_0, x_0, θ_0), such that

$$J(b, \psi)J(a, \phi) = \mathrm{Op}_h(j_1), \quad J(a, \phi)J(b, \psi) = \mathrm{Op}_h(j_2), \qquad (A.13)$$

where $j_1, j_2 \in S^0(\mathbf{R}^{2n}, 1)$, and $j_1 - 1, j_2 - 1$ are of class $S^{-\infty}$ in a neighborhood of (y_0, η_0) and (x_0, ξ_0) respectively. We say that $J(b, \psi)$ is a microlocal inverse of $J(a, \phi)$.

A semi-classical version of Egorov's theorem says that if $J(a, \phi)$, $J(b, \psi)$ are as above, and if $P = \mathrm{Op}_h(p)$ with $p \in S^0(1)$, then $Q := J(b, \psi)PJ(a, \phi)$ is of the form $\mathrm{Op}_h(q)$ with $q \in S^0(1)$ and moreover,

$$q \equiv p \circ \kappa \bmod S^{-1}(1) \text{ in a neighborhood of } (y_0, \eta_0). \qquad (A.14)$$

We leave the proof as an exercise or refer for instance to [GrSj].

Notes

Theorem 11.1 is due to Petkov–Robert [PeRo] and Ivrii [I1]. For the proof of (11.22) we have followed [I2,3]. For the computation of the coefficient, α, of the expression (11.22) we have used the idea of [PeRo]. When the measure of periodic trajectories is non-zero clustering of eigenvalues may appear, see Petkov–Popov [PePo] for a very general result in this direction, corresponding to earlier results of Guriev–Safarov and Safarov.

In the case of a compact manifold without boundary, Duistermaat–Guillemin [DuGu] proved two term asymptotics supposing that the measure of periodic bicharacteristics equals zero. Under a similar condition on the periodic trajectories, this result was obtained by Ivrii [I4] for second order differential operators in compact manifold with boundary. See also Zelditch [Zel].

12. A more general study of the trace

In many situations, it is impossible to construct parametrices for small times for the Schrödinger equation associated with an h-pseudor. This happens for instance when we are interested in systems or in boundary value problems. For such problems, Ivrii (see [I1]) made the beautiful observation that one can nevertheless get full asymptotic expansions for the trace. By suitable scaling arguments and hyperbolic energy estimates, he showed that the trace in the region $h^\delta \leq |t| \leq 1/C$ is $\mathcal{O}(h^\infty)$, for $0 < \delta < 1$. Then in the region $|t| \leq h^\delta$, he observed that very rough parametrix constructions suffice to get the full asymptotics of the trace. In this chapter, we shall not follow the original approach by Ivrii, but rather develop a corresponding stationary version. Many ideas of Ivrii are still present in this version. A new amusing feature is the very systematic use of almost analytic extensions, and a corresponding absence of propagation estimates. An advantage of this presentation is that we can also treat situations where the spectral parameter is implicit, and where there is no really natural associated evolution equation, and we get a flexible tool which can be combined with Grushin reductions and effective Hamiltonians (see Chapter 13). For the transparency of the presentation, however, we shall mainly be concerned with the resolvent. We now formulate the assumptions and the results.

Let $\alpha < \beta$ be two real numbers, and let $P(h)$ be an h-pseudor with symbol $p(x,\xi;h) \sim \sum_{j \geq 0} h^j p_j(x,\xi)$ in $S^0(\mathbf{R}^{2n}, 1; \mathcal{L}(\mathbf{C}^m, \mathbf{C}^m))$ and denote the scalar products on $\mathbf{R}^{2n}$ and $\mathbf{C}^m$ by $\langle\,,\,\rangle$ and $(\,,\,)$ respectively. We assume:

(H1) $P(h)$ is symmetric for every $h \in\,]0, h_0]$. In particular $p_j(x,\xi)$ is a Hermitian matrix for every j and every $(x,\xi) \in \mathbf{R}^{2n}$.

(H2) Let $\lambda_1(x,\xi) \leq \ldots \leq \lambda_m(x,\xi)$ be the eigenvalues of $p_0(x,\xi)$ arranged in increasing order. Then

$$\liminf_{|(x,\xi)| \to \infty} d(\lambda_j(x,\xi), [\alpha,\beta]) > 0$$

for all j, where d denotes the standard distance on $\mathbf{R}$.

Definition 12.1. We say that $p_0(x,\xi)$ is microhyperbolic at $(\overline{x}, \overline{\xi})$ in the direction $T \in \mathbf{R}^{2n}$, if there are constants $C_0, C_1, C_2 > 0$ such that

$$(\langle dp_0(x,\xi), T\rangle\omega, \omega) \geq \frac{1}{C_0}\|\omega\|^2 - C_1\|p_0(x,\xi)\omega\|^2,$$

for all $(x,\xi) \in \mathbf{R}^{2n}$ with $\|(x,\xi) - (\overline{x},\overline{\xi})\| \leq \frac{1}{C_2}$ and for all $\omega \in \mathbf{C}^m$. (In this definition, we may replace p_0 by any other smooth function with values in the Hermitian matrices.) If for some constants $C_0, C_1 > 0$ the above estimate holds for all $(x, \xi) \in \mathbf{R}^{2n}$, $\omega \in \mathbf{C}^m$, we say that p_0 is uniformly microhyperbolic on $\mathbf{R}^{2n}$ in the direction T.

For $\tau \in \mathbf{R}$, we put $\Sigma_\tau := \{(x, \xi) \in \mathbf{R}^{2n}; \det(p_0 - \tau) = 0\}$. We introduce the following assumption:

(H3) $(p_0 - \tau)$ is microhyperbolic (in some direction) at every point of Σ_τ, when $\tau \in \{\alpha, \beta\}$.

The assumptions (H1), (H2) imply that if $\eta > 0$ is small enough, then (for $h > 0$ small enough) the spectrum of $P(h)$ in $[\alpha - \eta, \beta + \eta]$ is discrete. We choose η small, so that (H3) still holds, when τ varies in $[\alpha - \eta, \alpha + \eta] \cup [\beta - \eta, \beta + \eta]$. Let $(\lambda_j(h))_{0 \leq j \leq N(h)}$ be the increasing sequence of eigenvalues of $P(h)$ in $[\alpha - \eta, \beta + \eta]$, repeated according to their multiplicities. Put

$$N_h(\alpha, \beta) = \#\{\lambda_j(h); \lambda_j(h) \in [\alpha, \beta]\}.$$

Let $\theta \in C_0^\infty(\mathbf{R})$ have its support in a small neighborhood of the origin and be equal to one near the origin. Let $f \in C_0^\infty(]\tau - \eta, \tau + \eta[)$, $\tau \in \{\alpha, \beta\}$. Under the assumptions (H1-3), we have,

Theorem 12.2. *For every integer $N \geq 1$, we have*

$$\mathrm{tr}\,((\mathcal{F}_h^{-1}\theta)(\lambda - P)f(P)) = (2\pi h)^{-n}(f(\lambda)\sum_{j=0}^{N-1} h^j \gamma_j(\lambda) + \mathcal{O}(h^N)),$$

uniformly for $\lambda \in [\tau - \eta, \tau + \eta]$. Here, γ_j are smooth functions of λ, independent of θ, f, and for $j = 0$, we have

$$\gamma_0(\lambda) = -\frac{1}{2\pi i}\int \mathrm{tr}\,((\lambda + i0 - p_0(x, \xi))^{-1} - (\lambda - i0 - p_0(x, \xi))^{-1})\chi(x, \xi)dxd\xi,$$

where $\chi \in C_0^\infty(\mathbf{R}^{2n})$ is equal to 1 in a neighborhood of

$$\Sigma_{[\tau - \eta, \tau + \eta]} := \cup_{\sigma \in [\tau - \eta, \tau + \eta]}\Sigma_\sigma.$$

From Theorem 12.2 and the arguments of Chapter 10, we will deduce

Theorem 12.3. *We have*

$$N_h(\alpha, \beta) = \frac{1}{(2\pi h)^n}\sum_{j=1}^{m}\iint_{\lambda_j^{-1}([\alpha, \beta])} dxd\xi + \mathcal{O}(h^{1-n}), \quad h \searrow 0.$$

The most essential step in the proof of Theorem 12.2 will be the following result:

Proposition 12.4. *There exists $C_0 > 0$, such that if $\theta \in C_0^\infty(]\frac{1}{2}, 1[)$, then for every $\delta \in]0, 1]$, we have*

$$\operatorname{tr}\left(f(P)\mathcal{F}_h^{-1}\theta_\epsilon(\lambda - P)\right) = \mathcal{O}(h^\infty), \tag{12.1}$$

uniformly for $\lambda \in]\tau - \eta, \tau + \eta[$, $\epsilon \in [h^{1-\delta}, \frac{1}{C_0}]$, where we have written $\theta_\epsilon(t) = \theta(\frac{t}{\epsilon})$.

Proof. Let $\tilde{f}$ be an almost analytic extension of f, such that:

$$\tilde{f} \in C_0^\infty(\mathbf{C}), \tag{12.2}$$

$$\tilde{f}(z) = f(z) \text{ for all } z \in \mathbf{R}, \tag{12.3}$$

$$\overline{\partial}\tilde{f}(z) = \mathcal{O}(|\operatorname{Im} z|^N) \text{ for all } N \in \mathbf{N}. \tag{12.4}$$

Let $\psi(t) \in C^\infty(\mathbf{R}; [0,1])$ be equal to 1 for $t \leq 1$ and equal to 0 for $t \geq 2$. Let $M > 0$ be a sufficiently large constant, to be fixed later, and put

$$\psi_{\frac{Mh}{\epsilon} \log \frac{1}{h}}(z) = \psi\left(\frac{\epsilon \operatorname{Im} z}{Mh \log \frac{1}{h}}\right).$$

We then get

$$\overline{\partial}(\tilde{f}\psi_{\frac{Mh}{\epsilon} \log \frac{1}{h}})$$

$$= \begin{cases} \mathcal{O}(|\operatorname{Im} z|^N), & \text{if } \operatorname{Im} z < 0 \\ \mathcal{O}(\psi_{\frac{Mh}{\epsilon} \log \frac{1}{h}}(z)|\operatorname{Im} z|^N + \frac{\epsilon}{Mh \log \frac{1}{h}} 1_{[1,2]}(\frac{\epsilon \operatorname{Im} z}{Mh \log \frac{1}{h}})), & \operatorname{Im} z \geq 0. \end{cases} \tag{12.5}$$

The starting point will be the Cauchy formula:

$$f(P)(\mathcal{F}_h^{-1}\theta_\epsilon)(\lambda - P) = -\frac{1}{\pi}\int \overline{\partial}(\tilde{f}\psi_{\frac{Mh}{\epsilon}\log\frac{1}{h}})(z)(\mathcal{F}_h^{-1}\theta_\epsilon)(\lambda - z)(z - P)^{-1}L(dz).$$

Let $\chi \in C_0^\infty(\mathbf{R}^{2n}; [0,1])$ be equal to 1 in a small neighborhood of $\Sigma_{[\tau-\eta,\tau+\eta]}$ and put $\tilde{P}(h) = P(h) + i\chi^w(x, hD_x)I$. By definition of χ, $(\tilde{P} - z)$ is elliptic for z in a complex neighborhood K of $[\tau - \eta, \tau + \eta]$. Let χ_1, χ_2 be two functions in $C_0^\infty(\mathbf{R}^{2n})$ which are equal to 1 in a neighborhood of supp χ. From now on we shall sometimes use the same symbol for an h-pseudor and for its Weyl-symbol. Replacing $(z - P)^{-1}$ in the three last terms of the identity:

$$(z - P)^{-1} = \chi_1(z - P)^{-1}\chi_2 + (1 - \chi_1)(z - P)^{-1}\chi_2 + \chi_1(z - P)^{-1}(1 - \chi_2)$$
$$+ (1 - \chi_1)(z - P)^{-1}(1 - \chi_2)$$

by

$$(z - P)^{-1} = (z - \widetilde{P})^{-1} - (z - \widetilde{P})^{-1}(P - \widetilde{P})(z - P)^{-1},$$

and using the fact that $(z - \widetilde{P})^{-1}$ is holomorphic in K, as well as the fact that $\operatorname{supp}(\widetilde{P} - P) \cap \operatorname{supp}(1 - \chi_j) = \emptyset$ for $j = 1, 2$, we get, using the cyclicity of the trace:

$$\operatorname{tr}\left(f(P)\mathcal{F}_h^{-1}\theta_\epsilon(\lambda - P)\right) =$$

$$-\operatorname{tr}\frac{1}{\pi}\int \overline{\partial}(\widetilde{f}\psi_{\frac{Mh}{\epsilon}\log\frac{1}{h}})(z)\mathcal{F}_h^{-1}\theta_\epsilon(\lambda - z)\chi_1(z - P)^{-1}\chi_2 L(dz) + \mathcal{O}(h^\infty)$$

$$= -\operatorname{tr}\frac{1}{\pi}\int_{\operatorname{Im} z < 0} \ldots L(dz) - \operatorname{tr}\frac{1}{\pi}\int_{\operatorname{Im} z > 0} \ldots L(dz) + \mathcal{O}(h^\infty). \qquad (12.6)$$

Here we also used the (Paley–Wiener) estimate,

$$\mathcal{F}_h^{-1}\theta_\epsilon(\lambda - z) = \begin{cases} \mathcal{O}(\frac{\epsilon}{h}e^{\frac{\epsilon \operatorname{Im} z}{h}}) & \text{for } \operatorname{Im} z > 0, \\ \mathcal{O}(\frac{\epsilon}{h})e^{\frac{\epsilon \operatorname{Im} z}{2h}} & \text{for } \operatorname{Im} z < 0. \end{cases} \qquad (12.7)$$

Remark 12.5. In the domain $\operatorname{Im} z > 0$, we have

$$\psi_{\frac{Mh}{\epsilon}\log\frac{1}{h}}|\operatorname{Im} z|^N \le (2\frac{Mh}{\epsilon}\log\frac{1}{h})^N \qquad (12.8)$$

for all $N \ge 0$, and using also (12.7), we will be able to neglect this contribution from (12.5) to (12.6).

(12.4) and (12.7) imply that the first term of the third member of (12.6) is $\mathcal{O}(h^\infty)$. In view of (12.6) and Remark 12.5, (12.1) will follow from

$$-\operatorname{tr}\frac{1}{\pi}\int_{\operatorname{Im} z > \frac{Mh}{\epsilon}\log\frac{1}{h}} \overline{\partial}(\widetilde{f}\psi_{\frac{Mh}{\epsilon}\log\frac{1}{h}})(z)\mathcal{F}_h^{-1}\theta_\epsilon(\lambda - z)\chi_1(z - P)^{-1}\chi_2 L(dz)$$

$$= \mathcal{O}(h^\infty). \qquad (12.9)$$

If we choose χ_1, χ_2 of the form $\sum_i \chi_i(x, \xi)$ where each χ_i has its support in a small neighborhood of some point in $\Sigma_{[\tau - \eta, \tau + \eta]}$, we see that it suffices to show:

Lemma 12.6. *For every* (i, j) *with* $\operatorname{supp}\chi_i \cap \operatorname{supp}\chi_j \ne \emptyset$, *and for every* $N \in \mathbf{N}$, *there exists* $M(N) \ge 0$, *such that for* $M \ge M(N)$:

$$-\frac{1}{\pi}\int_{\operatorname{Im} z > \frac{Mh}{\epsilon}\log\frac{1}{h}} \overline{\partial}(\widetilde{f}\psi_{\frac{Mh}{\epsilon}\log\frac{1}{h}})(z)\mathcal{F}_h^{-1}\theta_\epsilon(\lambda - z)\operatorname{tr}(\chi_i(z - P)^{-1}\chi_j)L(dz)$$

$$= \mathcal{O}(h^N). \qquad (12.10)$$

In fact, by the cyclicity of the trace, it is easy to see that the corresponding integrals with $\operatorname{supp}\chi_j \cap \operatorname{supp}\chi_k = \emptyset$, are $\mathcal{O}(h^\infty)$. In the following, we fix (i,j) as in the lemma. To prove Lemma 12.6, we need

Lemma 12.7. *Let* $\widetilde{P} = \widetilde{p}(x, hD_x) + P(h) - p_0(x, hD_x)$ *be selfadjoint, where* $\widetilde{p} \in S(\mathbf{R}^{2n}, 1; \mathcal{L}(\mathbf{C}^m, \mathbf{C}^m))$ *and* $\widetilde{p}(x,\xi) = p_0(x,\xi)$ *in a small neighborhood of* $\operatorname{supp}\chi_i \cup \operatorname{supp}\chi_j$. *Then for every* $N \in \mathbf{N}$, *there exists* $M(N) \geq 0$, *such that for* $M \geq M(N)$:

$$\operatorname{tr} \frac{1}{\pi} \int_{\operatorname{Im} z > \frac{Mh}{\epsilon} \log \frac{1}{h}} \overline{\partial}(\widetilde{f}\psi_{\frac{Mh}{\epsilon} \log \frac{1}{h}})(z) \mathcal{F}_h^{-1}\theta_\epsilon(\lambda - z)(\chi_i(z - P)^{-1}\chi_j)L(dz) -$$

$$\operatorname{tr} \frac{1}{\pi} \int_{\operatorname{Im} z > \frac{Mh}{\epsilon} \log \frac{1}{h}} \overline{\partial}(\widetilde{f}\psi_{\frac{Mh}{\epsilon} \log \frac{1}{h}})(z) \mathcal{F}_h^{-1}\theta_\epsilon(\lambda - z)(\chi_i(z - \widetilde{P})^{-1}\chi_j)L(dz)$$

$$= \mathcal{O}(h^N). \tag{12.11}$$

Proof of Lemma 12.7. Let $\widetilde{\chi} \in C_0^\infty(\mathbf{R}^{2n})$ be equal to one in a small neighborhood of $\operatorname{supp}\chi_j \cup \operatorname{supp}\chi_j$ and have its support contained in the interior of an h-independent set where the symbols of $P(h)$ and $\widetilde{P}(h)$ coincide. The identity

$$(z - P)\widetilde{\chi}(z - \widetilde{P})^{-1}\chi_j =$$
$$\chi_j + \widetilde{\chi}(\widetilde{P} - P)(z - \widetilde{P})^{-1}\chi_j - [P, \widetilde{\chi}](z - \widetilde{P})^{-1}\chi_j + \mathcal{O}(h^\infty), \quad (12.12)$$

and the fact that $\chi_j \widetilde{\chi} = \chi_j + \mathcal{O}(h^\infty)$ in trace norm, imply

$$\chi_i(z - \widetilde{P})^{-1}\chi_j - \chi_i(z - P)^{-1}\chi_j =$$
$$\chi_i(z - P)^{-1}\widetilde{\chi}(\widetilde{P} - P)(z - \widetilde{P})^{-1}\chi_j - \chi_i(z - P)^{-1}[P, \widetilde{\chi}](z - \widetilde{P})^{-1}\chi_j$$
$$+ \mathcal{O}(\frac{h^\infty}{|\operatorname{Im} z|}). \tag{12.13}$$

Here the first term of the second member is $\mathcal{O}(h^\infty|\operatorname{Im} z|^{-2})$ in trace norm. For the second term, we shall use the fact that modulo $\mathcal{O}(h^\infty)$ in trace norm $[P, \widetilde{\chi}]$ has a symbol with support in a compact set $\widetilde{K}$ such that $\operatorname{supp}\chi_i \cap \widetilde{K} = \emptyset$.

Let $G_0 \in \dot{C}_0^\infty(\mathbf{R}^{2n})$ be real-valued and such that

$$G_0 = 1 \text{ near } \operatorname{supp}\chi_i, \quad G_0 = 0 \text{ near } \widetilde{K}.$$

Put $G = \alpha G_0$, $\alpha > 0$. We notice that the symbol $a = e^{G \log \frac{1}{h}}$ is of class $S_\delta^\delta(1)$ for every $\delta > 0$. By $e^{G \log \frac{1}{h}}$ we also denote the corresponding h-pseudor, which is elliptic and has an inverse operator $(e^{G \log \frac{1}{h}})^{-1}$ with symbol in the same classes. It is clear that for some $k \in \mathbf{N}$:

$$e^{G \log \frac{1}{h}}(z - P)(e^{G \log \frac{1}{h}})^{-1} = z - P + \mathcal{O}(\alpha h \log \frac{1}{h})\|\nabla G_0\|_{C^k}, \qquad (12.14)$$

in operator norm, for $h \leq h(\alpha)$, where $h(\alpha) > 0$ is some continuous function. Using the fact that $G = \alpha$ near $\operatorname{supp} \chi_i$ and that $G = 0$ near $\widetilde{K}$, we get in trace norm:

$$e^{\alpha \log \frac{1}{h}}\chi_i(z - P)^{-1}[P, \widetilde{\chi}]$$
$$= \chi_i e^{G \log \frac{1}{h}}(z - P)^{-1}(e^{G \log \frac{1}{h}})^{-1}[P, \widetilde{\chi}] + \mathcal{O}(\frac{h^\infty}{|\operatorname{Im} z|})$$
$$= \chi_i(z - e^{G \log \frac{1}{h}}P(e^{G \log \frac{1}{h}})^{-1})^{-1}[P, \widetilde{\chi}] + \mathcal{O}(\frac{h^\infty}{|\operatorname{Im} z|}). \qquad (12.15)$$

We choose G with
$$\alpha = \min(\frac{\operatorname{Im} z}{\widetilde{C} h \log \frac{1}{h}}, \mathcal{O}(1)),$$

where $\mathcal{O}(1)$ is some arbitrarily large and fixed constant and $\widetilde{C}$ is sufficiently large. Then the expression (12.15) is $\mathcal{O}((h^n |\operatorname{Im} z|)^{-1})$ in trace norm, and we get with a new constant $C_1 > 0$:

$$\chi_i(z - P)^{-1}[P, \widetilde{\chi}] = \mathcal{O}(h^{-n}|\operatorname{Im} z|^{-1}e^{-\alpha \log \frac{1}{h}})$$
$$= \mathcal{O}(h^{-n}|\operatorname{Im} z|^{-1} \max(h^{\mathcal{O}(1)}, e^{-\frac{\operatorname{Im} z}{C_1 h}})).$$

For $\frac{Mh}{\epsilon} \log \frac{1}{h} \leq \operatorname{Im} z \leq 2\frac{Mh}{\epsilon} \log \frac{1}{h}$, we get, by using also (12.7):

$$(\mathcal{F}_h^{-1}\theta_\epsilon)(\lambda - z)\chi_i(z - P)^{-1}[P, \widetilde{\chi}] =$$
$$\mathcal{O}(\epsilon h^{-n-1}|\operatorname{Im} z|^{-1} \max(h^{\mathcal{O}(1)}, e^{-\frac{1}{h}\operatorname{Im} z(\frac{1}{C_1} - \frac{1}{C_0})})), \qquad (12.16)$$

where we recall that $\epsilon \leq \frac{1}{C_0}$. We choose $C_0 > C_1$. Then the LHS of (12.11) is

$$\mathcal{O}(h^\infty) + \int_{\substack{\frac{Mh}{\epsilon} \log \frac{1}{h} \leq \operatorname{Im} z \leq 2\frac{Mh}{\epsilon} \log \frac{1}{h} \\ |\operatorname{Re} z| \leq \text{const.}}} \mathcal{O}(1)(\frac{\epsilon}{Mh \log \frac{1}{h}})^2 \epsilon h^{-n-1} \max(h^{\mathcal{O}(1)}, e^{-\frac{1}{h}\operatorname{Im} z(\frac{1}{C_1} - \frac{1}{C_0})}) L(dz)$$

$$= \mathcal{O}(h^\infty) + \mathcal{O}(1)\frac{\epsilon}{Mh \log \frac{1}{h}}\epsilon h^{-n-1} \max(h^{\mathcal{O}(1)}, e^{-C_0 M(\log \frac{1}{h})(\frac{1}{C_1} - \frac{1}{C_0})}).$$

Choosing first M sufficiently large and then the exponent in $h^{\mathcal{O}(1)}$ sufficiently large, we see that this is $\mathcal{O}(h^N)$ for any given N.   #

Proof of Lemma 12.6. Combining Lemma 12.7 with the result of the appendix to this chapter, we see that we may assume (after changing the symbol away from supp $\chi_i \cup$ supp χ_j), that the leading symbol p has the property that for $\tau \in \{\alpha, \beta\}$, $(\tau - p)$ is globally and uniformly microhyperbolic with respect to H_ϕ, where ϕ is a linear function which may depend on τ. After a linear canonical transformation (see the appendix to Chapter 7), we may assume that $\phi = x_1$.

For $t \in \mathbf{R}$, we put

$$\chi_{k,t} = e^{-it\phi/h}\chi_k e^{it\phi/h}, \ k = i, j,$$
$$P_t = e^{-it\phi/h}Pe^{it\phi/h}.$$

Notice that the symbol of P_t is obtained from that of P by translating by $-t$ in the ξ_1-direction. Introducing almost analytic extensions of the symbols χ_i, χ_j, P, we can define $\chi_{j,t}, \chi_{k,t}, P_t$ for $t \in \mathbf{C}$, such that

$$\frac{\partial}{\partial \bar{t}}Q = \mathcal{O}(|\mathrm{Im}\,t|^\infty) \text{ in } S^0,$$

where Q_t denotes one of $\chi_{i,t}, \chi_{j,t}, P_t$, and such that $\chi_{j,t}$ have uniformly compact support in (x, ξ) whenever t is restricted to a compact set. We put

$$\tilde{Q}_t = e^{-i\mathrm{Re}\,t\phi/h}Q_{i\mathrm{Im}\,t}e^{i\mathrm{Re}\,t\phi/h}.$$

Lemma 12.8. *We have in* S^0:

$$\frac{\partial}{\partial \bar{t}}\tilde{Q} = \mathcal{O}(|\mathrm{Im}\,t|^\infty), \tag{12.17}$$

$$\nabla_t^k \tilde{Q}_t = \mathcal{O}(1), \ k \in \mathbf{N}, \tag{12.18}$$

$$\tilde{Q}_t - Q_t = \mathcal{O}(|\mathrm{Im}\,t|^\infty). \tag{12.19}$$

Proof. Q_t satisfies:

$$(*) \quad \begin{cases} hD_tQ_t + [\phi, Q_t] = 0 \text{ for } t \text{ real} \\ \frac{\partial}{\partial \bar{t}}Q_t = \mathcal{O}(|\mathrm{Im}\,t|^\infty), \\ \nabla_t^k Q_t = \mathcal{O}(1) \ \forall k \in \mathbf{N}. \end{cases}$$

and this implies that $hD_tQ_t + [\phi, Q_t] = \mathcal{O}(h|\mathrm{Im}\,t|^\infty)$. Also if $\hat{Q}_t$ is a C^∞ function of t with values in S^0 satisfying $(*)$ and $\hat{Q}_0 = Q_0$, then

$$\hat{Q}_t - Q_t = \mathcal{O}(|\mathrm{Im}\,t|^\infty) \text{ in } S^0.$$

Let us verify that $\widetilde{Q}_t$ is such a solution: We have:

$$h\frac{\partial}{\partial \mathrm{Re}\, t}\widetilde{Q}_t = -ie^{-i\mathrm{Re}\, t\phi/h}[\phi, Q_{i\mathrm{Im}\, t}]e^{i\mathrm{Re}\, t\phi/h},$$

$$h\frac{\partial}{\partial \mathrm{Im}\, t}\widetilde{Q}_t = e^{-i\mathrm{Re}\, t\phi/h}h\frac{\partial}{\partial \mathrm{Im}\, t}(Q_{i\mathrm{Im}\, t})e^{i\mathrm{Re}\, t\phi/h}.$$

From the construction of Q_t, we have

$$[\phi, Q_{i\mathrm{Im}\, t}] = -\frac{h}{i}(\frac{\partial}{\partial t}Q)_{i\mathrm{Im}\, t} + h\mathcal{O}(|\mathrm{Im}\, t|^\infty)$$

$$= -\frac{h}{i}\frac{1}{i}\frac{\partial}{\partial \mathrm{Im}\, t}(Q_{i\mathrm{Im}\, t}) + h\mathcal{O}(|\mathrm{Im}\, t|^\infty),$$

and it follows that $\frac{\partial}{\partial t}\widetilde{Q}_t = \mathcal{O}(|\mathrm{Im}\, t|^\infty)$, so $\widetilde{Q}_t$ is a solution of $(*)$. Then (12.19) follows.   #

We put

$$k(t, z) = \mathrm{tr}\,(\widetilde{\chi}_{i,t}(z - \widetilde{P}_t)^{-1}\widetilde{\chi}_{j,t}).$$

k is well defined in the zone $|\mathrm{Im}\, t| \leq \frac{1}{\mathcal{O}(1)}|\mathrm{Im}\, z|$ and has the following properties:

$$k(t, z) = \mathcal{O}(h^{-n}|\mathrm{Im}\, z|^{-1}),$$

$$\frac{\partial}{\partial t}k(t, z) = \mathcal{O}(h^{-n}|\mathrm{Im}\, t|^\infty|\mathrm{Im}\, z|^{-2}).$$

Since $k(t, z)$ only depends on $\mathrm{Im}\, t$, it follows that

$$k(i\mathrm{Im}\, t, z) = k(0, z) + \mathcal{O}(\frac{|\mathrm{Im}\, t|^\infty}{h^n|\mathrm{Im}\, z|^2}).$$

We can then replace $\mathrm{tr}\,(\chi_i(z - P)^{-1}\chi_j)$ in the integral in the first member of (12.10) by $k(i\frac{Mh}{C_1\epsilon}\log\frac{1}{h}, z)$, where C_1 does not depend on the other constants.

By construction of P_t we see that the leading symbol of $z - \widetilde{P}_{i\mathrm{Im}\, t}$ is of the form

$$z - p_{i\mathrm{Im}\, t}(x, \xi) = z - p_0(x, \xi) - i\mathrm{Im}\, t\frac{\partial p_0(x, \xi)}{\partial \xi_1} + \mathcal{O}(|\mathrm{Im}\, t|^2),$$

writing $p_{i\mathrm{Im}\, t}$ instead of $p_{0,i\mathrm{Im}\, t}$ for short. Put $\mu = \mathrm{Im}\, t = \frac{Mh}{C_1\epsilon}\log(\frac{1}{h})$ and let $-\frac{\mu}{2} < \mathrm{Im}\, z < 0$. Using the global and uniform microhyperbolicity (Definition 12.1) with $T = (0; (-1, 0, 0, \dots, 0))$, we obtain for $|\mathrm{Re}\, z - \tau| < \eta$:

$$\mathrm{Im}\,(z - p_{i\mu}(x, \xi)) \geq (2\mu + \mathrm{Im}\, z)I - \mathcal{O}(1)\mu(\mathrm{Re}\, z - p(x, \xi))^2$$

in the sense of Hermitian matrices, where we assumed for simplicity that the constant C_0 in Definition 12.1 is equal to $\frac{1}{2}$. Then, if h is small enough, there exists a constant $C_2 > 0$ such that,

$$\operatorname{Im}(z - p_{i\mu}(x,\xi)) + C_2\mu(z - p_{i\mu}(x,\xi))^*(z - p_{i\mu}(x,\xi)) \geq (\mu + \operatorname{Im} z)I,$$
(12.20)

where the star indicates that we take the usual complex adjoint. Now the semi-classical version of the sharp Gårding inequality (see Theorem 7.12) implies,

$$\operatorname{Im}(\operatorname{Op}_h^w(z - p_{i\mu})u|u) + C_2\mu\|\operatorname{Op}_h^w(z - p_{i\mu})u\|^2$$

$$\geq (\operatorname{Op}_h^w(\operatorname{Im}(z - p_{i\mu}) + C_2\mu(z - p_{i\mu})^*(z - p_{i\mu}))u|u) - \mathcal{O}(1)h\mu\|u\|^2$$

$$\geq (\mu + \operatorname{Im} z)\|u\|^2 - \mathcal{O}(1)(h + \mu h)\|u\|^2 \geq \frac{\mu}{3}\|u\|^2, \qquad (12.21)$$

for all $u \in L^2(\mathbf{R}^n; \mathbf{C}^m)$, for h small enough. (12.21) combined with the inequality $ab \leq \frac{\mu}{6}a^2 + \frac{3}{2\mu}b^2$ gives

$$\frac{\mu}{3}\|u\|^2 \leq \|\operatorname{Op}_h^w(z - p_{i\mu})u\|\|u\| + C_1\mu\|\operatorname{Op}_h^w(z - p_{i\mu})u\|^2$$

$$\leq \frac{\mu}{6}\|u\|^2 + (\frac{3}{2\mu} + C_1\mu)\|\operatorname{Op}_h^w(z - p_{i\mu})u\|^2.$$

Then there exists $C_2 > 0$ such that,

$$\frac{\mu}{C_2}\|u\| \leq \|\operatorname{Op}_h^w(z - p_{i\mu})u\|, \qquad (12.22)$$

for all $u \in L^2(\mathbf{R}^n; \mathbf{C}^m)$. We conclude that $(z - P_{\frac{iMh}{C_1\epsilon} \log \frac{1}{h}})^{-1}$ extends as a holomorphic function of z to the zone $\operatorname{Im} z \geq -\frac{Mh}{2C_1^2\epsilon} \log \frac{1}{h}$ and consequently the LHS of (12.10) is equal to

$$\frac{1}{\pi} \int_{\operatorname{Im} z < 0} \bar{\partial}(\tilde{f}\psi_{\frac{Mh}{\epsilon} \log \frac{1}{h}, \frac{Mh}{C_1^2} \log \frac{1}{h}})(z)\mathcal{F}_h^{-1}\theta_\epsilon(\lambda - z)k(i\frac{Mh}{C_1\epsilon} \log \frac{1}{h}, z)L(dz)$$

$$+\mathcal{O}(h^N),$$

where we define $\psi_{\alpha,\beta} = \psi(\frac{-\operatorname{Im} z}{\beta})\psi(\frac{\operatorname{Im} z}{\alpha})$. Using Remark 12.5, which extends to the zone $\operatorname{Im} z < 0$, and the fact that

$$k(i\frac{Mh}{C_1\epsilon} \log \frac{1}{h}, z) = \mathcal{O}(1)(h^{-n}\epsilon/Mh \log \frac{1}{h})$$

in the zone $-\frac{Mh}{2C_1^2\epsilon} \log \frac{1}{h} \leq \operatorname{Im} z < 0$, we get

$$\frac{1}{\pi} \int_{\operatorname{Im} z < 0} \tilde{f}\bar{\partial}(\psi_{\frac{Mh}{\epsilon} \log \frac{1}{h}, \frac{Mh}{C_1^2} \log \frac{1}{h}})\mathcal{F}_h^{-1}\theta_\epsilon(\lambda - z)k(i\frac{Mh}{C_1\epsilon} \log \frac{1}{h}, z)L(dz)$$

$$= \mathcal{O}(1)\frac{h^{-n}}{(\frac{Mh}{\epsilon} \log \frac{1}{h})^2} \int_{\substack{\operatorname{Re} z \in]\tau-\eta, \tau+\eta[, \\ -2\frac{Mh}{C_1^2} \log \frac{1}{h} \leq \operatorname{Im} z \leq -\frac{Mh}{C_1^2} \log \frac{1}{h}}} \frac{\epsilon}{h}e^{\frac{\epsilon|\operatorname{Im} z|}{2h}}L(dz)$$

$$= \mathcal{O}(1)\frac{\epsilon^2 h^{-n-2}}{M \log \frac{1}{h}}e^{-\frac{M}{2C_1^2} \log \frac{1}{h}} = \mathcal{O}(1)\frac{\epsilon^2 h^{-n-2+\frac{M}{2C_1^2}}}{M \log \frac{1}{h}}.$$

To finish the proof, it suffices to choose M sufficiently large. This also ends the proof of Proposition 12.4.   # #

Remark 12.9. Proposition 12.4 remains valid if $\operatorname{supp}\theta \subset]-1, -\tfrac{1}{2}[$.

Proof of Theorem 12.2. In the following we let C be a large fixed constant and δ a constant in $]0, \tfrac{1}{2}[$. According to the preceding remark and Proposition 12.1, we have:

$$\operatorname{tr}(\mathcal{F}_h^{-1}\theta(\lambda - P)f(P)) =$$
$$-\operatorname{tr}\frac{1}{\pi}\int \overline{\partial}(\psi_{h^\delta}\widetilde{f})(z)\mathcal{F}_h^{-1}\theta_\epsilon(\lambda - z)(z - P)^{-1}\chi L(dz) + \mathcal{O}(h^\infty). \quad (12.23)$$

Here χ, θ are the functions given in Theorem 12.2, $\epsilon = h^{1-\delta}$ and $\psi_{h^\delta} = \psi(\frac{\operatorname{Im} z}{h^\delta})$, where $\psi \in C_0^\infty(]-2, 2[)$ is equal to 1 on $[-1, 1]$. In fact, we can represent $(\theta - \theta_\epsilon)$ as a sum of functions of the type appearing in Proposition 2.4, Remark 12.9, which leads to (12.23) without $\chi = \chi^w(x, hD_x)$. The insertion of χ can then be justified in the same way as the insertion of χ_1, χ_2 in (12.6). Since $\overline{\partial}(\psi_{h^\delta}\widetilde{f}) = \psi_{h^\delta}\overline{\partial}\widetilde{f} + \widetilde{f}\overline{\partial}(\psi_{h^\delta})$, we can again use Remark 12.5 to reduce the domain of integration in (12.22) to $K_\delta = \{z \in \mathbf{C}; |\operatorname{Im} z| \geq h^\delta\}$. Here we also use the fact that $\mathcal{F}_h^{-1}\theta_\epsilon(\lambda - z) = \mathcal{O}(\frac{\epsilon}{h}e^{\frac{2\epsilon}{h}|\operatorname{Im} z|})$ by Payley–Wiener. According to Proposition 8.6, the symbol of $(z - P)^{-1}$ is given for z in K_δ by:

$$E_0(x, \xi) + hE_1(x, \xi, z) + \ldots + h^N E_N(x, \xi, z) + \mathcal{O}(h^{N(1-2\delta)}). \quad (12.24)$$

Here $E_j(x, \xi, z)$ is a finite sum of terms of the form

$$(z - p_0(x, \xi))^{-1}b_1(x, \xi, z)(z - p_0(x, \xi))^{-1} \cdot \ldots \cdot b_{k-1}(x, \xi, z)(z - p_0)^{-1}$$

with $k \leq 2j + 1$ and $b_k \in S^0(\mathbf{R}^{2n}, 1; \mathcal{L}(\mathbf{C}^m, \mathbf{C}^m))$ uniformly in z. We then get:

$$\operatorname{tr}((\mathcal{F}_h^{-1}\theta)(\lambda - P)f(P)) = \frac{1}{(2\pi h)^n}\sum_{j=0}^N h^j a_j + \mathcal{O}(h^{N(1-2\delta)-n}), \quad (12.25)$$

where

$$a_j = -\frac{1}{\pi}\int \operatorname{tr}\iint \overline{\partial}(\psi_{h^\delta}\widetilde{f})(z)\mathcal{F}_h^{-1}\theta_\epsilon(\lambda - z)E_j(x, \xi, z)\chi(x, \xi)dxd\xi L(dz).$$
$$(12.26)$$

Lemma 12.10. *The functions $e_j(z) = \iint \operatorname{tr} E_j(x, \xi, z)\chi(x, \xi)dxd\xi$, defined on $]\tau - \eta, \tau + \eta[\pm i]0, \alpha[$ for some constant $\alpha > 0$, extend to C^∞-functions on $]\tau - \eta, \tau + \eta[\pm i[0, \alpha[$.*

Proof. For the simplicity of the notation, we only treat the case $j = 0$. Decomposing χ by a partition of unity, we may also assume that $\chi(x, \xi)$ has its support in some small neighborhood of some point in $\Sigma_{[\tau-\eta, \tau+\eta]}$. Let U be a constant direction of microhyperbolicity for $p_0 - \lambda$ in a neighborhood of supp χ.

Let $\tilde{p}_0$ be an almost analytic extension of p_0. As with the formula (12.22), we show that $(z - \tilde{p}_0((x, \xi) - itU))^{-1}$ exists and is of norm $\mathcal{O}(\frac{1}{t})$ for Re $z \in]\tau - \eta, \tau + \eta[$, Im $z \geq 0$, $t \geq 0$, Im $z + t > 0$. Stokes' formula gives for Im $z > 0$:

$$
\begin{aligned}
G(z) &= \iint (z - p_0(x, \xi))^{-1} \chi(x, \xi) dx d\xi \\
&= \iint_{\mathbf{R}^{2n} - it_0 U} (z - \tilde{p}_0)^{-1} \tilde{\chi}(x, \xi) dx d\xi \\
&\quad + \iiint_{\substack{0 \leq t \leq t_0, \\ (x, \xi) \in \mathbf{R}^{2n}}} \Gamma^*(((z - \tilde{p}_0)^{-1} \overline{\partial} p_0 (z - \tilde{p}_0)^{-1} \tilde{\chi} + (z - p_0)^{-1} \overline{\partial} \tilde{\chi}) \wedge dx d\xi),
\end{aligned}
$$

where $\Gamma : (t, x, \xi) \mapsto (x, \xi) - itU$ and $\tilde{\chi}$ is an almost analytic extension of χ. Here the first term of the last member extends to a C^∞ function on $]\tau - \eta, \tau + \eta[+ i]0, \alpha]$. The same is true for the last term, since the integrand is $\mathcal{O}(t^\infty)$ as well as all its derivatives with respect to z. We have then proved that e_0 extends to a smooth function on $]\tau - \eta, \tau + \eta[+ i]0, \alpha[$. Using the Hermitian property we also get the extension to $]\tau - \eta, \tau + \eta[- i]0, \alpha[$.   #

From (12.26), Lemma 12.10 and Stokes' formula, we get

$$
a_j = -\frac{1}{2\pi i} \int \tilde{f}(\lambda') \mathcal{F}_h^{-1} \theta_\epsilon(\lambda - \lambda')(e_j(\lambda' + i0) - e_j(\lambda' - i0)) d\lambda'. \quad (12.27)
$$

Here

$$
\mathcal{F}_h^{-1} \theta_\epsilon = \mathcal{F}_{h/\epsilon}^{-1} \theta = \mathcal{F}^{-1}(\theta(\frac{h}{\epsilon}\cdot)) = \delta_0 + \mathcal{O}(h^\infty)
$$

in the sense of distributions, so (12.27) gives

$$
a_j = -\frac{1}{2\pi i} f(\lambda)(e_j(\lambda + i0) - e_j(\lambda - i0)) + \mathcal{O}(h^\infty), \quad (12.28)
$$

and we get Theorem 12.2 with

$$
\gamma_j(\lambda) = -\frac{1}{2\pi i}(e_j(\lambda + i0) - e_j(\lambda - i0)). \quad (12.29)
$$

#

150   *Spectral Asymptotics in the Semi-Classical Limit*

Notice that for $f \in C_0^\infty(]\tau - \eta, \tau + \eta[)$, $(x, \xi) \in \operatorname{supp} \chi$:

$$-\frac{1}{2\pi i} \int f(\lambda)(\operatorname{tr}(\lambda + i0 - p_0(x,\xi))^{-1} - \operatorname{tr}(\lambda - i0 - p_0(x,\xi))^{-1})d\lambda$$
$$= \sum_{j=1}^{m} f(\lambda_j(x,\xi)).$$

Consequently,

$$\int f(\lambda)\gamma_0(\lambda)d\lambda = \sum_{j=1}^{m} \iint f(\lambda_j(x,\xi))\chi(x,\xi)dxd\xi = \sum_{j=1}^{m} \iint f(\lambda_j(x,\xi))dxd\xi.$$
(12.30)

Proof of Theorem 12.3. Let $f_1 \in C_0^\infty(]\alpha - \eta, \alpha + \eta[; [0,1])$, $f_2 \in C_0^\infty(]\alpha + \frac{\eta}{2}, \beta - \frac{\eta}{2}[; [0,1])$ $f_3 \in C_0^\infty(]\beta - \eta, \beta + \eta[; [0,1])$ satisfy $f_1 + f_2 + f_3 = 1$ on $[\alpha - \frac{\eta}{2}, \beta + \frac{\eta}{2}]$. Let $\lambda_0(h) \le \lambda_1(h) \le \ldots \le \lambda_{N(h)}$ be the eigenvalues of $P(h)$ (counted with their multiplicity) in $[\alpha - \eta, \beta + \eta]$. We have

$$N_h(\alpha, \beta) = \sum_{\alpha \le \lambda_j(h) \le \beta} (f_1 + f_2 + f_3)(\lambda_j(h))$$
$$= \sum_{\alpha \le \lambda_j(h)} f_1(\lambda_j(h)) + \sum f_2(\lambda_j(h)) + \sum_{\lambda_j(h) \le \beta} f_3(\lambda_j(h))$$
$$= \sum_{\alpha \le \lambda_j(h)} f_1(\lambda_j(h)) + \operatorname{tr} f_2(P(h)) + \sum_{\lambda_j(h) \le \beta} f_3(\lambda_j(h)). (12.31)$$

According to Theorem 9.6, we have:

$$\operatorname{tr} f_2(P(h)) = (2\pi h)^{-n} \sum_{j=1}^{m} \iint f_2(\lambda_j(x,\xi))dxd\xi + \mathcal{O}(h^{1-n}).$$
(12.32)

Using the argument of Chapter 10 (Theorem 10.1), we have:

$$\sum_{\alpha \le \lambda_j(h)} f_1(\lambda_j(h)) = (2\pi h)^{-n} \int_\alpha^{+\infty} f_1(\lambda)\gamma_0(\lambda)d\lambda + \mathcal{O}(h^{1-n}),$$
(12.33)

$$\sum_{\lambda_j(h) \le \beta} f_3(\lambda_j(h)) = (2\pi h)^{-n} \int_{-\infty}^\beta f_3(\lambda)\gamma_0(\lambda)d\lambda + \mathcal{O}(h^{1-n}).$$
(12.34)

Approaching $1_{[\alpha,\infty[}f_1, f_{]-\infty,\beta]}f_3$ from above and from below by C_0^∞-functions and using the fact that $\gamma_0(\lambda)$ is continuous, we get:

$$\int_{-\infty}^\beta f_3(\lambda)\gamma_0(\lambda)d\lambda = \sum_{j=1}^{m} \iint (f_3 1_{]-\infty,\beta]})(\lambda_j(x,\xi))dxd\xi$$
(12.35)

$$\int_{\alpha}^{\infty} f_1(\lambda)\gamma_0(\lambda)d\lambda = \sum_{j=1}^{m} \iint (f_1 1_{[\alpha,\infty[})(\lambda_j(x,\xi))dxd\xi, \qquad (12.36)$$

and combining this with (12.31), (12.32), we get Theorem 12.3. #

Appendix: Extension of a microhyperbolic function

Let $p(x)$ be smooth in a neighborhood of $0 \in \mathbf{R}^n$ with values in the Hermitian $m \times m$-matrices, and assume that $p(x)$ is microhyperbolic in the constant direction $t \in \mathbf{R}^n$, so that

$$t(\partial_x)p := \langle t, \partial_x p(x) \rangle \geq \frac{1}{C_0} - C_1(p(x))^2, \qquad (A.1)$$

for some $C_0 > 0$ in the sense of Hermitian matrices. After a conjugation with a constant unitary transformation, we may assume that p takes the block matrix form

$$p(x) = \begin{pmatrix} p_{11}(x) & p_{21}(x) \\ p_{12}(x) & p_{22}(x) \end{pmatrix}, \qquad (A.2)$$

with $p_{11}(0) = 0$, $p_{21}(0) = 0$, $p_{12}(0) = 0$ and with $p_{22}(0)$ bijective. Put

$$p_0(x) = \begin{pmatrix} \langle dp_{11}(0), x \rangle & 0 \\ 0 & p_{22}(0) \end{pmatrix}.$$

Lemma A.1. $p_0(x)$ *is uniformly microhyperbolic on* $\mathbf{R}^n$ *in the direction* t.

Proof. Let

$$\pi_0 = \begin{pmatrix} 1 & 0 \\ 0 & 0 \end{pmatrix}.$$

Then for every $\delta \in]0,1]$,

$$t(\partial_x)p_0 = \langle t, dp(0) \rangle + \begin{pmatrix} 0 & \mathcal{O}(1) \\ \mathcal{O}(1) & \mathcal{O}(1) \end{pmatrix}$$

$$\geq \frac{1}{C_0} - C_1(p(0))^2 - \mathcal{O}(1)(\delta\pi_0^2 + \frac{1}{\delta}(I - \pi_0)^2).$$

Choosing $\delta > 0$ small enough, we get

$$t(\partial_x)p_0 \geq \frac{1}{2C_0} - \mathcal{O}(1)(p_0(x))^2, \qquad (A.3)$$

uniformly with respect to x. #

Let $\chi \in C_0^\infty(\mathbf{R}^n; [0,1])$ be equal to 1 near 0 and put with a (new) sufficiently small $\delta > 0$:

$$p_\delta(x) = \chi(\frac{x}{\delta})(p(x) - p_0(x)) + p_0(x). \tag{A.4}$$

If $|x| = \mathcal{O}(\delta)$, we have

$$p_\delta(x) = \begin{pmatrix} 0 & 0 \\ 0 & f(x) \end{pmatrix} + \mathcal{O}(\delta),$$

where $f(x)$, $f(x)^{-1} = \mathcal{O}(1)$, uniformly with respect to x. Hence,

$$p_\delta(x)^2 = \begin{pmatrix} \mathcal{O}(\delta^2) & \mathcal{O}(\delta) \\ \mathcal{O}(\delta) & f(x)^2 + \mathcal{O}(\delta) \end{pmatrix} \geq \begin{pmatrix} -\mathcal{O}(\delta^2) & 0 \\ 0 & \frac{1}{\mathcal{O}(1)} \end{pmatrix}. \tag{A.5}$$

Using the fact that $p_\delta(x) = p_0(x)$ for $|x| \geq \mathcal{O}(\delta)$, we see that (A.5) extends uniformly to all $x \in \mathbf{R}^n$.

For δ small enough, we have

$$t(\partial_x)p_\delta = \begin{pmatrix} \langle t, p_{11}(0) \rangle + \mathcal{O}(\delta) & \mathcal{O}(1) \\ \mathcal{O}(1) & \mathcal{O}(1) \end{pmatrix},$$

and the argument that gave (A.3) now shows that

$$t(\partial_x)p_\delta \geq \frac{1}{3C_0}I - \begin{pmatrix} 0 & 0 \\ 0 & \mathcal{O}(1) \end{pmatrix}.$$

Thanks to (A.5), we get for δ small enough,

$$t(\partial_x)p_\delta \geq \frac{1}{4C_0}I - \mathcal{O}(1)p_\delta(x)^2. \tag{A.6}$$

We have then proved

Lemma A.2. *If $\delta > 0$ is small enough, then p_δ is uniformly microhyperbolic in the direction t.*

In the following, we let δ be a small fixed constant. We need to further modify p_δ away from $x = 0$ in order to become of class $S(1)$. Notice that all the derivatives of p_δ are uniformly bounded on $\mathbf{R}^n$. If $f \in C_0^\infty(\mathbf{R})$, we know from Chapter 8, that

$$f(p_\delta(x)) = -\frac{1}{\pi} \int \overline{\partial}\tilde{f}(z)(z - p_\delta(x))^{-1} L(dz) \tag{A.7}$$

is in $S(1)$, and if we let f vary in a class of $C_0^\infty(\mathbf{R})$ functions with uniform bounds on the diameter of the supports and on the supremum of every derivative, then $f(p_\delta)$ varies in a bounded set in $S(1)$.

If $f \in S(\mathbf{R}, 1)$, then we can find such a class such that for every $x_0 \in \mathbf{R}^n$, we can find $f_1, \ldots, f_m$ in the class such that

$$f(p_\delta(x)) = f_1(p_\delta(x)) + \ldots + f_m(p_\delta(x))$$

for x in some neighborhood of x_0. It follows that $f(p_\delta) \in S(\mathbf{R}^n, 1)$. Now, choose $f \in S(\mathbf{R}, 1)$ real-valued, such that $f(t) = t$ for $|t| \leq 1$, $f(t) \geq 1$ for $t \geq 1$, $f(t) \leq -1$ for $t \leq -1$, and put $\tilde{p}(x) = f(p_\delta(x))$. Without any loss of generality, we may assume that the norm of $p_\delta(x)$ is smaller than 1 for x in some neighborhood of 0.

Proposition A.3.

(i) $\tilde{p}(x) = p(x)$ in a small neighborhood of 0,

(ii) $\tilde{p} \in S(\mathbf{R}^n, 1; \mathcal{L}(\mathbf{C}^m, \mathbf{C}^m))$,

(iii) $\tilde{p}$ is globally and uniformly microhyperbolic in the direction t.

Proof. *(i)* follows from the construction and we have verified *(ii)*.

For every $x_0 \in \mathbf{R}^n$, we can find $s \in [\frac{1}{\mathcal{O}(1)}, \frac{1}{2}]$, such that $\pm[s, s + \frac{1}{\mathcal{O}(1)}]$ are disjoint from the spectrum of $p_\delta(x)$ for x in some neighborhood W_{x_0} of x_0. (Here the $\mathcal{O}(1)$ is uniform with respect to x_0.) Then $\pi = \pi(x) = 1_{[-s,s]}(p_\delta(x))$ is C^∞ and

$$\tilde{p} = \pi p_\delta \pi + (1 - \pi)a(1 - \pi), \quad [a, \pi] = 0, \qquad (A.8)$$

where $0 < a$, $\partial_x^\alpha a$, a^{-1}, $\partial_x^\alpha \pi$ are $\mathcal{O}_\alpha(1)$ for $x \in W_{x_0}$, uniformly with respect to x_0.

Let A, B be Hermitian matrices. Then for every $\alpha > 0$, we have:

$$AB + BA \geq -\alpha A^2 - \frac{1}{\alpha}B^2. \qquad (A.9)$$

From (A.8), we get, writing $t(\partial_x) = t$,

$$t(\tilde{p}) = t(\pi)p_\delta\pi + \pi p_\delta t(\pi) + \pi t(p_\delta)\pi - $$
$$t(\pi)a(1 - \pi) - (1 - \pi)at(\pi) - (1 - \pi)t(a)(1 - \pi) \quad (A.10)$$

Using (A.6), as well as the fact that $\pi^2 = \pi$, we get with a new constants $C_0, C_1 > 0$,

$$\pi t(p_\delta)\pi \geq \frac{\pi}{C_0} - C_1 \pi p_\delta^2 \pi. \qquad (A.11)$$

From (A.8) and the properties of a, we obtain

$$\pi t(p_\delta)\pi \geq \frac{1}{C_0} - \mathcal{O}(1)\widetilde{p}^2. \tag{A.12}$$

The argument used above, combined with (A.9), gives for arbitrary $\alpha > 0$,

$$\pi p_\delta t(\pi) + t(\pi)p_\delta\pi \geq -\alpha - \frac{\mathcal{O}(1)}{\alpha}\widetilde{p}^2,$$

$$-t(\pi)a(1-\pi) - (1-\pi)at(\pi) \geq -\alpha - \frac{\mathcal{O}(1)}{\alpha}\widetilde{p}^2,$$

$$(1-\pi)t(a)(1-\pi) \geq -\mathcal{O}(1)\widetilde{p}^2.$$

Choosing α sufficiently small, we obtain

$$t(\partial_x)\widetilde{p} \geq \frac{1}{2C_0}I - \mathcal{O}(1)\widetilde{p}^2. \tag{A.13}$$

#

Notes

Trace formulae have been studied and used by many authors, see for instance [CdV2], [DuGu]. In the semi-classical regime, a trace formula has been studied in detail by Chazarain for the Schrödinger operator and by Helffer–Robert [HeRo1] and Ivrii [I1] for a general class of h-pseudors. We mention also the papers of Brummelhuis–Uribe [BrUr] and Petkov–Popov [PePo]. A trace formula for several commuting operators was established by Colin de Verdière [CdV3], Uribe and Zelditch [UrZe]. See also the recent paper of Charbonnel–Popov [CharPo]. The presentation of this chapter follows a paper of [DiSj]. The present work is generalized to the case where the dependence on spectral parameter is non-linear in [Di3]. Applications for the periodic Schrödinger operator with slowly and strong varying perturbations are treated in [Di3].

13. Spectral theory for perturbed periodic problems

Let $\Gamma = \oplus_{j=1}^{n} \mathbf{Z} e_j$ be the lattice generated by some basis $e_1, \ldots, e_n$ in $\mathbf{R}^n$. Consider the Schrödinger operator

$$P_{A,\varphi} = \sum_{j=1}^{n} (D_{y_j} + A_j(hy))^2 + V(y) + \varphi(hy), \quad (h > 0, h \to 0),$$

where V is Γ-periodic: $V(x + \gamma) = V(x), \forall \gamma \in \Gamma$, and φ is bounded with all its derivatives. $A(x) = (A_1(x), \ldots, A_n(x))$ is a magnetic potential such that all derivatives of non-vanishing order are bounded. In solid state physics, the Hamiltonian $P_{A,\varphi}$ describes the motion of an electron in a periodic crystal with external electric and magnetic fields. Such problems arise naturally in the investigation of impurity levels in the one-electron model of solids, and in particular in the theory of the colour of crystals. We refer the reader to [DH]. Let $\Gamma^* = \{\gamma^* \in \mathbf{R}^n; \gamma^* \cdot \gamma \in 2\pi\mathbf{Z}, \forall \gamma \in \Gamma\}$ be the dual lattice so that $\Gamma^* = \oplus_{i=1}^{n} \mathbf{Z} e_i^*$, where e_i^* is the dual basis, $e_i^* \cdot e_k = \delta_{ik} 2\pi$. For a fixed $\xi \in \mathbf{R}^n / \Gamma^*$, let $\lambda_1(\xi) \leq \lambda_2(\xi) \leq \ldots$ be the eigenvalues of the operator $(D_y + \xi)^2 + V(y) : L^2(\mathbf{R}^n / \Gamma) \to L^2(\mathbf{R}^n / \Gamma)$. It is well-known (see [ReSi]) that the spectrum of the non-perturbed periodic Schrödinger operator, $P_0 = -\Delta + V(y)$, consists of the closed intervals $J_1 = \lambda_1(\mathbf{R}^n / \Gamma^*)$, $J_2 = \lambda_2(\mathbf{R}^n / \Gamma^*), \ldots$ There are many papers dealing with different aspects of the spectral theory of $P_{A,\varphi}$ (see [ADH], [ReSi] and [Bi1,2]). To study the spectrum of $P_{A,\varphi}$ we use the method of the effective Hamiltonian. This method was introduced in solid state physics and has subsequently been used by many people: Buslaev [Bu1], Guillot–Ralston–Trubowitz [GRT], Nenciu [Ne1,2], Helffer–Sjöstrand [HeSj6], Gérard–Martinez–Sjöstrand [GMS] etc. The effective Hamiltonian approximation is to replace, for h small, $P_{A,\varphi}$ by the collection of h-pseudors:

$$\lambda_j(hD_x + A(x)) + \varphi(x) \text{ for } j \in \mathbf{N}, \tag{13.0}$$

where $\lambda_j(\xi)$ are the Bloch eigenvalues described above. In the case of Schrödinger operators with constant magnetic fields and no external electric field (i.e. when A_j are linear and $\varphi = 0$), rigorous reductions from $P_A = P_{A,0}$ to (13.0) have been given by Nenciu [Ne1], and Helffer–Sjöstrand [HeSj6]. To construct asymptotic solutions, $\tilde{u}$, of $P_{A,\varphi}\tilde{u} = \lambda_0\tilde{u}$ near some energy level λ_0, Buslaev (also Guillot–Ralston–Trubowitz) uses the following idea: if $u(x,y) \in \mathcal{D}'(\mathbf{R}_x^n \times \mathbf{R}_y^n)$ is a solution of

$$P(x, y, hD_x + D_y + A(x))u$$

$$= (\sum_{j=1}^{n} (hD_{x_j} + D_{y_j} + A_j(x))^2 + V(y) + \varphi(x))u = \lambda_0 u,$$

which is Γ–periodic in y, then $\tilde{u} = u(hy, y)$ satisfies $P_{A,\varphi}\tilde{u} = \lambda_0\tilde{u}$. They construct u by considering $P(x, y, hD_x + D_y + A(x))$ as an h-(pseudo)differential operator in x with operator valued symbol. This in turn is related to the study of operators of the form (13.0). Gérard–Martinez–Sjöstrand [GMS] use this idea to relate the spectrum of $P_{A,\varphi}$ near an energy level λ_0 to that of $P = P(x, y, hD_x + D_y + A(x))$. They obtain an $N \times N$ system of h-pseudos of order 0, $E^w_{-+}(x, hD_x + A(x), \lambda_0; h)$ (which is the effective Hamiltonian) such that the symbol $E_{-+}(x, \xi, \lambda_0; h)$ is Γ^*-periodic in ξ and has an asymptotic expansion in powers of h in $S^0(\mathbf{R}^{2n}, \mathcal{L}(\mathbf{C}^N, \mathbf{C}^N))$. Here and in the following, the order function will be 1 when nothing else is indicated, and we then write $S^0(\mathbf{R}^{2n}) = S^0(\mathbf{R}^{2n}, 1)$. The spectra of $P_{A,\varphi}$ and P are related by:

$$\lambda \in \sigma(P_{A,\varphi}) \Longleftrightarrow 0 \in \sigma(E^w_{-+}(x, hD_x + A(x), \lambda; h)),$$

for λ in a neighborhood of λ_0, where this last operator is considered as a bounded operator: $V_0^N \to V_0^N$ and V_0 is the Hilbert space $\{u = \sum_{\gamma \in \Gamma} c_\gamma \delta(x - h\gamma); (c_\gamma)_{\gamma \in \Gamma} \in l^2(\Gamma)\}$. One can use the effective Hamiltonian to prove that a gap in the spectrum of P_0 is stable under small perturbations of the magnetic field (see Proposition 13.27), a result due to Avron–Simon [AvSi] and Nenciu [Ne2]. This reduction from $(P_{A,\varphi} - \lambda)$ to $E^w_{-+}(x, hD_x + A(x), \lambda_0; h)$ is used by Dimassi [Di1] to study the discrete spectrum of $P_{A,\varphi}$. He obtains an asymptotic expansion in powers of h of $\operatorname{tr} f(P_{A,\varphi})$ for $f \in C^\infty_0(I)$, where I is an interval disjoint from the essential spectrum. In this chapter, we will develop the method of the effective Hamiltonian to the study of the spectrum of $P = P(hy, y, D_y + A(hy))$ on $\mathbf{R}^n$ ($h > 0, h \to 0$), where $P(x, y, \eta)$ is elliptic, periodic in y and has smooth bounded coefficients in (x, y). $A(x)$ is a magnetic potential with bounded derivatives. We follow essentially the papers of [GMS] and [Di1].

Let $P(x, y, \eta) \in C^\infty(\mathbf{R}^{3n})$ be real valued and have the following properties:

(H1) $P(x, y, \eta) = \sum_{|\alpha| \le m} a_\alpha(x, y)\eta^\alpha$,

(H2) $a_\alpha(x, y) = a_\alpha(x, y + \gamma), \forall |\alpha| \le m, \forall \gamma \in \Gamma$,

where Γ is a lattice $\oplus_{i=1}^n \mathbf{Z}e_i$ for a basis $(e_1, e_2, \ldots, e_n)$ of $\mathbf{R}^n$.

(H3) $|\partial_x^\gamma \partial_y^\beta a_\alpha(x, y)| \le C_{\alpha,\beta,\gamma} \quad \forall \alpha, \beta, \gamma \in \mathbf{N}^n, |\alpha| \le m$.

(H4) $p_m(x, y, \eta) := \sum_{|\alpha|=m} a_\alpha(x, y)\eta^\alpha \ge \frac{1}{C_0}|\eta|^m$ for some $C_0 > 0$.

Let $A(x) = (A_1(x), A_2(x), \ldots, A_n(x)) \in C^\infty(\mathbf{R}^n; \mathbf{R}^n)$. We assume that:

(H5) $\forall \alpha \in \mathbf{N}^n \setminus \{0\}$ there exists C_α such that $|\partial_x^\alpha A(x)| \le C_\alpha$.

For $m \in \mathbf{N}$, we put

$$H_A^m = \{u \in L^2(\mathbf{R}^n); (hD_x + A(x))^\alpha u \in L^2(\mathbf{R}^n), \forall |\alpha| \le m\},$$

$$\mathcal{E}_m = \{u \in L^2(\mathbf{R}^{2n}); (D_y + hD_x + A(x))^\alpha u \in L^2(\mathbf{R}^{2n}), \forall |\alpha| \le m\},$$

which are Hilbert spaces with the natural norm.

Notice that the commutator

$$[D_{x_j} + A_j, D_{x_k} + A_k] = \frac{1}{i}\left(\frac{\partial A_k}{\partial x_j} - \frac{\partial A_j}{\partial x_k}\right),$$

is a C^∞ function which is bounded with all its derivatives. Hence, if we change the order of the factors $(hD_{x_j} + A_j(x))$ in the definition of H_A^m, the space H_A^m (as a vector space) does not change and only its norm changes into an equivalent norm. Let $\chi \in C_0^\infty(B(0,2))$ with $\chi(x) = 1$ for $|x| \le 1$, where $B(x, r) = \{y \in \mathbf{R}^n; |y - x| < r\}$. Put $\chi_j(x) = \chi(x/j)$. For $u \in H_A^m$ and $|\alpha| \le m$ we have:

$$[(hD_x + A(x))^\alpha, \chi_j]u = \sum_{\beta \neq 0, |\beta| \le |\alpha|} h^{|\beta|} C_{\alpha,\beta}(hD_x^\beta \chi_j)(hD_x + A(x))^{\alpha-\beta}u,$$

where $C_{\alpha,\beta}$ are constants. As $\operatorname{supp}(D_x^\beta \chi_j) \subset \{x \in \mathbf{R}^n; (j \le |x| \le 2j)\}$ and $|(D_x^\beta \chi_j)| \le C_\beta j^{-|\beta|}$ for all $\beta \in \mathbf{N}^n \setminus \{0\}$ with C_β independent of j, the right hand side of the last equality tends to zero in $L^2(\mathbf{R}^n)$. Hence, every u in H_A^m can be approximated by elements with compact support, and each u in H_A^m with compact support can be approximated by C_0^∞-functions by means of a standard regularization. Then we have proved:

Proposition 13.1. H_A^m *is a Hilbert space in which* C_0^∞ *is dense.*

Suppose that P is independent of y, and put:

$$\operatorname{Op}_A^w(P) = P^w(x, hD_x + A(x)).$$

Proposition 13.2. *If we consider* $\operatorname{Op}_A^w(P)$ *as a symmetric operator on* $L^2(\mathbf{R}^n)$ *with domain* $C_0^\infty(\mathbf{R}^n)$*, then* $\operatorname{Op}_A^w(P)$ *is essentially selfadjoint and the domain of the selfadjoint extension is* $H_A^m(\mathbf{R}^n)$*.*

Proof. Here it is enough to treat the case of a fixed value of h, say 1. Using the composition formula (7.14) for Weyl quantization and the fact that $A(x)$

has bounded derivatives as well as the fact that for any $\alpha \in \mathbf{N}^n$, $\beta \in \mathbf{N}^{2n}$, $|\beta| \geq 1$, we have

$$D_{x,\xi}^\beta ((\xi + A(x))^\alpha) = \sum_{|\gamma| < |\alpha|} a_{\alpha,\beta,\gamma}(x)(\xi + A(x))^\gamma,$$

where the $a_{\alpha,\beta,\gamma}(x)$ and all their derivatives are bounded functions on $\mathbf{R}^n$, we see that for any $\alpha \in \mathbf{N}^n$ and any function $a(x)$ bounded with all its derivatives,

$$a(x)(D_x + A(x))^\alpha = [a(x)(\xi + A(x))^\alpha]^w + \sum_{|\beta| < |\alpha|} [b_{\alpha,\beta}(x)(\xi + A(x))^\beta]^w,$$

$$(13.1)$$

where the $b_{\alpha,\beta}$ and all their derivatives are bounded.

From (13.1) we deduce

$$[a(x)(\xi + A(x))^\alpha]^w = a(x)(D_x + A(x))^\alpha + \sum_{|\beta| < |\alpha|} c_{\alpha,\beta}(x)(D_x + A(x))^\beta,$$

$$(13.2)$$

where the $c_{\alpha,\beta}$ have the same property as the $b_{\alpha,\beta}$.

Then $\mathrm{Op}_A^w(P)$ can be written:

$$\mathrm{Op}_A^w(P) = \sum_{|\alpha| \leq m} b_\alpha(x)(D_x + A(x))^\alpha, \qquad (13.3)$$

where b_α and all their derivatives are bounded on $\mathbf{R}^n$.

If we fix $x_0 \in \mathbf{R}^n$ and put $\xi_0 = A(x_0)$, we have

$$e^{ix\xi_0} \mathrm{Op}_A^w(P) e^{-ix\xi_0} = \mathrm{Op}_{A-\xi_0}^w(P) = \sum_{|\alpha| \leq m} b_\alpha(x)(D_x + A(x) - A(x_0))^\alpha,$$

$$(13.4)$$

which is a bounded operator: $H_{A-\xi_0}^{m+k} \to H_{A-\xi_0}^k$.

If Ω is an open set in $\mathbf{R}^n$ we define:

$$\|u\|_{H_A^k(\Omega)}^2 = \sum_{|\alpha| \leq k} \|(D_x + A(x))^\alpha u\|_{L^2(\Omega)}^2.$$

The operator (13.4) can be written

$$\sum_{|\alpha| \leq m} c_\alpha(x, x_0) D_x^\alpha, \qquad (13.5)$$

with $c_\alpha(x, x_0) = a_\alpha(x)$ for $|\alpha| = m$, $|\partial_x^\beta c_\alpha(x, x_0)| \leq C_\beta$ for all $\beta \in \mathbf{N}^n$ and all $x \in B(x_0, 2)$ with C_β independent of x_0, as follows from assumption (H5).

The norms $\| \ \|_{H_{A-\varepsilon_0}^k(B(x_0,2))}$ and $\| \ \|_{H^k(B(x_0,2))}$ are equivalent uniformly with respect to x_0. Standard a priori estimates for elliptic operators give for any $u \in H_A^{m+k}(B(x_0, 2))$,

$$\|u\|_{H_A^{k+m}(B(x_0,1))} \leq c_k \big[\|\mathrm{Op}_A^w(P)u\|_{H_A^k(B(x_0,2))} + \|u\|_{L^2(B(x_0,2))}\big], \quad (13.6)$$

where c_k is independent of x_0.

Considering a covering of $\mathbf{R}^n$ by a family of balls $B(x_0, 2)$ with x_0 in a lattice, we obtain from (13.6):

Lemma 13.3. *If* $u \in L^2(\mathbf{R}^n)$ *and* $\mathrm{Op}_A^w(P)u \in H_A^k(\mathbf{R}^n)$, *then* $u \in H_A^{m+k}(\mathbf{R}^n)$ *and we have:*

$$\|u\|_{H_A^{k+m}} \leq C_k(\|\mathrm{Op}_A^w(P)u\|_{H_A^k} + \|u\|_{L^2}), \quad (13.7)$$

where C_k *is independent of* u .

Using this lemma with $k = 0$, we get Proposition 13.2.   #

Remark 13.4. If we replace $(D_x + A(x))$ which appears in the definitions of H_A^{k+m}, H_A^k and $\mathrm{Op}_A^w(P)$ by $(D_x + A(x+y))$ with y parameter in $\mathbf{R}^n$, then (13.7) remains true with C independent of y.

We now drop the assumption that P is independent of y and consider the operator $P = P^w(x, y, hD_x + D_y + A(x))$, which is the Weyl quantization of $P(x, y, h\xi + \eta + A(x))$. The change of variable

$$(x, y) \to (\tilde{x}, \tilde{y}) = (x - hy, y), \quad (13.8)$$

transforms P into:

$$\tilde{P} = P^w(\tilde{x} + h\tilde{y}, \tilde{y}, D_{\tilde{y}} + A(\tilde{x} + h\tilde{y})).$$

Lemma 13.3 and Remark 13.4 show that $\tilde{P}$ is essentially selfadjoint with domain

$$\tilde{\mathcal{E}}_m = \{u \in L^2(\mathbf{R}^{2n}); (D_{\tilde{y}} + A(\tilde{x} + h\tilde{y}))^\alpha u \in L^2(\mathbf{R}^{2n}), \forall\, |\alpha| \leq m\}.$$

Now the change of variables (13.8) (which can be realized as a unitary transformation) shows that P is essentially selfadjoint with domain $\mathcal{E}_m$.

To study P we will use the Floquet–Bloch reduction in the y variable. Let Γ^* be the dual lattice of Γ, given by

$$\Gamma^* = \{\gamma^* \in \mathbf{R}^n; \gamma^* \cdot \gamma \in 2\pi\mathbf{Z}, \; \forall\gamma \in \Gamma\}. \tag{13.9}$$

For $u \in \mathcal{S}(\mathbf{R}^{2n})$ and $\theta \in \mathbf{R}^n$, we put

$$\mathcal{U}u(x,y;\theta) = \sum_{\gamma\in\Gamma} e^{i\gamma\cdot\theta} u(x, y - \gamma). \tag{13.10}$$

We notice that $\mathcal{U}u(x,y,\theta)$ only depends on θ modulo the dual lattice Γ^*, so we may view $Uu(x,y,\theta)$ as a smooth function on $\mathbf{R}^{2n} \times (\mathbf{R}^n/\Gamma^*)$. For $\theta \in \mathbf{R}^n/\Gamma^*$, we put

$$\mathcal{D}'_\theta = \{u \in \mathcal{D}'(\mathbf{R}^{2n}); u(x, y + \gamma) = e^{i\gamma\cdot\theta} u(x,y)\}$$

and

$$\mathcal{H}_\theta = \{L^2_{\text{loc}}(\mathbf{R}^{2n}) \cap \mathcal{D}'_\theta; \int_{\mathbf{R}^n_x \times E} |u(x,y)|^2 dxdy < +\infty\},$$

where E is a fundamental domain of Γ. When $\theta = 0$ we write simply $L^2(\mathbf{R}^n_x \times \mathbf{R}^n_y/\Gamma)$. From (13.10) we see that

$$\mathcal{U}u(.,\theta) \in \mathcal{H}_\theta, \tag{13.11}$$

and if we view $\mathcal{H}_\theta$ as a bundle over $\mathbf{R}^n/\Gamma^*$, we can view $\mathcal{U}u$ as a smooth section of this bundle (thinking of $\mathcal{H}_\theta$ as a subspace of $L^2_{\text{loc}}(\mathbf{R}^{2n})$). We write $\mathcal{U}u \in C^\infty(\mathbf{R}^n/\Gamma^*; \mathcal{H}_\theta)$.

Every $v \in L^2(\mathbf{R}^n/\Gamma^*)$ can be expanded in a Fourier series:

$$v(\theta) = \sum_{\gamma\in\Gamma} \widehat{v}(\gamma) e^{i\gamma\theta},$$

where $\widehat{v}(\gamma) = \text{vol}\,(E^*)^{-1} \int_{E^*} e^{-i\gamma\theta} v(\theta) d\theta$. We have

$$\text{vol}\,(E^*)^{-1} \int_{E^*} |v(\theta)|^2 d\theta = \sum_{\gamma\in\Gamma} |\widehat{v}(\gamma)|^2. \tag{13.12}$$

For every x, y we can view (13.10) as the Fourier series expansion of $\mathcal{U}u(x,y;\cdot)$, so (13.12) gives:

$$\text{vol}\,(E^*)^{-1} \int_{E^*} |\mathcal{U}u(x,y;\theta)|^2 d\theta = \sum_{\gamma\in\Gamma} |u(x,y-\gamma)|^2. \tag{13.13}$$

Integrating this over $\mathbf{R}_x^n \times E$, we obtain

$$(\text{vol}\,(E^*))^{-1} \iiint_{\mathbf{R}_x^n \times E \times E^*} |\mathcal{U}u(x,y;\theta)|^2 dx dy d\theta)^{1/2}$$

$$= (\iint_{\mathbf{R}_x^n \times \mathbf{R}_y^n} |u(x,y)|^2 dx dy)^{1/2}. \tag{13.14}$$

This means that $\mathcal{U}$ can be extended to an isometry: $L^2(\mathbf{R}^{2n}) \to L^2(\mathbf{R}^n/\Gamma^*; \mathcal{H}_\theta)$, if we equip the latter Hilbert space (which by definition is the space of $v \in L_{\text{loc}}^2(\mathbf{R}^{3n})$ such that $v(x, y + \gamma; \theta) = e^{i\gamma\theta} v(x, y; \theta)$ and $v(x, y; \theta + \gamma^*) = v(x, y; \theta)$ for all $\gamma \in \Gamma$ and all $\gamma^* \in \Gamma^*$) with the norm given by the left hand side of (13.14).

Let $v(x, y; \theta) \in C^\infty(\mathbf{R}^n/\Gamma^*; \mathcal{H}_\theta)$, and write the Fourier series expansion with respect to θ:

$$v(x, y; \theta) = \sum_{\gamma \in \Gamma} \widehat{v}_\gamma(x, y) e^{i\gamma\theta}, \tag{13.15}$$

$$\widehat{v}_\gamma(x, y) = \text{vol}\,(E^*)^{-1} \int_{E^*} e^{-i\gamma\theta} v(x, y; \theta) d\theta. \tag{13.16}$$

Using the property $v(x, y - \gamma; \theta) = e^{-i\gamma \cdot \theta} v(x, y; \theta)$, we see that

$$\widehat{v}_\gamma(x, y) = \widehat{v}_0(x, y - \gamma). \tag{13.17}$$

Hence when v is smooth we have $v = \mathcal{U}\mathcal{W}v$, where:

$$\mathcal{W}v(x, y) = \text{vol}\,(E^*)^{-1} \int_{E^*} v(x, y; \theta) d\theta = \widehat{v}_0(x, y). \tag{13.18}$$

(13.18) combined with (13.15) and (13.17) gives

$$\|\mathcal{W}v\|_{L^2(\mathbf{R}_x^n \times \mathbf{R}_y^n)}^2 = \sum_{\gamma \in \Gamma} \iint_{\mathbf{R}_x^n \times E} |\widehat{v}_0(x, y - \gamma)|^2 dx dy$$

$$= \sum_{\gamma \in \Gamma} \iint_{\mathbf{R}_x^n \times E} |\widehat{v}_\gamma(x, y)|^2 dx dy = \|v\|_{L^2(\mathbf{R}^n/\Gamma^*, \mathcal{H}_\theta)}^2. \tag{13.19}$$

(13.19) shows that $\mathcal{W}$ is an isometry: $L^2(\mathbf{R}^n/\Gamma^*; \mathcal{H}_\theta) \to L^2(\mathbf{R}^n)$, so $\mathcal{W}$ is a bounded right inverse of $\mathcal{U}$. Since $\mathcal{U}$ is also an isometry we conclude that $\mathcal{U}$ is unitary and that $\mathcal{W} = \mathcal{U}^{-1}$.

Notice that $\mathcal{U}$ commutes formally with $(hD_x + D_y + A(x))^\alpha$, so for every $k \in \mathbf{N}$, $\mathcal{U}$ is unitary from $\mathcal{E}_k$ into $L^2(\mathbf{R}^n/\Gamma^*; \mathcal{H}_\theta^k)$, where $\mathcal{H}_\theta^k = \{u \in \mathcal{H}_\theta; (hD_x + D_y + A(x))^\alpha u \in \mathcal{H}_\theta, \forall |\alpha| \le k\}$. Let P_θ denote the selfadjoint operator $P^w(x, y, hD_x + D_y + A(x))$ acting in the sense of distributions on the space $\mathcal{H}_\theta$ with domain $\mathcal{H}_\theta^m$. Thanks to (13.3) and using the fact that $(hD_x + D_y + A(x))^\alpha$ commutes with $\mathcal{U}$, we see that

$$\mathcal{U}P\mathcal{U}^{-1} = \int^{\oplus} P_\theta d\theta, \qquad (13.20)$$

where by definition the last expression is the selfadjoint operator Q on $L^2(\mathbf{R}^n/\Gamma^*; \mathcal{H}_\theta)$ with domain $L^2(\mathbf{R}^n/\Gamma^*; \mathcal{H}_\theta^m)$ given by

$$Qv(x, y; \theta) = (P_\theta v(\cdot, \cdot; \theta))(x, y).$$

For $\theta \in \mathbf{R}^n/\Gamma^*$, P_θ is unitarily equivalent to the operator

$$\widetilde{P}_\theta = e^{i(x/h-y)\cdot\theta} P_\theta e^{-i(x/h-y)\cdot\theta},$$

acting on

$$\mathcal{K}_0 = \{u \in L^2_{\mathrm{loc}}(\mathbf{R}^{2n}); u(x, y + \gamma) = u(x, y), \iint_{\mathbf{R}_x^n \times \mathbf{R}_y^n/\Gamma} |u(x, y)|^2 dx dy < +\infty\}$$

with domain $\mathcal{K}_m = \{u \in \mathcal{K}_0; (hD_x + D_y + A(x))^\alpha u \in \mathcal{K}_0, \forall\, |\alpha| \leq m\}$. Thanks to the definition of P_θ and using the fact that P has the form (13.3) as well as the fact that $(hD_x + D_y + A(x))^\alpha$ commutes with $e^{i(x/h-y)\theta}$, we see that $\widetilde{P}_\theta$ is the same differential operator P acting on $\mathcal{K}_0$ with domain $\mathcal{K}_m$. Hence the spectrum of P_θ is θ-independent. Using (13.20) we get:

Proposition 13.5. *The spectrum of P acting on $L^2(\mathbf{R}^{2n})$ with domain $\mathcal{E}_m$ is the same as the spectrum of P acting on $\mathcal{K}_0$ with domain $\mathcal{K}_m$.*

Remark 13.6. From (13.1), (13.2) we have

$$\mathcal{K}_m = \{u \in \mathcal{K}_0; [(h\xi + \eta + A(x))^\alpha]^w u \in \mathcal{K}_0, \forall |\alpha| \leq m\}.$$

Now we consider in more detail the operator P. We will give a reduction of the study of $\sigma(P)$ by considering P as a h-pseudor in the x variables with an operator-valued symbol $P(x, \xi + A(x)) = P^w(x, y, D_y + \xi + A(x))$ as an element of $\mathcal{L}(\mathcal{K}_{m,\xi+A(x)}, \mathcal{K}_0)$, and by introducing a suitable auxiliary (so-called Grushin) problem, which will permit a reduction to an h-pseudor; an effective Hamiltonian.

Here

$$K_0 = L^2(\mathbf{R}^n/\Gamma),$$

$$K_{m,\xi} = \{u \in K_0; (D_y + \xi)^\alpha u \in K_0, \forall |\alpha| \le m\}.$$

Notice that only the norm of $K_{m,\xi}$ depends on ξ, not the space itself, and under the hypothesis (H5) we have:

$$\|u\|_{K_{m,\xi+A(x)}} \le C((\langle \xi - \zeta \rangle + \langle x - z \rangle)^m \|u\|_{K_{m,\zeta+A(z)}}$$

$$\forall u \in K_{m,0} \quad (x, z, \xi, \zeta) \in \mathbf{R}^{4n}, \tag{13.21}$$

$$\|\partial_x^\alpha \partial_\xi^\beta P(x, \xi + A(x))\|_{\mathcal{L}(K_{m,\xi+A(x)}, K_0)} \le C_{\alpha,\beta}. \tag{13.22}$$

We need some basic results about operator valued pseudors. Our main reference is here the unpublished work of Balazard–Konlein [BaKo]. We shall consider a family of Hilbert spaces $\mathcal{A}_X$, $X = (x, \xi) \in \mathbf{R}^{2n}$ satisfying:

$$\mathcal{A}_X = \mathcal{A}_Y \text{ as vector spaces for all } X, Y \in \mathbf{R}^{2n}, \tag{13.23}$$

there exist $N_0 \ge 0$ and $C \ge 0$ such that

$$\|u\|_{\mathcal{A}_X} \le C\langle X - Y \rangle^{N_0} \|u\|_{\mathcal{A}_Y} \text{ for all } u \in \mathcal{A}_0, X, Y \in \mathbf{R}^{2n}. \tag{13.24}$$

Let $\mathcal{B}_X, X \in \mathbf{R}^{2n}$ be a second family with the same properties. We say that $p \in C^\infty(\mathbf{R}^{2n}; \mathcal{L}(\mathcal{A}_0, \mathcal{B}_0))$ belongs to $S^0(\mathbf{R}^{2n}; \mathcal{L}(\mathcal{A}_X, \mathcal{B}_X))$ if for every $\alpha \in \mathbf{N}^{2n}$, there is a constant C_α such that

$$\|\partial_X^\alpha p\|_{\mathcal{L}(\mathcal{A}_X; \mathcal{B}_X)} \le C_\alpha, \text{ for all } X \in \mathbf{R}^{2n}. \tag{13.25}$$

We can then associate with p the operator $p^w(x, hD_x)$, and similarly to Proposition 7.7, Lemma 7.8 and Theorem 7.11 we have:

Proposition 13.7. *Let $p \in S^0(\mathbf{R}^{2n}; \mathcal{L}(\mathcal{A}_X, \mathcal{B}_X))$, where $\mathcal{A}_X, \mathcal{B}_X$ satisfy (13.23),(13.24). Then $\mathrm{Op}_h^w(p) = p^w(x, hD_x)$ is uniformly continuous $\mathcal{S}(\mathbf{R}^n; \mathcal{A}_0) \to \mathcal{S}(\mathbf{R}^n; \mathcal{B}_0)$.*

Proposition 13.8. *Assume $\mathcal{A}_X = \mathcal{A}_0$, $\mathcal{B}_X = \mathcal{B}_0, \forall X \in \mathbf{R}^{2n}$. If $p \in S^0(\mathbf{R}^{2n}; \mathcal{L}(\mathcal{A}_0, \mathcal{B}_0))$ (i.e. $\|\partial_X^\alpha p\|_{\mathcal{L}(\mathcal{A}_0; \mathcal{B}_0)} \le C_\alpha$, for all $X \in \mathbf{R}^{2n}$) then $\mathrm{Op}_h^w(p)$ is uniformly bounded*

$$L^2(\mathbf{R}^n; \mathcal{A}_0) \to L^2(\mathbf{R}^n; \mathcal{B}_0).$$

Let $\mathcal{C}_X$ be a third family of Hilbert spaces which also satisfies (13.23), (13.24).

Proposition 13.9. *Let $p \in S^0(\mathbf{R}^{2n}; \mathcal{L}(\mathcal{B}_X, \mathcal{C}_X)), q \in S^0(\mathbf{R}^{2n}; \mathcal{L}(\mathcal{A}_X, \mathcal{B}_X))$. Then $\mathrm{Op}_h^w(p) \circ \mathrm{Op}_h^w(q) = \mathrm{Op}_h^w(r)$, where $r \in S^0(\mathbf{R}^{2n}; \mathcal{L}(\mathcal{A}_X, \mathcal{C}_X))$ is given by*

$$r = \exp\left(\frac{ih}{2}\sigma(D_x, D_\xi; D_y, D_\eta)\right)(p(x, \xi)q(y, \eta))\big|_{x=y, \xi=\eta},$$

where σ is the usual symplectic 2form. We have the asymptotic formula:

$$r \sim \sum_{k=0}^{\infty} \frac{1}{k!} (\frac{ih}{2}\sigma(D_x, D_\xi; D_y, D_\eta))^k p(x,\xi)q(y,\eta)|_{x=y,\xi=\eta}$$

in the sense of Chapter 7.

Denote by $\mathcal{F}_{0,\xi}$ the space $\{u \in L^2_{loc}(\mathbf{R}^n_y); u(y+\gamma) = e^{i\gamma\cdot\xi}u(y)\}$ for $\xi \in \mathbf{R}^n/\Gamma^*$, and by $\mathcal{F}_{m,\xi}$ the space $\{u \in \mathcal{F}_{0,\xi}; D^\alpha_y u \in \mathcal{F}_{0,\xi}, |\alpha| \leq m\}$. We will fix some energy level $\lambda_0 \in \mathbf{R}^+$.

Proposition 13.10. *There exist $N \in \mathbf{N}$, a complex neighborhood ϑ of λ_0, and functions $\varphi_j(x,\xi,y) \in C^\infty(\mathbf{R}^{2n}_{x,\xi}; K_{m,\xi}) \cap C^\infty(\mathbf{R}^n_x \times \mathbf{R}^n_\xi \times \mathbf{R}^n_y)$ for $1 \leq j \leq N$, such that for each $(x,\xi) \in \mathbf{R}^{2n}$ and each $\lambda \in \vartheta$ the operator:*

$$\mathcal{P}(x,\xi,\lambda) \begin{pmatrix} u \\ u^- \end{pmatrix} = \begin{pmatrix} (P^w(x,y,D_y+\xi) - \lambda)u + R_-(x,\xi)u^- \\ R_+(x,\xi)u \end{pmatrix} \qquad (13.26)$$

is invertible from $K_{m,\xi} \times \mathbf{C}^N$ into $K_0 \times \mathbf{C}^N$ with an inverse $\mathcal{E}_0(x,\xi,\lambda) = \begin{pmatrix} E^0 & E^0_+ \\ E^0_- & E^0_{-+} \end{pmatrix}$ uniformly bounded with respect to (x,ξ,λ) together with all derivatives in $\mathcal{L}(K_0 \times \mathbf{C}^N, K_{m,\xi} \times \mathbf{C}^N)$ for $(x,\xi) \in \mathbf{R}^n, \lambda \in \vartheta$. Here $(R_+(x,\xi)u)_j = \langle u, \varphi_j(x,\xi,\cdot)\rangle_{\mathcal{F}_{0,0}}$ and $R_-(x,\xi)u^- = \sum_{j=1}^N u_j^- \varphi_j(x,\xi,\cdot)$.

Moreover, the functions φ_j satisfy the estimates:

$$\begin{cases} \|\partial^\alpha_x \partial^\beta_\xi \varphi_j\|_{K_{m,\xi}} \leq C_{\alpha,\beta}, \ \alpha, \beta \in \mathbf{N}^n, x, \xi \in \mathbf{R}^n \\ \forall \gamma^* \in \Gamma^*, \varphi_j(x, \xi+\gamma^*, y) = e^{-iy\gamma^*}\varphi_j(x,\xi,y) \end{cases} . \qquad (13.27)$$

Proof. By the appendix to this chapter, for a given $\lambda_1 \in \mathbf{R}$, there exist N analytic functions $\psi_j : (\mathbf{R}^{n*}/\Gamma^*)_\xi \to \mathcal{F}_{0,\xi}$, linearly independent for every ξ, and $C_0 > 0$ such that for all $u \in \text{Vect}(\psi_1, \ldots, \psi_N)^\perp$:

$$(((-\Delta)^{m/2} - \lambda_1)u, u)_{\mathcal{F}_{0,\xi}} \geq \frac{1}{C_0}\|u\|^2_{\mathcal{F}_{0,\xi}}. \qquad (13.28)$$

Using hypothesis (H3–4), we get

$$(P^w(x,y,D_y)u, u)_{\mathcal{F}_{0,\xi}} \geq \frac{1}{C_1}((-\Delta)^{m/2}u, u)_{\mathcal{F}_{0,\xi}} - C_1\|u\|^2_{\mathcal{F}_{0,\xi}}, \qquad (13.29)$$

uniformly for $x \in \mathbf{R}^n, \xi \in \mathbf{R}^n/\Gamma^*, u \in \mathcal{F}_{m,\xi}$.

Combining (13.28), (13.29), we obtain:

$$\operatorname{Re}\left((P^w(x,y,D_y) - \lambda)u, u\right)_{\mathcal{F}_{0,\xi}} \geq \left(\frac{1}{C_0 C_1} + \frac{\lambda_1}{C_1} - \operatorname{Re}\lambda - C_1\right)\|u\|^2_{\mathcal{F}_{0,\xi}},$$

(13.30)

and taking λ_1 large enough we get

$$\operatorname{Re}\left((P^w(x,y,D_y) - \lambda)u, u\right)_{\mathcal{F}_{0,\xi}} \geq \frac{1}{C_2}\|u\|^2_{\mathcal{F}_{0,\xi}},$$

(13.31)

with $C_2 > 0$, uniformly for $x \in \mathbf{R}^n, \lambda \in \vartheta, \xi \in \mathbf{R}^n/\Gamma^*$ and $u \in \mathcal{F}_{m,\xi} \cap$ Vect $(\psi_1, \ldots, \psi_N)^\perp$.

We define $\widetilde{R}_+(\xi)$, $\widetilde{R}_-(\xi)$ by: $(\widetilde{R}_+(\xi)u)_j = (u, \psi_j(\cdot, \xi))_{\mathcal{F}_{0,\xi}}$ and $\widetilde{R}_-(\xi)(u_1^-, u_2^-, \ldots, u_N^-) = \sum_{j=1}^{N} u_j^- \psi_j(\cdot, \xi)$. As in the appendix, (13.31) implies that the Grushin problem

$$\begin{cases} (P^w(x,y,D_y) - \lambda)u + \widetilde{R}_-(\xi)u^- = v \\ \widetilde{R}_+(\xi)u = v_+ \end{cases}$$

is bijective from $\mathcal{F}_{m,\xi} \times \mathbf{C}^N$ into $\mathcal{F}_{0,\xi} \times \mathbf{C}^N$ with an inverse uniformly bounded for $x \in \mathbf{R}^n, \lambda \in \vartheta, \xi \in \mathbf{R}^n/\Gamma^*$. With $\varphi_j(y,\xi) = e^{-iy\cdot\xi}\psi_j(y,\xi)$ we define $\mathcal{P}(x,\xi,\lambda)$ as in the proposition, noticing also that $u \to e^{-iy\cdot\xi}u$ is unitary from $\mathcal{F}_{m,\xi}$ into $K_{m,\xi}$. Then $\mathcal{P}(x,\xi,\lambda)$ is invertible with a uniformly bounded inverse $\mathcal{E}_0$ in $\mathcal{L}(K_0 \times \mathbf{C}^N, K_{m,\xi} \times \mathbf{C}^N)$. From $\mathcal{P} \circ \mathcal{E}_0 = I$ and $\mathcal{E}_0 \circ \mathcal{P} = I$ we get $\partial^\alpha_{x,\xi}\mathcal{E}_0 = -\mathcal{E}_0 \circ \partial^\alpha_{x,\xi}\mathcal{P} \circ \mathcal{E}_0$ for $|\alpha| = 1$. As $\partial^\alpha_{x,\xi}\mathcal{P} = \mathcal{O}(1)$ in $\mathcal{L}(K_{m,\xi} \times \mathbf{C}^N, K_0 \times \mathbf{C}^N)$, then $\partial^\alpha_{x,\xi}\mathcal{E}_0 = \mathcal{O}(1)$ in $\mathcal{L}(K_0 \times \mathbf{C}^N, K_{m,\xi} \times \mathbf{C}^N)$. Continuing to take higher and higher order derivatives of $\mathcal{P} \circ \mathcal{E}_0$ and using the fact that $\partial^\alpha_{x,\xi}\mathcal{P}$ is bounded in $\mathcal{L}(K_{m,\xi} \times \mathbf{C}^N, K_0 \times \mathbf{C}^N)$ for all α, we see that $\partial^\alpha_{x,\xi}\mathcal{E}_0 = \mathcal{O}(1)$ in $\mathcal{L}(K_0 \times \mathbf{C}^N, K_{m,\xi} \times \mathbf{C}^N)$.   #

Proposition 13.11. *The operator* $\mathcal{E}_0^w(x, hD_x + A(x), \lambda)$ *is continuous from* $\mathcal{S}(\mathbf{R}^n; K_0 \times \mathbf{C}^N)$ *into* $\mathcal{S}(\mathbf{R}^n; K_{m,0} \times \mathbf{C}^N)$, *from* $\mathcal{S}'(\mathbf{R}^n; K_0 \times \mathbf{C}^N)$ *into* $\mathcal{S}'(\mathbf{R}^n; K_{m,0} \times \mathbf{C}^N)$ *and uniformly bounded from* $\mathcal{K}_0 \times L^2(\mathbf{R}^n_x; \mathbf{C}^N)$ *into* $\mathcal{K}_m \times L^2(\mathbf{R}^n_x; \mathbf{C}^N)$ *for* $\lambda \in \vartheta$. *Moreover, we have:*

$$\mathcal{P}^w(x, hD_x + A(x), \lambda) \circ \mathcal{E}_0^w(x, hD_x + A(x), \lambda)$$
$$= 1 + h\mathcal{R}^w(x, hD_x + A(x), \lambda; h),$$

(13.32)

where $\mathcal{R}(x,\xi,\lambda;h) \sim \sum_{j=0}^{\infty} \mathcal{R}_j(x,\xi,\lambda)h^j$ *in* $S^0(\mathbf{R}^{2n}; \mathcal{L}(K_0 \times \mathbf{C}^N))$ *and* $\mathcal{R}, \mathcal{R}_j$ *depend holomorphically on* λ.

Proof. The continuity of $\mathcal{E}_0^w(x, hD_x + A(x), \lambda)$ in $\mathcal{S}$ and $\mathcal{S}'$ follows from the properties of $\mathcal{E}_0$ (Proposition 13.10) and Proposition 13.7. Using Remark

13.6, to show the L^2-boundedness, it suffices to show that for all $|\alpha| \leq m$

$$\begin{pmatrix} ((hD_x + D_y + A(x))^\alpha)^w & 0 \\ 0 & 1 \end{pmatrix} \circ \mathcal{E}_0^w \text{ is } \mathcal{O}(1) \text{ in}$$

$$\mathcal{L}(\mathcal{K}_0 \times L^2(\mathbf{R}_x^n; \mathbf{C}^N); \mathcal{K}_0 \times L^2(\mathbf{R}_x^n; \mathbf{C}^N)). \tag{13.33}$$

The fact that $((h\xi + \eta + A(x))^\alpha)^w$ can be viewed as an h-pseudor with operator valued symbol $(\xi + A(x) + D_y)^\alpha : K_{m,\xi + A(x)} \to K_0$ and Proposition 13.9 show that

$$\begin{pmatrix} (\xi + D_y + A(x))^\alpha & 0 \\ 0 & 1 \end{pmatrix} \circ \mathcal{E}_0^w(x, \xi + A(x), \lambda) \in$$

$$S^0(\mathbf{R}^{2n}; \mathcal{L}(K_0 \times \mathbf{C}^N, K_0 \times \mathbf{C}^N)),$$

combining this with Proposition 13.8 we get (13.33). The formula (13.32) follows from Proposition 13.9.   #

Proposition 13.12. *Assume (H1) to (H5). For h sufficiently small and $\lambda \in \vartheta$, $\mathcal{P}^w(x, hD_x + A(x), \lambda)$ has a uniformly bounded inverse of the form $\mathcal{E}^w(x, hD_x + A(x), \lambda; h)$, where*

$$\mathcal{E}(x, \xi, \lambda; h) \in S^0(\mathbf{R}^{2n}; \mathcal{L}(K_0 \times \mathbf{C}^N, K_{m,\xi} \times \mathbf{C}^N))$$

has an asymptotic expansion

$$\mathcal{E}^w(x, \xi, \lambda, h) \sim \sum_{j=0}^\infty \mathcal{E}_j(x, \xi, \lambda)h^j, \mathcal{E}_0 = \mathcal{P}(x, \xi, \lambda)^{-1}$$

as above. This inverse has the same continuity properties as $\mathcal{E}_0(x, hD_x + A(x), \lambda)$.

Proof. The Calderon–Vaillancourt result (Proposition 13.8) and Proposition 13.11 show that for h small enough

$$\|h\mathcal{R}^w(x, hD_x + A(x), \lambda; h)\|_{\mathcal{L}(L^2(\mathbf{R}^n \times \mathbf{R}^n/\Gamma) \times L^2(\mathbf{R}^n; \mathbf{C}^N))} \leq 1/2.$$

Then for h sufficiently small $(1 + h\mathcal{R}^w)^{-1}$ exists in $\mathcal{L}(L^2(\mathbf{R}^n \times \mathbf{R}^n/\Gamma) \times L^2(\mathbf{R}^n; \mathbf{C}^N))$. Consequently $\mathcal{P}^w(x, hD_x + A(x))$ has a right inverse $\mathcal{E}^w(x, hD_x + A(x), \lambda; h) = \mathcal{E}_0(x, hD_x + A(x), \lambda) \circ (1 + h\mathcal{R}^w)^{-1}$. We know from Chapter 8 that if $p(x, \xi; h)$ is an elliptic scalar valued symbol in $S^0(\mathbf{R}^{2n})$ then there exists $q(x, \xi, h) \in S^0(\mathbf{R}^{2n})$ such that $p^w(x, hD_x; h) \circ q^w(x, hD_x; h) = q^w(x, hD_x; h) \circ p^w(x, hD_x; h) = I$. We also know that $q(x, \xi; h) \sim \sum_{j \geq 0} q_j(x, \xi)h^j$ when $p(x, \xi; h) \sim \sum_{j \geq 0} p_j(x, \xi)h^j$ in $S^0(\mathbf{R}^{2n})$. Using the fact that this remains true in the case of operators with operator valued symbol, we see that for h small enough

$$(1 + h\mathcal{R}^w)^{-1} = 1 + h\widetilde{\mathcal{R}}^w,$$

where $\widetilde{\mathcal{R}}^w$ is an h-pseudor with the same properties as $\mathcal{R}^w$. Consequently $\mathcal{E}(x, \xi, \lambda; h)$ has an asymptotic expansion $\mathcal{E}^w(x, \xi, \lambda; h) \sim \sum_{j=0}^{\infty} \mathcal{E}_j(x, \xi, \lambda) h^j$ in $S^0(\mathbf{R}^{2n}; \mathcal{L}(K_0 \times \mathbf{C}^N, K_{m,\xi} \times \mathbf{C}^N))$. For $\lambda \in \vartheta \cap \mathbf{R}$, $\mathcal{P}^w(x, hD_x + A(x))$ is selfadjoint and then $\mathcal{E}^w(x, hD_x + A(x), \lambda; h)$ is also a left inverse for $\lambda \in \vartheta \cap \mathbf{R}$, and also for $\lambda \in \vartheta$ by analytic continuation.   #

Remark 13.13.

(1) Proposition 13.10 remains true if we replace λ_0 by a compact interval.

(2) Let $\mathcal{P}(\lambda) = \begin{pmatrix} P - \lambda & R_- \\ R_+ & 0 \end{pmatrix}$ be the operator constructed in Proposition 13.10 and $\begin{pmatrix} E(\lambda) & E_+(\lambda) \\ E_-(\lambda) & E_{-+}(\lambda) \end{pmatrix}$ the matrix of the inverse $\mathcal{E}(\lambda) = \mathcal{E}^w(x, hD_x + A(x), \lambda; h)$. ($E_{-+}(\lambda)$ is called the effective Hamiltonian associated with $(P - \lambda)$ and is selfadjoint when λ is real.) We have the following properties:

(i) $\mathcal{E}(\lambda)$ is analytic for $\lambda \in \vartheta$.

(ii)

$$(P - \lambda)^{-1} = E(\lambda) - E_+(\lambda)(E_{-+}(\lambda))^{-1} E_-(\lambda), \qquad (13.34)$$

$$E_{-+}(\lambda)^{-1} = -R_+(\lambda)(P - \lambda)^{-1} R_-(\lambda). \qquad (13.35)$$

(13.34), (13.35) follow from the identities: $\mathcal{P}(\lambda) \circ \mathcal{E}(\lambda) = I, \mathcal{E}(\lambda) \circ \mathcal{P}(\lambda) = I$ and can be used to show that many properties of $(P - \lambda)$ are the same as for $E_{-+}(\lambda)$.

(iii) Using the fact that R_+, R_- are independent of λ we get

$$\partial_\lambda E_{-+}(\lambda) = E_-(\lambda) E_+(\lambda). \qquad (13.36)$$

Proposition 13.14. *Under assumptions (H1) to (H5), for $\lambda \in \vartheta, h$ small enough, one has the following equivalence:*

$$\lambda \in \sigma(P) \iff 0 \in \sigma(E_{-+}^w(x, hD_x + A(x), \lambda; h)).$$

Proof. This follows from Remark 13.13 (formulas (13.34), (13.35)) and the continuity property of $\mathcal{E}$ obtained in Proposition 13.12.   #

Remark 13.15. Because of (13.27), we have:

$$\begin{cases} R_-(x, \xi + \gamma^*) = e^{-iy\gamma^*} R_-(x, \xi) \\ R_+(x, \xi + \gamma^*) = R_+(x, \xi) e^{iy\gamma^*} \end{cases}$$

On the operator level we get:

$$\begin{cases} e^{-ix\gamma^*/h} R_-^w(x, hD_x + A(x)) e^{ix\gamma^*/h} = e^{-iy\gamma^*} R_-^w(x, hD_x + A(x)) \\ e^{-ix\gamma^*/h} R_+^w(x, hD_x + A(x)) e^{ix\gamma^*/h} = R_+^w(x, hD_x + A(x)) e^{iy\gamma^*}. \end{cases}$$

Combining this with the fact that

$$P^w(x, y, \xi + \gamma^* + D_y) = e^{-iy\gamma^*} P^w(x, y, \xi + D_y) e^{iy\gamma^*},$$

we get

$$[\mathcal{P}^w(x, hD_x + A(x)), \begin{pmatrix} e^{i(x/h-y)\gamma^*} & 0 \\ 0 & e^{ix\gamma^*/h} \end{pmatrix}] = 0. \qquad (13.37)$$

Let us consider $P_0 = P^w(hy, y, D_y + A(hy))$ with domain $\widetilde{H}_{m,A} = \{u \in L^2(\mathbf{R}^n); (D_y + A(hy))^\alpha u \in L^2(\mathbf{R}^n), \forall |\alpha| \leq m\}$. From Proposition 13.2 P_0 is essentially selfadjoint on $C_0^\infty(\mathbf{R}^n)$ with domain $\widetilde{H}_{m,A}$. To reduce the study of P_0, we apply the same method as for P. We get

Proposition 13.16. *The spectrum of* $P_0 = P^w(hy, y, D_y, +A(hy))$ *acting on* $L^2(\mathbf{R}^n)$ *with domain* $\widetilde{H}_{m,A}$ *is the same as the spectrum of* $P = P^w(x, y, hD_x + D_y + A(x))$ *acting on* $L_0 := \{\sum_{\gamma \in \Gamma} v(x)\delta(x - hy + h\gamma); h^{-n/2}v \in L^2(\mathbf{R}_x^n)\}$ *with domain* $L_m := \{u \in L_0; (hD_x + D_y + A(x))^\alpha u \in L_0, |\alpha| \leq m\}$. *Here* L_0, L_m *are equipped with the natural norms.*

Remark 13.17.

(1) The nature of the spectrum of the two operators P and P_0 may conceivably differ.

(2) Using (13.1) and the fact that $((h\xi + \eta + A(x))^\alpha)^w$ can be viewed as an h-pseudor with operator valued symbol $(\xi + A(x) + D_y)^\alpha : K_{m,\xi+A(x)} \to K_0$, we see that $L_m = \{u \in L_0; ((h\xi + \eta + A(x))^\alpha)^w u \in L_0, |\alpha| \leq m\}$.

It follows from the last result that we can reduce the study of the spectrum of P_0 to that of P acting on L_0 with domain L_m. Let $\mathcal{S}(\mathbf{R}_x^n \times \mathbf{R}_y^n/\Gamma) = \{\varphi \in$

$C^\infty(\mathbf{R}^{2n}); \langle x \rangle^N \partial_{x,y}^\alpha \varphi \in L^2(\mathbf{R}_x^n \times \mathbf{R}_y^n/\Gamma), \forall N, \alpha\}$. For $\varphi \in \mathcal{S}(\mathbf{R}_x^n \times (\mathbf{R}_y^n/\Gamma))$, we have

$$\varphi(x,y) = \sum_{\gamma^* \in \Gamma^*} c_{\gamma^*}(x) e^{i\gamma^* y}, \qquad (13.38)$$

with

$$c_{\gamma^*}(x) = \mathrm{vol}\,(E)^{-1} \int_E e^{-i\gamma^* \cdot \tilde{y}} \varphi(x, \tilde{y}) d\tilde{y}. \qquad (13.39)$$

Integrating by parts in (13.39) and using the Schwartz inequality, we get:

$$|c_{\gamma^*}(x)| \le C \sum_{|\alpha| \le N} \left(\int_E |\partial_y^\alpha \varphi(x, \tilde{y})|^2 d\tilde{y} \right)^{1/2} (1 + |\gamma^*|)^{-N}, \qquad (13.40)$$

with C independent of x in $\mathbf{R}^n$. Combining (13.38) and (13.40) we get (with a new constant C) for every fixed $N > n$,

$$\int_{\mathbf{R}^n} |\varphi(hy, y)|^2 dy \le C h^{-n} \sum_{|\alpha| \le N} \iint_{\mathbf{R}_x^n \times E} |\partial_y \varphi^\alpha(x, \tilde{y})|^2 d\tilde{y} dx. \qquad (13.41)$$

A distribution $u = \sum_\Gamma v(x) \delta(x - h(y - \gamma))$ in L_0 can be written $u = \sum_\Gamma v(h(y - \gamma)) \delta(x - h(y - \gamma))$. If $\varphi \in \mathcal{S}(\mathbf{R}_x^n \times (\mathbf{R}_y^n/\Gamma))$, then

$$\langle u, \varphi \rangle := \iint_{(x,y) \in \mathbf{R}_x^n \times E} u(x,y) \varphi(x,y) dx dy$$

$$= \sum_{\gamma \in \Gamma} \int_E v(h(y - \gamma)) \varphi(h(y - \gamma), y) dy = \int_{\mathbf{R}^n} v(hy) \varphi(hy, y) dy. (13.42)$$

(13.41) and (13.42) show that u can be viewed as an element of $\mathcal{S}'(\mathbf{R}_x^n \times (\mathbf{R}_y^n/\Gamma))$. Hence we can hope to adapt Proposition 13.12 to L_0. We will characterize more precisely the space L_0 and L_m. Let us denote by V_0 the subspace of $\mathcal{S}'(\mathbf{R}_x^n)$ consisting of distributions of the form:

$$w(x) = \sum_{\gamma \in \Gamma} f_\gamma \delta(x - h\gamma)$$

with $(f_\gamma)_{\gamma \in \Gamma} \in l^2(\Gamma)$. V_0 is equipped with its natural Hilbert space structure, and we first study this space in more detail. Let τ_{γ^*} be the operator of multiplication by $e^{ix\gamma^*/h}$.

Lemma 13.18. *Let* $\chi \in C_0^\infty(\mathbf{R}^n)$ *satisfy* $\sum_{\gamma^* \in \Gamma^*} \chi(\xi + \gamma^*) = 1$. *Then for* $u, v \in V_0$, *the quantity* $(\chi^w(hD_x)u, v)_{L^2(\mathbf{R}^n)}$ *is well defined and independent of the choice of* χ. *Moreover;*

$$(\chi^w(hD_x)u, v)_{L^2(\mathbf{R}^n)} = \frac{1}{h^n \mathrm{vol}\,(\mathbf{R}^n/\Gamma)} (u, v)_{V_0}.$$

Proof. If $u = \sum_{\gamma \in \Gamma} f_u(\gamma)\delta(\cdot - h\gamma) \in V_0$, then $\mathcal{F}_h u(\xi) = \sum_{\gamma \in \Gamma} f_u(-\gamma)e^{i\gamma\xi} \in L^2(\mathbf{R}^n/\Gamma^*))$ (where $\mathcal{F}_h u(\xi) = \int_{\mathbf{R}^n} e^{-ix\xi/h}u(x)dx$). If $v = \sum_{\gamma \in \Gamma} f_v(\gamma)\delta(\cdot - h\gamma) \in V_0$, we have

$$(\mathcal{F}_h u, \mathcal{F}_h v)_{L^2(\mathbf{R}^n/\Gamma^*)} = \text{vol}\,(\mathbf{R}^n/\Gamma^*)(u, v)_{V_0}.$$

$(\chi^w(hD_x)u, v)$ can be defined by the Parseval formula:

$$(\chi^w(hD_x)u, v)_{L^2(\mathbf{R}^n)} = (2\pi h)^{-n}(\mathcal{F}_h(\chi^w(hD_x)u), \mathcal{F}_h v)_{L^2(\mathbf{R}^n)}$$

$$= (2\pi h)^{-n}\int_{\mathbf{R}^n}\chi(\xi)\mathcal{F}_h u(\xi)\overline{\mathcal{F}_h v(\xi)}d\xi$$

$$= (2\pi h)^{-n}\sum_{\gamma^* \in \Gamma^*}\int_{E^*}\chi(\xi+\gamma^*)\mathcal{F}_h u(\xi+\gamma^*)\overline{\mathcal{F}_h v(\xi+\gamma^*)}d\xi.$$

Since $\sum_{\gamma^* \in \Gamma^*}\chi(\xi+\gamma^*) = 1$ and $\mathcal{F}_h u(\xi+\gamma^*) = \mathcal{F}_h u(\xi)$, $\mathcal{F}_h v(\xi+\gamma^*) = \mathcal{F}_h v(\xi)$ for all γ^* in Γ^*, we get

$$(\chi^w(hD_x)u, v)_{L^2(\mathbf{R}^n)} = (2\pi h)^{-n}(\mathcal{F}_h u, \mathcal{F}_h v)_{L^2(\mathbf{R}^n/\Gamma^*)}$$

$$= \frac{1}{h^n\text{vol}\,(\mathbf{R}^n/\Gamma)}\langle u, v\rangle_{V_0}, \tag{13.43}$$

where we also used the fact that $\text{vol}\,(\mathbf{R}^n/\Gamma^*) \times \text{vol}\,(\mathbf{R}^n/\Gamma) = (2\pi)^n$.   #

Proposition 13.19. *For every* $u \in V_0$ *there exists* $u_0 \in L^2(\mathbf{R}^n)$ *such that with* $u_{\gamma^*} = \tau_{\gamma^*}u_0$:

$$u = \sum_{\gamma^* \in \Gamma^*}u_{\gamma^*} \quad \text{with convergence in } \mathcal{S}'(\mathbf{R}^n). \tag{13.44}$$

Moreover, $\exists C > 0$ *independent of* u *for which* u_0 *can be chosen such that for every bounded set* $\mathcal{B} \subset S^0(\mathbf{R}^{2n})$ *and every* $N \in \mathbf{N}$, *there exists a constant* $C_N > 0$ *such that* $\forall A \in \mathcal{B}$ *with* $\text{dist}\,(\text{supp}\,A, \mathbf{R}_x^n \times \{0\}) \geq C$, *we have*

$$\|A^w(x, hD_x)u_0\|_{L^2(\mathbf{R}^n)} \leq C_N h^{-\frac{n}{2}}(\text{dist}\,(\text{supp}\,A, \mathbf{R}^n \times \{0\}))^{-N} \tag{13.45}$$

uniformly for $h > 0$ *small enough. The constants* C_N *can be taken* $\mathcal{O}(\|u\|_{V_0})$ *uniformly with respect to* u.

Conversely, if $u_0 \in L^2(\mathbf{R}^n)$ *satisfies (13.45) and if* $u_{\gamma^*} = \tau_{\gamma^*}u_0$, *then the sum* $\sum_{\gamma^* \in \Gamma^*}u_{\gamma^*}$ *converges in* $\mathcal{S}'(\mathbf{R}^n)$ *towards an element* u *of* V_0, *with* $\|u\|_{V_0}$ *bounded by a constant times the sum of* $h^{\frac{n}{2}}\|u_0\|_{L^2}$ *and a finite number of the* C_N *in (13.45)* .

Proof. If $u \in V_0$, we take

$$u_0 = \chi(hD_x)u = (\mathcal{F}_h^{-1}\chi) * u = (2\pi h)^{-n} \sum_{\gamma \in \Gamma} f_u(\gamma)\hat{\chi}(\gamma - \frac{x}{h})$$

with $\chi \in C_0^\infty(\mathbf{R}^n, [0,1])$, $\sum_{\gamma^* \in \Gamma} \chi(\xi + \gamma^*) = 1$. Then $u_0 \in H^s$ for every $s \in \mathbf{R}$ and satisfies

$$\|(hD_x)^\alpha u_0\|_{L^2} \leq C_\alpha h^{-\frac{n}{2}} \|u\|_{V_0} \tag{13.46}$$

By integrating by parts in the oscillatory integral which gives $A^w(x, hD_x)u_0$, we obtain for every $k \in \mathbf{N}$,

$$A^w(x, hD_x)u_0 = \sum_{|\gamma+\beta|=2k} h^{2k-|\beta|}b_\gamma^w(x, hD_x)(h^{|\beta|}\partial_x^\beta(u_0(x))),$$

with $b_\gamma(x, \xi) = \partial_x^\gamma A(x, \xi)(\xi^2)^{-k}\mathcal{O}_{k,\gamma}(1)$, the $\mathcal{O}_{k,\gamma}(1)$ factor having bounded derivatives to all orders. Using Theorem 7.11 we get:

$$\|A^w(x, hD_x)u_0\| \leq \mathcal{O}_k(1)(\mathrm{dist}\,(\mathrm{supp}\,A, \mathbf{R}^n \times \{0\}))^{-2k} \sum_{|\beta|\leq 2k} \|(hD_x)^\beta u_0\|,$$
$$\tag{13.47}$$

with $\mathcal{O}_k(1)$ independent of h and u_0. Now (13.45) follows from (13.46) and (13.47).

To prove (13.44) we only remark that for $u \in V_0$ we have $\tau_{\gamma^*}u = u$ and $u_{\gamma^*} = \chi(hD_x - \gamma^*)u$.

Let us prove the converse statement. Let $A \in C_0^\infty(\mathbf{R}^n)$ be real valued and such that $A(\xi) = 1$ in a sufficiently large bounded neighborhood of 0. If u_0 satisfies (13.45) then for every $N \in \mathbf{N}$, we have ;

$$\|(1 - A^w(hD_x))\langle hD_x\rangle^N u_0\|_{L^2(\mathbf{R}^n)} = C_N h^{-n/2},$$

where $\langle x \rangle = (1 + |x|^2)^{1/2}$. By Theorem 7.11 we have

$$\|A^w(hD_x)\langle hD_x\rangle^N u_0\|_{L^2(\mathbf{R}^n)} = \mathcal{O}_N(1)\|u_0\|.$$

Consequently,

$$(2\pi h)^{-n/2}\|\langle \xi\rangle^N \mathcal{F}_h u_0\|_{L^2} = \|\langle hD_x\rangle^N u_0\|_{L^2} = \mathcal{O}_N(1)(\|u_0\| + C_N h^{-n/2}).$$
$$\tag{13.48}$$

Put $u = \sum_{\gamma^* \in \Gamma^*} \tau_{\gamma^*}u_0$. From (13.48) we deduce that

$$\mathcal{F}_h(u)(\xi) = \sum_{\gamma^* \in \Gamma^*} (\mathcal{F}_h u_0)(\xi + \gamma^*) \in L^2(\mathbf{R}^n/\Gamma^*)$$

172   *Spectral Asymptotics in the Semi-Classical Limit*

and $\|\mathcal{F}_h(u)(\xi)\|_{L^2(\mathbf{R}^n/\Gamma^*)} = \mathcal{O}_N(1)(h^{n/2}\|u_0\| + C_N)$. Now (13.43) gives the result.   #

Proposition 13.20. *Let* $B(x,\xi) \in S^0(\mathbf{R}^{2n})$ *with* $B(x,\xi + \gamma^*) = B(x,\xi)$ *for any* $\gamma^* \in \Gamma^*$. *Then* $B^w(x, hD_x)$ *is bounded on* V_0, *uniformly with respect to* h *small enough.*

Proof. Let $u \in V_0$ and let us write $u = \sum_{\gamma^*} u_{\gamma^*}$ as in Proposition 13.19. Then $B^w(x, hD_x)u = \sum_{\gamma^*} v_{\gamma^*}$, where

$$v_{\gamma^*} = B^w(x, hD_x)u_{\gamma^*} = \tau_{\gamma^*}B^w(x, hD_x + \gamma^*)u_0.$$

Since $B(x,\xi + \gamma^*) = B(x,\xi)$, we get $v_{\gamma^*} = \tau_{\gamma^*}v_0$, where $v_0 = B^w(x, hD_x)u_0$. Using the same argument as in (13.47), we see that v_0 satisfies (13.45) with constants C_N estimated by similar constants for u_0. Then Proposition 13.19 shows that $v \in V_0$ and $\|v\|_{V_0} \le C_0(h^{\frac{n}{2}}\|B^w(x, hD_x)u_0\|_{L^2} + \|u\|_{V_0}) \le C(h^{\frac{n}{2}}\|u_0\|_{L^2} + \|u\|_{V_0})$. Thanks to (13.46) we obtain the result.   #

Proposition 13.21. *Let* $Q_1(x,\xi), Q_2(x,\xi), B(x,\xi) \in S^0(\mathbf{R}^{2n})$. *We assume that* Q_1, Q_2 *and* B *are* Γ^*-*periodic in* ξ *and* $\Pi_x \operatorname{supp} Q_1 = \{x \in \mathbf{R}^n; (x,\xi) \in \operatorname{supp} Q_1 \text{ for some } \xi \in \mathbf{R}^n\}$ *is compact . We also assume that* $\operatorname{dist}(\Pi_x \operatorname{supp} Q_1, \Pi_x \operatorname{supp} Q_2) \ge \epsilon_0 > 0$ *for some fixed* $\epsilon_0 > 0$. *Then for every* $N \in \mathbf{N}$ *there exists a constant* C_N *such that*

$$\|Q_1^w(x, hD_x) \circ B^w(x, hD_x) \circ Q_2^w(x, hD_x)\|_{\mathcal{L}(V_0)} \le C_N h^N.$$

Proof. Let $u \in V_0$ and let us write $u = \sum_{\gamma^*} u_{\gamma^*}$ as in Proposition 13.19. Then $Q_1^w(x, hD_x)B^w(x, hD_x)Q_2^w(x, hD_x)u = \sum_{\gamma^*} v_{\gamma^*}$, where

$$v_{\gamma^*} = Q_1^w(x, hD_x) \circ B^w(x, hD_x) \circ Q_2^w(x, hD_x)u_{\gamma^*}.$$

As in the proof of the preceding proposition, $v_{\gamma^*} = \tau_{\gamma^*}v_0$, with

$$v_0 = Q_1^w(x, hD_x)B^w(x, hD_x)Q_2^w(x, hD_x)u_0.$$

From the assumption on the support of Q_1, Q_2, we see by the Weyl calculus (Chapter 7) that for every $N \in \mathbf{N}$

$$\|v_0\|_{L^2(\mathbf{R}^n)} = \mathcal{O}_N(h^N)\|u_0\|.$$

Let $A \in S^0(\mathbf{R}^{2n})$ be as in Proposition 13.19. By the argument used in (13.47), we see that for every integer $N \in \mathbf{N}$,

$$\|A^w(x, hD_x)v_0\|_{L^2(\mathbf{R}^{2n})} \le C_N h^N(1 + \operatorname{dist}(\operatorname{supp} A, \mathbf{R}^n \times \{0\}))^{-N}, \quad (13.49)$$

with $C_N = \mathcal{O}(\|u\|_{V_0})$. Now by Proposition 13.19 we can conclude.   #

We will generalize the above results to the space L_0. If

$$u = \sum_{\gamma \in \Gamma} v(x)\delta(x - hy + h\gamma) \in L_0,$$

we have

$$\|u\|_{L_0}^2 = h^{-n}\|v\|_{L^2}^2 = \int_{\mathbf{R}^n/\Gamma^*} \|v(hy + h\cdot)\|_{l^2(\Gamma)}^2 dy.$$

Let χ be as in Proposition 13.18 and consider u as an element of $\tau_{hy}V_0$ with y as a parameter. Let $w = \sum_{\gamma \in \Gamma} a(x)\delta(x - h(y - \gamma)) \in L_0$. Similarly to (13.43) we have

$$(\chi^w(hD_x)u_y, w_y)_{L^2(\mathbf{R}_x^n)}$$
$$= \frac{1}{(2\pi h)^n} \sum_{\alpha,\beta \in \Gamma} v(h(y+\alpha))\overline{a(h(y+\beta))} \int_{\mathbf{R}^n} e^{i(\alpha-\beta)\cdot\xi}\chi(\xi)d\xi.$$

Since $\sum_{\gamma^* \in \Gamma^*} \chi(\xi + \gamma^*) = 1$, then $\widehat{\chi}(\gamma) = \int_{\mathbf{R}^n} e^{-i\gamma\xi}\chi(\xi)d\xi = 0$ for all $\gamma \in \Gamma$ and $\widehat{\chi}(0) = \mathrm{vol}\,(\mathbf{R}^n/\Gamma^*)$. Consequently

$$(\chi^w(hD_x)u_y, w_y)_{L^2(\mathbf{R}^n)} = \frac{\mathrm{vol}\,(\mathbf{R}^n/\Gamma^*)}{(2\pi h)^n} \sum_{\alpha \in \Gamma} u(h(y+\alpha))\overline{a(h(y+\alpha))}.$$

Integrating this over $\mathbf{R}_y^n/\Gamma$, we get

$$(\chi^w(hD_x)u, w)_{L^2(\mathbf{R}_x^n \times \mathbf{R}_y^n/\Gamma^*)} = \frac{1}{\mathrm{vol}\,(\mathbf{R}^n/\Gamma)h^n}h^{-n}(v, a)_{L^2(\mathbf{R}^n)}.$$

Then the analogue of Proposition 13.18 holds for L_0. Specifically:

$$(\chi^w(hD_x)u, u)_{L^2(\mathbf{R}_x^n \times \mathbf{R}_y^n/\Gamma^*)} = \frac{1}{\mathrm{vol}\,(\mathbf{R}^n/\Gamma)h^n}\|u\|_{L_0}^2.$$

We will now prove the analogue of Proposition 13.19 for L_0. Let T_{γ^*} be the operator of multiplication by $e^{i(x/h-y)\gamma^*}$ and notice that $T_{\gamma^*}u = u$ for $u \in L_0$ and $\gamma^* \in \Gamma^*$.

Proposition 13.22. *For any $u \in L_0$ there exists $u_0 \in L^2(\mathbf{R}_x^n \times \mathbf{R}_y^n/\Gamma)$ such that with $u_{\gamma^*} = T_{\gamma^*}u_0$: $u = \sum_{\gamma^* \in \Gamma^*} u_{\gamma^*}$ with convergence in $\mathcal{S}'(\mathbf{R}_x^n \times \mathbf{R}_y^n/\Gamma)$. Moreover, $\exists C > 0$ independent of u_0 such that u_0 can be chosen such that for every bounded set $\mathcal{B} \subset S^0(\mathbf{R}^{2n})$ and every $N \in \mathbf{N}$, there exists a constant*

$C_N > 0$ *such that* $\forall A \in \mathcal{B}$ *(independent of* (y, η)*) with* $\operatorname{dist}(\operatorname{supp} A, \mathbf{R}_x^n \times \{0\}) \geq C$, *we have*

$$\|A^w(x, hD_x)u_0\|_{L^2(\mathbf{R}_x^n \times \mathbf{R}_y^n / \Gamma)} \leq C_N h^{-\frac{n}{2}} (\operatorname{dist}(\operatorname{supp} A, \mathbf{R}^n \times \{0\}))^{-N} \quad (*)$$

uniformly for $h > 0$ *small enough. The constants* C_N *are* $\mathcal{O}(\|u\|_{L_0})$.

Conversely, if $u_0 \in L^2(\mathbf{R}_x^n \times \mathbf{R}_y^n / \Gamma)$ *satisfies* $(*)$ *and if* $u_{\gamma^*} = T_{\gamma^*} u_0$, *then the series* $\sum_{\gamma^* \in \Gamma^*} u_{\gamma^*}$ *converges in* $\mathcal{S}'(\mathbf{R}_x^n \times \mathbf{R}_y^n / \Gamma)$ *towards a distribution* u *in* L_0, *with* $\|u\|_{L_0}$ *bounded by a constant times the sum of* $h^{\frac{n}{2}}\|u_0\|_{L^2(\mathbf{R}_x^n \times \mathbf{R}_y^n / \Gamma)}$ *and a finite number of the* C_N *in* $(*)$.

The proof is similar to that of Proposition 13.19 and we omit it.

Theorem 13.23. *Assume (H1) to (H5). Then for* $h > 0$ *small enough and* $\lambda \in \vartheta$, $\mathcal{P}^w(x, hD_x + A(x), \lambda)$ *is uniformly bounded from* $L_m \times (V_0)^N$ *to* $L_0 \times (V_0)^N$ *and has the uniformly bounded two sided inverse* $\mathcal{E}^w(x, hD_x + A(x), \lambda, h)$ *from* $L_0 \times (V_0)^N$ *to* $L_m \times (V_0)^N$.

Proof. We know that $P^w(x, y, hD_x + D_y + A(x))$ is uniformly bounded from L_m to L_0. It then remains to show in view of Proposition 13.12 that the following operators are uniformly bounded:

$$R_+^w(x, hD_x + A(x)) : L_0 \to V_0^N, \tag{13.50}$$

$$R_-^w(x, hD_x + A(x)) : V_0^N \to L_m, \tag{13.51}$$

$$E_+^w(x, hD_x + A(x), \lambda; h) : V_0^N \to L_m, \tag{13.52}$$

$$E_-^w(x, hD_x + A(x), \lambda; h) : L_0 \to V_0^N, \tag{13.53}$$

$$E^w(x, hD_x + A(x), \lambda; h) : L_0 \to L_m, \tag{13.54}$$

$$E_{-+}^w(x, hD_x + A(x), \lambda; h) : V_0^N \to V_0^N. \tag{13.55}$$

Here (13.55) is a consequence of Proposition 13.20 and the same proof gives (13.50) and (13.53). Let us show (13.52) ((13.51) and (13.54) can be proved similarly). From Remark 13.17 it suffices to show that for all $\alpha \in \mathbf{N}^n$ with $|\alpha| \leq m$, $((h\xi + \eta + A(x))^\alpha)^w \circ E_+^w(x, hD_x + A(x), \lambda; h)$ is uniformly bounded from V_0^N to L_0. This composed operator is of the form $C^w(x, hD_x + A(x), \lambda; h)$ with $C(x, \xi, \lambda; h)$ in $S^0(\mathbf{R}^{2n}; \mathcal{L}(\mathbf{C}^N; L^2(\mathbf{R}^n / \Gamma)))$ and

$$T_{\gamma^*} C^w(x, hD_x + A(x), \lambda; h) = C^w(x, hD_x + A(x), \lambda; h)\tau_{\gamma^*}, \tag{13.56}$$

for all $\gamma^* \in \Gamma^*$ ((13.56) follows from (13.37) with $\mathcal{P}$ replaced by $\mathcal{E}$).

Let $u \in V_0^N$ and decompose $u = \sum_{\gamma^*} u_{\gamma^*}$ as in Proposition 13.21. Then we have $C^w(x, hD_x + A(x), \lambda; h)u = \sum v_\gamma^*$ with

$$v_{\gamma^*} = C^w(x, hD_x + A(x), \lambda; h)\tau_{\gamma^*}u_0 = T_{\gamma^*}C^w(x, hD_x + A(x), \lambda; h)u_0,$$

and thus we are in the situation where Proposition 13.21 applies and gives the result.   #

Using Theorem 13.23 and Remark 13.13 we get:

Corollary 13.24. *For $\lambda \in \vartheta$ and $h > 0$ sufficiently small, we have:*

$$\lambda \in \sigma(P^w(hy, y, D_y + A(hy))$$

(where the operator is equipped with the domain $\tilde{H}_A^m$) if and only if

$$0 \in \sigma(E_{-+}^w(x, hD_x + A(x), \lambda; h)),$$

where the last operator is considered as a bounded operator: $V_0^N \to V_0^N$.

It is of some interest to see what kind of Grushin problem we obtain for the original operator $P_0 = P^w(hy, y, D_y + A(hy))$ if we compose the Grushin problem of Theorem 13.23 with the earlier identifications. We recall that we have the unitary map $L^2(\mathbf{R}^n) \ni u \mapsto f = \sum_{\gamma \in \Gamma} v(x)\delta(x - h(y - \gamma)) \in L_0$, defined by $v(hy) = u(y)$ (see Proposition 13.16). We shall compute the jth component of $R_+^w(x, hD_x + A(x))f$ for some fixed j, $1 \leq j \leq N$ in terms of u. We recall that $\varphi_j(x, \xi, y)$ is Γ-periodic in y, and $\varphi_j(x, \xi + \gamma^*, y) = e^{-iy\gamma^*}\varphi_j(x, \xi, y)$ for all γ^* in Γ^*. We also recall that

$$(R_+(x, \xi)u)_j = \int_E u(y)\overline{\varphi_j(x, \xi, y)}dy, \tag{13.57}$$

where E is a fundamental domain of Γ (φ_j and $R_+(x, \xi)$ are given by Proposition 13.10). For simplicity we will drop the index j:

$$(2\pi h)^n R_+^w(x, hD_x + A(x))f(x)$$

$$= \iint e^{i(x-\tilde{x})\xi/h}R_+(\frac{x+\tilde{x}}{2}, \xi + A(\frac{x+\tilde{x}}{2}))f(\tilde{x}, .)d\tilde{x}d\xi$$

$$= \iint e^{i(x-\tilde{x})\xi/h}\int_E \overline{\varphi(\frac{x+\tilde{x}}{2}, \xi + A(\frac{x+\tilde{x}}{2}), y)}f(\tilde{x}, y)dyd\tilde{x}d\xi$$

$$= \sum_{\gamma \in \Gamma}\int_{\mathbf{R}_\xi^n}\int_{\mathbf{R}^n}\int_{E_y} \delta(\tilde{x} - h\gamma - hy)e^{i(x-\tilde{x})\xi/h}v(\tilde{x})$$

$$\overline{\varphi(\frac{x+\tilde{x}}{2}, \xi + A(\frac{x+\tilde{x}}{2}), y)}dyd\tilde{x}d\xi$$

$$= \sum_{\gamma \in \Gamma}\int_{\mathbf{R}_\xi^n}\int_E e^{i(x-h(y+\gamma))\xi/h}v(h(y+\gamma))$$

$$\overline{\varphi(\frac{x+h(y+\gamma)}{2}, \xi + A(\frac{x+h(y+\gamma)}{2}), y)}dyd\xi. \tag{13.58}$$

Using the Γ-periodicity in y of φ, we get

$$R_+^w(x, hD_x + A(x))f(x)$$
$$= (2\pi h)^{-n} \int_{\mathbf{R}_\xi^n} \int_{\mathbf{R}_y^n} e^{i(x-hy)\xi/h} u(y) \overline{\varphi(\frac{x+hy}{2}, \xi + A(\frac{x+hy}{2}), y)} dy d\xi,$$

(13.59)

recalling that $v(hy) = u(y)$. Introducing the function $\psi(x, \xi, y) = e^{iy\xi}\varphi(x, \xi, y)$, we obtain

$$(R_+^w(x, hD_x + A(x))f)(x)$$
$$= (2\pi h)^{-n} \int_{\mathbf{R}_y^n} e^{iA(\frac{x+hy}{2})y} u(y) \int_{\mathbf{R}_\xi^n} e^{i(x\xi/h)} \overline{\psi(\frac{x+hy}{2}, \xi + A(\frac{x+hy}{2}), y)} d\xi dy,$$

$$=: (\widetilde{R}_+ u)(x). \tag{13.60}$$

So if we make the change of variables $\eta = \xi + A(\frac{x+hy}{2})$, we get

$$(R_+^w(x, hD_x + A(x))f)(x)$$
$$= (2\pi h)^{-n} \int_{\mathbf{R}_y^n} e^{iA(\frac{x+hy}{2})(y-x/h)} u(y) \int_{\mathbf{R}_\eta^n} e^{i(x\eta/h)} \overline{\psi(\frac{x+hy}{2}, \eta, y)} d\eta dy.$$

(13.61)

In general, if $f(\eta)$ is a Γ^*-periodic function, then

$$(2\pi h)^{-n} \int_{\mathbf{R}^n} e^{iy\eta/h} f(\eta) d\eta = \sum_{\gamma \in \Gamma} a_\gamma \delta(y - h\gamma), \tag{13.62}$$

with $a_\gamma = (\text{vol}\,(E^*))^{-1} \int_{E^*} e^{i\gamma\eta} f(\eta) d\eta$. Since $\psi(x, \eta, y)$ is Γ-periodic in η and $\psi(x, \eta, y + \gamma) = e^{i\gamma\eta}\psi(x, \eta, y)$ for all $\gamma \in \Gamma$, we get from (13.61) and (13.62):

$$(\widetilde{R}_+ u)(x)$$
$$= \sum_{\gamma \in \Gamma} (\int_{\mathbf{R}_y^n} e^{iA(\frac{x+hy}{2})(y-x/h)} u(y) \int_{E^*} e^{i\gamma\eta} \overline{\psi(\frac{x+hy}{2}, \eta, y)} \frac{d\eta}{\text{vol}\,(E^*)} dy) \delta(x - h\gamma)$$
$$= \sum_{\gamma \in \Gamma} (\int_{\mathbf{R}_y^n} e^{iA(\frac{h\gamma+hy}{2})(y-\gamma)} u(y) \int_{E^*} \overline{\psi(\frac{h\gamma + hy}{2}, \eta, y - \gamma)} \frac{d\eta}{\text{vol}\,(E^*)} dy) \delta(x - h\gamma)$$
$$= \sum_{\gamma \in \gamma} (\int_{\mathbf{R}_y^n} u(y) \overline{W_A(h\frac{\gamma + y}{2}, y - \gamma)} dy) \delta(x - h\gamma), \tag{13.63}$$

with $W_A(x, y) = \frac{1}{\text{vol}(E^*)} e^{-iy \cdot A(x)} \int_{E^*} \psi(x, \eta, y) d\eta$. Taking into account the index j, we get: $(\widetilde{R}_+ u)_j(x) = \sum_{\gamma \in \Gamma} (\widehat{R}_+ u)_j(\gamma)\delta(x - h\gamma)$, where $(\widehat{R}_+ u)_j \in l^2(\Gamma)$ is given by:

$$(\widehat{R}_+ u)_j(\gamma) = \int_{\mathbf{R}^n} u(y) \overline{W_{A,j}(h\frac{\gamma + y}{2}, y - \gamma)} dy,$$

with $W_{A,j}(x,y) = \frac{1}{\mathrm{vol}(E^*)} e^{-iy \cdot A(x)} \int_{E^*} \psi_j(x,\eta,y) d\eta$, for all $1 \leq j \leq N$.
Since the various identifications in our computation are unitary and since
$R_-^w(x, hD_x + A(x)) = R_+^w(x, hD_x + A(x))^*$, it is clear that this operator is
naturally identified with $\widehat{R}_- = \widehat{R}_+^*$. Summing up, we have proved:

Corollary 13.25. *For λ in a neighborhood of λ_0, the operator*

$$\begin{pmatrix} P^w(hy, y, D_y + A(hy)) - \lambda & \widehat{R}_- \\ \widehat{R}_+ & 0 \end{pmatrix} : H_{m,A} \times l^2(\Gamma; \mathbf{C}^N) \to L^2 \times l^2(\Gamma; \mathbf{C}^N)$$

is bijective with bounded inverse $\begin{pmatrix} \widehat{E} & \widehat{E}_+ \\ \widehat{E}_- & \widehat{E}_{-+} \end{pmatrix}$. *The matrix of $\widehat{E}_{-+}$ is equal
to the matrix of $E_{-+}^w(x, hD_x + A(x), \lambda, h)$ acting on V_0^N, if we identify the
latter space with $l^2(\Gamma; \mathbf{C}^N)$ in the natural way.*

We end this chapter by discussing Schrödinger operators with slowly varying
perturbations. Let $V(y), \varphi(x) \in C^\infty(\mathbf{R}^n, \mathbf{R})$, where $V(y)$ is Γ-periodic and
φ is bounded with all its derivatives and tends to zero at infinity. We are
interested in the operator

$$P_{A,\varphi} = \sum_{j=1}^n (D_{y_j} + A_j(hy))^2 + V(y) + \varphi(hy) = P^w(hy, y, D_y + A(hy)),$$

where $P(x, y, \eta) = \eta^2 + V(y) + \varphi(x)$.

Let $I \subset\subset \mathbf{R}$ be an open interval and put $\widetilde{I} = I - \varphi(\mathbf{R}^n)$. In the appendix we
construct $\psi_1(\xi, y), \psi_2(\xi, y), \ldots, \psi_N(\xi, y)$ smooth in all variables, Γ^*-periodic
in ξ and with $\psi_j(\xi, y + \gamma) = e^{i\gamma \xi} \psi_j(\xi, y)$, such that the problem:

$$\widetilde{\mathcal{P}}(\xi, \widetilde{z}) = \begin{pmatrix} -\Delta + V(y) - \widetilde{z} & \widetilde{R}_-(\xi) \\ \widetilde{R}_+(\xi) & 0 \end{pmatrix} : \mathcal{F}_{2,\xi} \times \mathbf{C}^N \to \mathcal{F}_\xi \times \mathbf{C}^N$$

is bijective for $\xi \in \mathbf{R}^n$ and $\widetilde{z}$ in a neighborhood of $\widetilde{I}$, with $\widetilde{R}_-(\xi) = (\widetilde{R}_+(\xi))^*$,
$(\widetilde{R}_+(\xi)u)_j = (u, \psi_j)$. Let $\begin{pmatrix} \widetilde{E}(\xi, \widetilde{z}) & \widetilde{E}_+(\xi, \widetilde{z}) \\ \widetilde{E}_-(\xi, \widetilde{z}) & \widetilde{E}_{-+}(\xi, \widetilde{z}) \end{pmatrix}$ be the inverse of $\widetilde{\mathcal{P}}(\xi, \widetilde{z})$.
Taking $\widetilde{z} = z - \varphi(x)$ we get an inverse

$$\begin{pmatrix} E^0(x, \xi, z) & E_+^0(x, \xi, z) \\ E_-^0(x, \xi, z) & E_{-+}^0(x, \xi, z) \end{pmatrix} = \begin{pmatrix} \widetilde{E}(\xi, z - \varphi(x)) & \widetilde{E}_+(\xi, z - \varphi(x)) \\ \widetilde{E}_-(\xi, z - \varphi(x)) & \widetilde{E}_{-+}(\xi, z - \varphi(x)) \end{pmatrix}$$

for the operator $\widetilde{\mathcal{P}}(\xi, z - \varphi(x))$. Putting $R_-(\xi) = e^{-iy\xi} \widetilde{R}_-(\xi)$, $R_+(\xi) =$
$\widetilde{R}_+(\xi) e^{iy\xi}$, $\mathcal{P}(x, \xi, z) = \begin{pmatrix} (D_y + \xi)^2 + V(y) - z & R_-(\xi) \\ R_+(\xi) & 0 \end{pmatrix}$, we know from
Proposition 13.11 that when h is small enough,

$$\mathcal{P}^w(x, hD_x + A(x), z) : \mathcal{K}_2 \times L^2(\mathbf{R}^n; \mathbf{C}^N) \to L^2(\mathbf{R}^n \times \mathbf{R}^n/\Gamma) \times L^2(\mathbf{R}^n; \mathbf{C}^N)$$

is bijective and has the uniformly bounded inverse $\mathcal{E}^w(x, hD_x + A(x), z; h)$. If $E_{-+}(x, \xi + A(x), z; h)$ is the $N \times N$ matrix which appears in the lower right corner of $\mathcal{E}(x, \xi + A(x), z; h)$ then $E_{-+}(x, \xi, z; h) \in S^0(\mathbf{R}^{2n}; \mathcal{L}(\mathbf{C}^N, \mathbf{C}^N))$ has a complete asymptotic expansion in powers of h and the leading term is $E^0_{-+}(x, \xi, z) = \tilde{E}_{-+}(\xi, z - \varphi(x))$. Moreover, E_{-+} is Γ^*-periodic with respect to ξ, and if $z \in I$, then $z \in \sigma(P_{A,\varphi})$ (as an operator acting on $L^2(\mathbf{R}^n)$) iff $0 \in \sigma(E^w_{-+}(x, hD_x + A(x), z, h))$, where E^w_{-+} now acts on the space V_0^N. In the following we denote by $\widehat{E}_{-+}$ the matrix of $E^w(x, hD_x + A(x), \lambda, h)$ acting on V_0^N, if we identify the latter space with $l^2(\Gamma; \mathbf{C}^N)$ in the natural way.

Remark 13.26.

(1) From Remark 13.13 and the construction above we have

$$\|(E^0_{-+}(x, \xi, z))^{-1}\|_{\mathcal{L}(\mathbf{C}^N, \mathbf{C}^N)} = \mathcal{O}(|\text{Im } z|^{-1}),$$
$$\det E^0_{-+}(x, \xi, z) = 0 \text{ iff } z \in \sigma(P_{\xi, \varphi}),$$
$$\dim \ker E^0_{-+}(x, \xi, z) = \dim \ker (P_{\xi, \varphi} - z), \qquad (13.64)$$

where $P_{\xi, \varphi} = (D_y + \xi)^2 + V(y) + \varphi(x)$ is considered as a non bounded-operator in $K_0 = L^2(\mathbf{R}^n/\Gamma)$. From simple general results on elliptic operators on compact manifolds, we know that $P_\xi = (D_y + \xi)^2 + V(y)$ has a discrete spectrum with eigenvalues counted with multiplicity: $\lambda_1(\xi) \leq \lambda_2(\xi) \leq \cdots$ with $\lambda_j(\xi) \to \infty, j \to \infty$. Ordinary perturbation theory shows that $\lambda_j(\xi)$ are continuous functions in ξ for every fixed j, and $\lambda_j(\xi)$ is even an analytic function of ξ near every point $\xi_0 \in \mathbf{R}^n/\Gamma^*$ where $\lambda_j(\xi_0)$ is a simple eigenvalue of P_{ξ_0}. The λ_j are called the Floquet eigenvalues. The sets $J_k = \lambda_k(\mathbf{R}^n/\Gamma^*)$ are closed intervals, and the spectrum $\sigma(P_0)$ of $P_0 = -\Delta + V(y)$ (as a non bounded operator in $L^2(\mathbf{R}^n)$) is given by $\sigma(P_0) = \sigma_{\text{ess}}(P_0) = \cup_{k=1}^\infty J_k$. Then we deduce:

$$\det E^0_{-+}(x, \xi, z) = 0 \text{ iff there exists } k \geq 1 \text{ such that } z = \lambda_k(\xi) + \varphi(x).$$
$$(13.65)$$

(2) Let $z_0 \in \mathbf{R}$, $d = \dim \text{Ker } E^0_{-+}(x, \xi, z)$ for a fixed (x, ξ). By ordinary perturbation theory (see Kato [Ka2]) we can reorder the eigenvalues $(\lambda_j(z))_{1 \leq j \leq N}$ of $E^0_{-+}(x, \xi, z)$ to be holomorphic in a neighborhood of $z_0 \in \mathbf{R}$ and $\lambda_1(z_0) = \cdots = \lambda_d(z_0) = 0$. Using (13.64) we see that

$$|\lambda_j(z)| \geq C_j |\text{Im } z|,$$

so $\lambda'_j(z_0) \neq 0$ for all $1 \leq j \leq N$. Hence, $z \mapsto \det E^0_{-+}(x, \xi, z)$ has a root z_0 of multiplicity d.

Proposition 13.27. *Suppose that $z_0 \notin \sigma(P_0)$. Then there exist $\epsilon > 0, h_0 > 0$ such that $z \notin \sigma(P_A)$ when $|z - z_0| < \epsilon, h \in]0, h_0[$. Here $P_A = \sum_{j=1}^{j=n} (D_{y_j} + A_j(hy))^2 + V(y)$.*

Proof. We have $E^0_{-+}(x, \xi + A(x), z) = \tilde{E}_{-+}(\xi + A(x), z)$, where $\tilde{E}_{-+}(\xi, z)$ is the effective Hamiltonian associated with $(D_y + \xi)^2 + V(y)$. The assumption $z_0 \notin \sigma(P_0)$ and (13.65) imply:

$$|\det E^0_{-+}(x, \xi, z_0)| \geq \frac{1}{C_0} \text{ with } C_0 \text{ is independent of } (x, \xi).$$

Theorem 8.3 shows that $(E^w_{-+}(x, hD_x + A(x), z; h))^{-1}$ exists for $|z - z_0| < \epsilon$, $0 < h < \tilde{h}_0$ small enough, and is equal to $\mathrm{Op}^w_h r(x, \xi + A(x), z; h)$, with $r \in S^0(\mathbf{R}^{2n}, \mathcal{L}(\mathbf{C}^N, \mathbf{C}^N))$ and $r(x, \xi, z; h) = r(x, \xi + \gamma^*, z; h)$. From Proposition 13.20 we conclude that $(E^w_{-+})^{-1}$ is bounded on V^N_0 for h small enough, and by Corollary 13.24 we see that $z \notin \sigma(P_A)$ when $|z - z_0| < \epsilon$.   #

Now we assume that

$$\sigma(-\Delta + V) \cap \bar{I} = \emptyset. \tag{$\tilde{\mathrm{H}}$}$$

Assumption $(\tilde{\mathrm{H}})$ and Proposition 13.27 imply that

$$\sigma(P_A) \cap \bar{I} = \emptyset,$$

for h small enough. Using the Weyl criterion (Chapter 4), and the fact that $\varphi(x)$ tends to zero at infinity, we see that for h small enough the spectrum of $P_{A,\varphi}$ in $\bar{I}$ is discrete. Let $f \in C^\infty_0(I)$. We have:

Theorem 13.28.

$$\mathrm{tr}\, f(P) \sim (2\pi h)^{-n} \sum_{j \geq 0} a_j h^j, \quad (h \searrow 0), \tag{13.66}$$

with

$$a_0 = \int_{E^*} \int_{\mathbf{R}^n_x} \sum_{k \geq 1} f(\varphi(x) + \lambda_k(\xi)) dx d\xi. \tag{13.67}$$

Proof. Let $\tilde{\varphi}(x) \in C^\infty(\mathbf{R}^n_x)$ be real valued, coincide with φ for large x and satisfy:

$$(\sigma(-\Delta + V) + \overline{\{\tilde{\varphi}(x), x \in \mathbf{R}^n_x\}}) \cap \bar{I} = \emptyset.$$

Let $\tilde{E}^0_{-+}(x, \xi, z)$ be the effective Hamiltonian associated with $\tilde{P}(x, y, \xi) = \xi^2 + V(y) + \tilde{\varphi}(x)$ and put

$$\tilde{E}_{-+}(x, \xi, z, h) = \tilde{E}^0_{-+}(x, \xi, z) + E_{-+}(x, \xi, z, h) - E^0_{-+}(x, \xi, z).$$

From the properties of $\tilde{\varphi}$ we have

$$E_{-+}(x, \xi, z, h) = \tilde{E}_{-+}(x, \xi, z, h) \text{ for large } x$$

and there exists $C > 0$ such that for h small enough

$$|\det \widetilde{E}_{-+}| \geq \frac{1}{C} \text{ uniformly in } x, \xi, z, h.$$

From Corollary 13.25 and Remark 13.13 we have:

$$(\widehat{E}_{-+})^{-1} = \widehat{R}_+ (z - P_{A,\varphi})^{-1} \widehat{R}_-$$

and

$$(z - P_{A,\varphi})^{-1} = -\widehat{E} + \widehat{E}_+ (\widehat{E}_{-+})^{-1} \widehat{E}_-. \tag{13.68}$$

As $\widehat{R}_-$ and $\widehat{R}_+$ are bounded,

$$\|\widehat{E}_{-+}^{-1}\|_{\mathcal{L}(l^2(\Gamma; \mathbf{C}^N))} = \mathcal{O}(|\mathrm{Im}\, z|^{-1}). \tag{13.69}$$

Let $\widetilde{f} \in C_0^\infty(\mathbf{C})$ be an almost analytic extension of f with support close to that of f such that for all $N \in \mathbf{N}$

$$\overline{\partial}_z \widetilde{f}(z) = \mathcal{O}(|\mathrm{Im}\, z|^N), \tag{13.70}$$

(see Chapter 8 for such a construction). By Theorem 8.1 we have

$$f(P_{A,\varphi}) = -\frac{1}{\pi} \int \frac{\partial \widetilde{f}}{\partial \overline{z}}(z)(z - P_{A,\varphi})^{-1} L(dz),$$

and the identity

$$\widehat{E}_{-+}^{-1} = \widehat{\widetilde{E}}_{-+}^{-1} - \widehat{E}_{-+}^{-1}(\widehat{E}_{-+} - \widehat{\widetilde{E}}_{-+})\widehat{\widetilde{E}}_{-+}^{-1},$$

with $\widehat{\widetilde{E}}_{-+}$ defined as in Corollary 13.25, combined with (13.68) and the fact that $\widehat{\widetilde{E}}_{-+}^{-1}, \widehat{E}$ are holomorphic in z near $\overline{I}$, give

$$f(P_{A,\varphi}) = -\frac{1}{\pi} \int \frac{\partial \widetilde{f}}{\partial \overline{z}}(z)(\widehat{E}_+ \widehat{E}_{-+}^{-1}(\widehat{\widetilde{E}}_{-+} - \widehat{E}_{-+})\widehat{\widetilde{E}}_{-+}^{-1} \widehat{E}_- L(dz). \tag{13.71}$$

Lemma 13.29. *Let* $Q(x, \xi) \in S^0(\mathbf{R}^{2n})$ *with* $Q(x, \xi + \gamma^*) = Q(x, \xi)$ *for any* $\gamma^* \in \Gamma^*$. *We assume that* $K = \Pi_x \operatorname{supp} Q$ *is compact. Then* $Q^w(x, hD_x)$ *is of trace class on* V_0 *and*

$$\operatorname{tr}(Q^w(x, hD_x)) = \frac{1}{(2\pi h)^n} \int_{E^*} \int_{\mathbf{R}_x^n} Q(x, \xi) dx d\xi + \mathcal{O}(h^\infty). \tag{13.72}$$

Proof. We denote by $A_{(\alpha,\gamma)}$ the coefficient of the matrix $\widehat{Q}$ when we identify V_0 with $l^2(\Gamma)$. Using the Γ^*-periodicity of Q on ξ we obtain, using (13.62):

$$A_{(\alpha,\gamma)} = \int_{E^*} e^{i(\alpha-\gamma)\xi}\, Q(\frac{h\alpha+h\gamma}{2},\xi)\frac{d\xi}{\operatorname{vol}(E^*)}.$$

By integrating by parts we see that

$$A_{(\alpha,\gamma)} = \mathcal{O}_N\left(\left(\frac{1}{1+\operatorname{dist}(\alpha,\frac{K}{h})}\right)^N\left(\frac{1}{1+\operatorname{dist}(\gamma,\frac{K}{h})}\right)^N\right) \text{ for every } N \in \mathbf{N}.$$

Consequently $Q^w(x,hD_x)$ is of trace class and $\|Q^w\|_{\mathrm{tr}} \leq \sum_{\alpha,\beta}|A_{\alpha,\beta}| = \mathcal{O}(h^{-n})$. To prove (13.72) it suffices to use the following lemma:

Lemma 13.30. *If $f \in \mathcal{S}(\mathbf{R}^n)$, then*

$$\int_{\mathbf{R}^n} f(x)dx = h^n\operatorname{vol}(E)\sum_{\gamma\in\Gamma} f(h\gamma) + \mathcal{O}(h^\infty), \quad h \to 0. \tag{13.73}$$

Proof. Define
$$F_h(x) = h^n\operatorname{vol}(E)\sum_{\gamma\in\Gamma} f(h(\gamma+x)).$$
F_h is C^∞, Γ-periodic and for all $\alpha \in \mathbf{N}^n$,

$$\partial_x^\alpha F_h(x) = \mathcal{O}_\alpha(h^{|\alpha|}), \tag{13.74}$$

uniformly on x in $\mathbf{R}^n$. Moreover, for every $x \in \mathbf{R}^n$

$$F_h(x) = \sum_{\gamma^*\in\Gamma^*} c_{\gamma^*} e^{i\gamma^* x}, \tag{13.75}$$

with

$$c_{\gamma^*} = (\operatorname{vol}(E))^{-1}\int_E e^{-i\gamma^* x}F_h(x)dx. \tag{13.76}$$

Integrating by parts in (13.76) and using (13.74) we get

$$c_{\gamma^*} = \mathcal{O}_N(1)h^N(1+|\gamma^*|)^{-N}, \tag{13.77}$$

for all $\gamma^* \neq 0$ and all $N \in \mathbf{N}$. (13.75) and (13.77) imply

$$F_h(x) = c_0 + \mathcal{O}(h^\infty) = \int_{\mathbf{R}^n} f(x)dx + \mathcal{O}(h^\infty),$$

and (13.73) follows as the special case $x = 0$. #

Lemma 13.29 follows.   #

By Lemma 13.27 $(\widehat{E}_{-+} - \widetilde{\widehat{E}}_{-+})$ is of trace class and we can take the trace and permute integration and the operator 'tr' in (13.71). The identity $\partial_z \widehat{E}_{-+} = \widehat{E}_- \widehat{E}_+$ shows that for $\mathrm{Im}\, z \neq 0$,

$$\mathrm{tr}\,(\widehat{E}_+ \widehat{E}_{-+}^{-1}(\widetilde{\widehat{E}}_{-+} - \widehat{E}_{-+})\widetilde{\widehat{E}}_{-+}^{-1}\widehat{E}_-) = \mathrm{tr}\,(\widehat{E}_{-+}^{-1}(\widetilde{\widehat{E}}_{-+} - \widehat{E}_{-+})\widetilde{\widehat{E}}_{-+}^{-1}\partial_z \widehat{E}_{-+}).$$
(13.78)

Let $\chi \in C_0^\infty(\mathbf{R}^n)$ be equal to 1 in a neigborhood of $\Pi_x(\mathrm{supp}\,(E_{-+}^0(x,\xi,z) - \widetilde{E}_{-+}^0(x,\xi,z)))$, and denote by $\widehat{\chi}$ the matrix associated with the operator of multiplication by $\chi(x)$ on V_0^N. Since $\Pi_x(\mathrm{supp}\,(E_{-+}^0(x,\xi,z) - \widetilde{E}_{-+}^0(x,\xi,z))) \cap \mathrm{supp}\,(1-\chi) = \emptyset$, (13.69) and Proposition 13.21 show that:

$$\|\widehat{E}_{-+}^{-1}(\widetilde{\widehat{E}}_{-+} - \widehat{E}_{-+})\widetilde{\widehat{E}}_{-+}^{-1}\partial_z \widehat{E}_{-+}(1-\widehat{\chi})\|_{\mathrm{tr}} = \mathcal{O}(h^\infty|\mathrm{Im}\, z|^{-1}),$$

so $\mathrm{tr}\, f(P_{A,\varphi}) = -\frac{1}{\pi}\mathrm{tr}\,\int \frac{\partial \widetilde{f}}{\partial \bar{z}}(z)\widehat{E}_{-+}^{-1}(\widetilde{\widehat{E}}_{-+} - \widehat{E}_{-+})\widetilde{\widehat{E}}_{-+}^{-1}\partial_z \widehat{E}_{-+}\widehat{\chi}L(dz) + \mathcal{O}(h^\infty)$.

Splitting the integral into two terms and using the fact that $\widetilde{\widehat{E}}_{-+}^{-1}\partial_z \widehat{E}_{-+}$ is holomorphic in z, we get

$$\mathrm{tr}\, f(P_{A,\varphi}) = -\frac{1}{\pi}\mathrm{tr}\,\int \frac{\partial \widetilde{f}}{\partial \bar{z}}(z)\widehat{E}_{-+}^{-1}\partial_z \widehat{E}_{-+}\widehat{\chi}L(dz) + \mathcal{O}(h^\infty).$$
(13.79)

Lemma 13.31. *There exists* $r(x,\xi;h) \in S^0(\mathbf{R}^{2n},\mathcal{L}(\mathbf{C}^N,\mathbf{C}^N))$ *such that* $r(x,\xi;h) \sim \sum_{j\geq 0} h^j r_j(x,\xi)$ *and*

$$\mathrm{Op}_h^w(r(x,\xi + A(x);h))$$
$$= -\frac{1}{\pi}\int_{|\mathrm{Im}\, z|\geq h^\delta} \frac{\partial \widetilde{f}}{\partial \bar{z}}(z)(E_{-+}^w(x, hD_x + A(x), z; h))^{-1}\partial_z E_{-+}^w L(dz).$$
(13.80)

Moreover, r_j *is* Γ^*-*periodic in* ξ *for all* $j \geq 0$ *and* $r_0(x,\xi)$ *is independent of* A *with:*

$$r_0(x,\xi) = -\frac{1}{\pi}\int \frac{\partial \widetilde{f}}{\partial \bar{z}}(z)(E_{-+}^0(x,\xi,z))^{-1}\partial_z E_{-+}^0(x,\xi,z)L(dz).$$
(13.81)

Proof. Let us recall that the results of Chapter 8 remain true in the case of operators with operator valued symbol. Let $l_1, l_2, \ldots$ be linear forms on $\mathbf{R}^{2n}$ and put $L_j = l_j(x, hD_x)$. From the identity $E_{-+}^w \circ (E_{-+}^w)^{-1} = I$ we have

$$\mathrm{ad}_{L_j}(E_{-+}^w)^{-1} = -(E_{-+}^w)^{-1} \circ \mathrm{ad}_{L_j} E_{-+}^w \circ (E_{-+}^w)^{-1},$$

where $\mathrm{ad}_{L_j} A$ denotes the commutator $[L_j, A]$. As $\mathrm{ad}_{L_j}(A \circ B) = (\mathrm{ad}_{L_j} A) \circ B + A \circ \mathrm{ad}_{L_j} B$,

$$\mathrm{ad}_{L_j}((E^w_{-+})^{-1} \partial_z E^w_{-+})$$
$$= -(E^w_{-+})^{-1} \circ \mathrm{ad}_{L_j} E^w_{-+} \circ (E^w_{-+})^{-1} \circ \partial_z E^w_{-+} + (E^w_{-+})^{-1} \circ \mathrm{ad}_{L_j} \partial_z E^w_{-+}.$$
$$(13.82)$$

Using (13.82), the fact that $\|(E^w_{-+})^{-1}\| = \mathcal{O}(|\mathrm{Im}\,z|^{-1})$ as in (13.69) and the fact that $E^w_{-+}, \partial_z E^w_{-+}$ are h-pseudors with symbol in $S^0(\mathbf{R}^{2n}, \mathcal{L}(\mathbf{C}^N, \mathbf{C}^N))$, we see that

$$\|\mathrm{ad}_{L_j}(E^w_{-+})^{-1} \circ \partial_z E^w_{-+}\| = \mathcal{O}(\frac{h}{|\mathrm{Im}\,z|^2}).$$

An easy induction (just indicated in the proof above) then shows that:

$$\|\mathrm{ad}_{L_j} \circ \ldots \circ \mathrm{ad}_{L_N}((E^w_{-+})^{-1} \circ \partial_z E^w_{-+})\| = \mathcal{O}(\frac{h^N}{|\mathrm{Im}\,z|^{N+1}}). \qquad (13.83)$$

Now by the Beals characterization of h-pseudors we can apply the same proof as for Theorem 8.7. The periodicity of r_j follows from that of $E^j_{-+}(x, \xi, z)$. #

If we restrict the integral in the right hand side of (13.79) to the domain $|\mathrm{Im}\,z| \leq h^\delta$ then we get a term $\mathcal{O}(h^\infty)$ in trace norm. If we restrict our attention to the domain $|\mathrm{Im}\,z| \geq h^\delta$ then by Lemma 13.29 and Lemma 13.31 we get (13.66). To finish let us compute a_0. We have

$$a_0 = \iint_{\mathbf{R}^n_x \times E^*} \mathrm{tr}\,[r_0(x, \xi + A(x))] dx d\xi = \iint_{\mathbf{R}^n_x \times E^*} \mathrm{tr}\,[r_0(x, \xi)] dx d\xi$$
$$= \iint_{\mathbf{R}^n_x \times E^*} (-\frac{1}{\pi} \int \frac{\partial \widetilde{f}}{\partial \overline{z}}(z) \mathrm{tr}\,(E^0_{-+}(x, \xi, z))^{-1} \partial_z E^0_{-+}(x, \xi, z) L(dz)) dx d\xi.$$
$$(13.84)$$

Thanks to Liouville's formula (i.e. $\mathrm{tr}\,(\partial_z A(z) A^{-1}(z)) = \frac{\partial_z \det A(z)}{\det A(z)}$ in the sense of matrices), we get

$$a_0 = \iint_{\mathbf{R}^n_x \times E^*} (-\frac{1}{\pi} \int \frac{\partial \widetilde{f}}{\partial \overline{z}}(z) \frac{\partial_z \det E^0_{-+}(x, \xi, z)}{\det E^0_{-+}(x, \xi, z)} L(dz)) dx d\xi.$$

To prove (13.67) we use Remark 13.64 and the following Lemma:

Lemma 13.32. *Let g be an analytic function. Let $(z_k)_{k \geq 1}$ be the roots (counted with their multiplicity) of g in $\mathrm{supp}\,(\widetilde{f})$. We have:*

$$\frac{-1}{\pi} \int \frac{\partial \widetilde{f}}{\partial \overline{z}} \frac{g'(z)}{g(z)} L(dz) = \sum_{k \geq 1} f(z_k).$$

Proof. This follows from the formula $\frac{1}{\pi}\bar{\partial}(\frac{1}{z-z_0}) = \delta(\cdot - z_0)$ and the fact that $\frac{g'(z)}{g(z)} = \sum_{k\geq 1}\frac{1}{z-z_k} + k(z)$, where k is holomorphic for z in a small neigborhood of supp $\tilde{f}$.

Appendix: Grushin problem

We will construct a suitable auxiliary (so-called Grushin) problem associated with the operator $(-\Delta + V(y) - \lambda_0)$, for some fixed energy level λ_0. The same proof applies to the operator $P_m = ((-\Delta)^{\frac{m}{2}} - \lambda_0)$.

Theorem A.1. *There exist N analytic functions $\varphi_j : \mathbf{R}^{n*}/\Gamma^* \to \mathcal{F}_{0,\xi}$, $1 \leq j \leq N$, such that for every $\xi \in \mathbf{R}^{n*}/\Gamma^*$ the Grushin problem,*

$$(P_\xi - \lambda_0)u + R_-(\xi)u^- = v, \ R_+(\xi)u = v^+, \tag{A.1}$$

has a unique solution $(u, u^-) \in \mathcal{F}_{2,\xi} \times \mathbf{C}^N$ for every $(v, v^+) \in \mathcal{F}_{0,\xi} \times \mathbf{C}^N$. Here we have put

$$P_\xi = D_y^2 + V(y), \ R_+(\xi)u(j) = (u, \varphi_j(\xi)), R_-(\xi)u^- = \sum_{1\leq j\leq N} u^-(j)\varphi_j(\xi).$$

In the next proposition we prove that if $\varphi_1(\xi), \ldots, \varphi_N(\xi)$ are linearly independent, then the well posedness of (A.1) only depends on the vector spaces $\mathcal{T}_\xi$ generated by $\varphi_1(\xi), \ldots, \varphi_N(\xi)$.

Proposition A.2. *Fix ξ and assume that $\varphi_1(\xi), \ldots, \varphi_N(\xi)$ are linearly independent. Then the problem (A.1) is well posed if and only if $(1-\pi)(P_\xi - \lambda_0)$ as an operator from $\mathcal{T}_\xi^\perp \cap \mathcal{F}_{2,\xi}$ to $\mathcal{T}_\xi^\perp$ is bijective. Here π denotes the orthogonal projection onto $\mathcal{T}_\xi$.*

Proof. In the case $v = 0, v^+ = 0$ the problem A.1 takes the form $(P_\xi - \lambda_0)u + R_-(\xi)u^- = 0, u \in \mathcal{T}_\xi^\perp \cap \mathcal{F}_{2,\xi}$. Then $(1 - \pi)(P_\xi - \lambda_0)u = 0$, so the injectivity of the restriction of $(1 - \pi)(P_\xi - \lambda_0)$ to $\mathcal{T}_\xi^\perp \cap \mathcal{F}_{2,\xi}$ and the fact that $\varphi_1(\xi), \ldots, \varphi_N(\xi)$ are linearly independent imply that we have uniqueness for the problem (A.1). If the restriction of $(1 - \pi)(P_\xi - \lambda_0)$ to $\mathcal{T}_\xi^\perp \cap \mathcal{F}_{2,\xi}$ is not injective, then there is $u \neq 0 \in \mathcal{T}_\xi^\perp \cap \mathcal{F}_{2,\xi}$ such that $(P_\xi - \lambda_0)u = \pi(P_\xi - \lambda_0)u \in$ Range of $R_-(\xi)$, and we see that there is u^- such that (A.1) holds with $v = 0, v^+ = 0$. We have then established the equivalence between uniqueness for (A.1) and injectivity of the restriction of $(1 - \pi)(P_\xi - \lambda_0)$ to $\mathcal{T}_\xi^\perp \cap \mathcal{F}_{2,\xi}$. Since $\varphi_1(\xi), \ldots, \varphi_N(\xi)$ are independent, $R_+(\xi)$ is surjective and the solvability of (A.1) is equivalent to the solvability of the first equation of (A.1), with $u \in T_\xi^\perp \cap \mathcal{F}_{2,\xi}$. Since the image of $R_-(\xi)$ is T_ξ, we see that solvability of (A.1) is equivalent to the surjectivity of the restriction of $(1 - \pi)(P_\xi - \lambda_0)$ to $T_\xi^\perp \cap \mathcal{F}_{2,\xi}$. #

Proposition A.3. *Let P be a second order elliptic selfadjoint operator on a compact manifold M with smooth coefficients. Let $\varphi_1, \ldots, \varphi_N \in L^2(M)$ be linearly independent functions and assume that that there is a constant $C_0 > 0$ such that*

$$(Pu, u) \geq C_0^{-1} \|u\|^2, u \in H^1(M) \cap [\varphi_1, \ldots, \varphi_N]^\perp, \qquad (A.2)$$

where $[\varphi_1, \ldots, \varphi_N]$ denotes the linear span of the functions $\varphi_1, \ldots, \varphi_N$, and $H^s(M)$ (for $s \in \mathbf{R}$) is the classical Sobolev space on M of order s. Then if $\widetilde{P}$ is another second order selfadjoint operator and $\widetilde{\varphi}_1, \ldots, \widetilde{\varphi}_N \in L^2(M)$, with $\|\widetilde{P} - P\|_{\mathcal{L}(H^1, H^{-1})}$ and $\|\widetilde{\varphi}_1 - \varphi_1\|, \ldots, \|\widetilde{\varphi}_N - \varphi_N\|$ small enough, there exists a constant $C_1 > 0$ such that

$$(\widetilde{P}u, u) \geq C_1^{-1} \|u\|_1^2 - C_1 \sum_1^N |(u, \widetilde{\varphi}_j)|^2, \qquad (A.3)$$

for all $u \in H^1(M)$.

Proof. Without loss of generality, we may assume that $\varphi_1, \ldots, \varphi_N$ is an orthonormal system. Choose $\psi_1, \ldots, \psi_N \in H^2$ with $(\psi_j, \varphi_k) = \delta_{j,k}$. For $u \in H^1$, we put $\widetilde{u} = u - \sum_1^N (u, \varphi_j) \psi_j \in H^1(M) \cap [\varphi_1, \ldots, \varphi_N]^\perp$, so we can apply (A.2) to $\widetilde{u}$: $(P\widetilde{u}, \widetilde{u}) \geq C_0^{-1} \|\widetilde{u}\|^2$. Now,

$$(P\widetilde{u}, \widetilde{u}) \leq (Pu, u) + C\|u\| \sum_1^N |(u, \varphi_j)| + C \sum_1^N |(u, \varphi_j)|^2,$$

$$\|u\|^2 \leq 2\|\widetilde{u}\|^2 + C \sum_1^N |(u, \varphi)|^2,$$

so with a new constant we get

$$(Pu, u) \geq C^{-1} \|u\|^2 - C \sum_1^N |(u, \varphi_j)|^2. \qquad (A.4)$$

Combining (A.4) with the Gårding inequality

$$(Pu, u) \geq C^{-1} \|u\|_1^2 - C\|u\|^2,$$

we obtain

$$(C^2 + 1)(Pu, u) \geq C^{-1}\|u\|_1^2 - C^3 \sum_1^N |(u, \varphi_j)|^2. \qquad \text{(A.5)}$$

From this we get (A.2) if $\|\tilde{P} - P\|_{\mathcal{L}(H^1, H^{-1})}, \|\tilde{\varphi}_1 - \varphi_1\|, \dots, \|\tilde{\varphi}_N - \varphi_N\|$ are small enough.   #

Proof of Theorem A.1. For a fixed $\xi_0 \in \mathbf{R}^{n*}/\Gamma^*$ we can find $\varphi_1(\cdot, \xi_0), \dots,$ $\varphi_N(\cdot, \xi_0)$ and a constant $C_0 > 0$ such that

$$((P_\xi - \lambda_0)u, u) \geq C_0^{-1}\|u\|^2 \text{ for all } u \in \mathcal{F}_{1, \xi_0} \cap [\varphi_1(\cdot, \xi_0), \dots, \varphi_N(\cdot, \xi_0)]^\perp$$
$$\text{(A.6)}$$

Let $E \subset \mathbf{R}^n$ be a fundamental domain of Γ. Modifying φ_j^0 by terms with small norm (which will not destroy (A.6)), we may assume that $\text{supp}(\varphi_j^0) \cap \partial E = \emptyset$, so that $\varphi_j(x, \xi_0) = \sum_\gamma \psi_j^0(x - \gamma)e^{i\xi_0\gamma}$, with $\psi_j^0 \in L^2(E) \cap \mathcal{E}'(\text{int}(E))$. Put $\varphi_j(x, \xi) = \sum_\gamma \psi_j^0(x - \gamma)e^{i\xi\gamma}$. Proposition A.3 shows that for ξ close to ξ_0 we have with a new constant $C_0 > 0$

$$((P_\xi - \lambda_0)u, u) \geq C_0^{-1}\|u\|^2 \text{ for all } u \in \mathcal{F}_{1, \xi_0} \cap [\varphi_1(\cdot, \xi), \dots, \varphi_N(\cdot, \xi)]^\perp$$
$$\text{(A.7)}$$

Clearly, if we add more functions to our system $\varphi_1, \dots, \varphi_N$ then (A-7) remains true for ξ in the same neighborhood of ξ_0 and with the same constant C_0. Varying the point ξ_0, and using the compactness of $\mathbf{R}^{n*}/\Gamma^*$, we obtain with a new N a system $\varphi_j(x, \xi)$, such that (A.7) holds for all $\xi \in \mathbf{R}^{n*}/\Gamma^*$ with a new constant $C_0 > 0$ which is independent of ξ. Without changing (A.7) we may eliminate successively all the ψ_j^0's which are linear combinations of the others (and make the corresponding elimination of φ_j). We then obtain (A.7) with $\varphi_1(\cdot, \xi), \dots, \varphi_N(\cdot, \xi)$ Γ^*-periodic and analytic in ξ and linearly independent for every $\xi \in \mathbf{R}^{n*}/\Gamma^*$. It is easy to show from (A.7) that $(1 - \pi_\xi)(P_\xi - \lambda_0) : \mathcal{F}_{2, \xi_0} \cap [\varphi_1(\cdot, \xi), \dots, \varphi_N(\cdot, \xi)]^\perp \to [\varphi_1(\cdot, \xi), \dots, \varphi_N(\cdot, \xi)]^\perp$ is bijective, and this completes the proof of Theorem A.1.   #

Notes

In this chapter we have only discussed one (semi-classical) aspect of a very wide subject. In addition to the papers, [GMS], [Di1], on which the chapter is based, we can mention [HeSj4], which gives precise information on the density of states of the periodic Schrödinger operator with magnetic field. The time-dependent periodic Schrödinger operator is discussed in the papers of Gérard [Gé] and Ralston [Ra]. In the one-dimensional case many results were obtained by Buslaev [Bu1,2,3] and Buslaev-Dimitrieva [BuDi].

The results of the present chapter can be applied to study the eigenvalues in gaps of the essential spectrum for certain perturbations with large coupling constant. Let A be a selfadjoint operator and consider the quantity $N(E, \lambda)$ (where $\lambda > 0$ and E is a regular point for A) defined as the number of eigenvalues of $A_t := A + tW$ crossing E as t increases from 0 to λ. Here W is a perturbation decaying at infinity. The behaviour of $N(E, \lambda)$ has been studied in detail for the periodic Schrödinger operator. Since the work of [ADH] we know that the behaviour of $N(E, \lambda)$ is dramatically different for non-negative and non-positive perturbations. Specifically, the leading term of the asymptotics for non-positive potentials is given by the classical Weyl formula and does not contain any information on the periodic background. See [ADH], [Bi1,2]. On the contrary, for non-negative perturbations the answer contains the density of states associated with the unperturbed operator and depends only on the asymptotics of the perturbation at infinity. An asymptotic expansion of $\operatorname{tr} f(A_t)$ for $f \in C_0^\infty(I)$, where I is an open interval disjoint from the essential spectrum, was obtained by [Di2]. In the one dimensional case more precise results were obtained by Sobolev [So]. The situation is more difficult if the perturbation is alternating. At present only partial results are known to this effect, see [ADH], [Hem1,2] and [Le]. The case when A is the Schrödinger operator with magnetic field was studied by Birman–Raikov [BiRa]. See also [GeSi].

14. Normal forms for some scalar pseudodifferential operators

In this chapter, we shall give local normal forms for classical pseudors valid near a non-degenerate minimum of the symbol. For a formal selfadjoint operator $P = p^w(x, hD_x)$ whose symbol admits a non-degenerate minimum at $(0, 0)$, we show that there exists a unitary fourior U such that the symbol of $U^* P U$ in a neighborhood of $(0, 0)$ is $\sim \sum_{j=0}^{\infty} \tilde{p}_j(x, \xi) h^j$, with $H_{p_0,2} \tilde{p}_j = \mathcal{O}((x, \xi)^{\infty})$, where $p_{0,2} = \sum_j \frac{\lambda_j}{2}(\xi_j^2 + x_j^2)$. Here λ_j are the eigenvalues of the matrix $(\partial_{x,\xi}^2 p)(0, 0)$. In particular, if $\mathcal{M}_\lambda := \{\alpha \in \mathbf{Z}^n; \lambda \cdot \alpha = 0\} = \{0\}$ then $\tilde{p}_j(x, \xi) = f_j(\frac{1}{2}(x_1^2 + \xi_1^2), \ldots, \frac{1}{2}(x_n^2 + \xi_n^2)) + \mathcal{O}((x, \xi)^{\infty})$, where $f_j(\tau_1, \ldots, \tau_n)$ is a real-valued smooth function defined in a neighborhood of $(0, \ldots, 0)$. As indicated at the end of Chapter 4, we apply this result to study more precisely the asymptotic behaviour of the lowest eigenvalues of an h-pseudor: $p^w(x, hD_x)$ when $p(x, \xi)$ has a non-degenerate minimum at $(0, 0)$, $p(x, \xi) > 0$ for $(x, \xi) \neq (0, 0)$ and $\liminf_{(x,\xi) \to \infty} p(x, \xi) > 0$. For any fixed $N > 0$, we get asymptotic formulas for the eigenvalues up to $\mathcal{O}(h^N)$. We now formulate the assumptions.

Let $\Omega \subset \mathbf{R}^{2n}$ be a neighborhood of the origin. We let $S_{cl}^0(\Omega)$ be the space of formal asymptotic sums $a(x, \xi; h) \sim \sum_{j=0}^{\infty} a_j(x, \xi) h^j$, with $a_j \in C^{\infty}(\Omega)$ (in the sense of Chapter 2 formula (2.10)). With such a symbol, we associate a formal h-Weyl quantization:

$$a^w(x, hD_x; h)u(x) = (2\pi h)^{-n} \iint e^{i(x-y)\xi/h} a\left(\frac{x+y}{2}, \xi; h\right) u(y) dy d\xi, \quad (14.1)$$

and if b is a second symbol of the same kind we have (formally)

$$a^w(x, hD_x; h) \circ b^w(x, hD_x; h) = c^w(x, hD_x; h),$$

with $c(x, \xi; h) \sim \sum_{j=0}^{\infty} c_j(x, \xi) h^j$ given by Proposition 7.7. We shall mostly only consider formal power series at $(x, \xi) = (0, 0)$, so if a_j, b_j are defined simply as formal power series at $(0, 0)$ then the composition formula still defines $c(x, \xi; h) \sim \sum_{j=0}^{\infty} c_j(x, \xi) h^j$ with c_j as formal power series at $(0, 0)$. Denote the corresponding symbol spaces by S_t^0. Put $S_{cl}^m(\Omega) = h^{-m} S_{cl}^0$, $S_t^m = h^{-m} S_t^0$.

Let $p(x, \xi; h) \sim \sum_{j=0}^{\infty} p_j(x, \xi) h^j \in S_{cl}^0$ be a real valued symbol, such that $p_0(x, \xi)$ has a non-degenerate minimum at $(0, 0)$. We shall use the following well-known fact (see for instance [HoZe] for a proof). If q is a positive definite quadratic form on $\mathbf{R}^{2n}$, then there exists a real linear canonical transformation κ and $\lambda_1, \ldots, \lambda_n > 0$, which can be invariantly defined in terms of the (linear) flow H_q, such that $q \circ \kappa = \sum \frac{1}{2}\lambda_j(x_j^2 + \xi_j^2)$. Applying this to the quadratic part of the Taylor expansion of p at $(0,0)$, and using Theorem A.2 of Chapter 7, we may assume that:

$$p_0(x,\xi) = \sum_{j=1}^{n} \frac{1}{2}\lambda_j(\xi_j^2 + x_j^2) + \mathcal{O}((x,\xi)^3). \tag{14.2}$$

In the following we denote by $p_{0,2}(x,\xi) = \sum_{j=1}^{n} \frac{1}{2}\lambda_j(x_j^2 + \xi_j^2)$ the quadratic part of $p_0(x,\xi)$, and by ρ the point (x,ξ). Let $\mathcal{M}_\lambda = \{\alpha \in \mathbf{Z}^n; \sum_{j=1}^{n} \alpha_j\lambda_j = 0\}$.

Definition 14.1. Put $\sqrt{2}y_j = \xi_j + ix_j, \sqrt{2}\eta_j = -i(\xi_j - ix_j)$. A resonant function is a smooth function with a Taylor series of the form

$$f \sim \sum_{\alpha,\beta;(\alpha-\beta)\in\mathcal{M}_\lambda} a_{\alpha,\beta}y^\alpha\eta^\beta.$$

Notice that $d\eta_j dy_j = d\xi_j dx_j$ so (y,η) are complex symplectic coordinates.

Remark 14.2.

(1) In the (y,η) variables we have

$$p_0 = \sum_{j=1}^{n} i\lambda_j y_j \eta_j + \mathcal{O}((y,\eta)^3), \tag{14.3}$$

and

$$p_{0,2} = \sum_{j=1}^{n} i\lambda_j y_j \eta_j, \quad H_{p_{0,2}} = \sum_{j=1}^{n} i\lambda_j\left(y_j \frac{\partial}{\partial y_j} - \eta_j \frac{\partial}{\partial \eta_j}\right).$$

Let f be a smooth function with Taylor series of the form $f \sim \sum_{\alpha,\beta\in\mathbf{N}} f_{\alpha,\beta}y^\alpha\eta^\beta$. Then

$$H_{p_{0,2}}f \sim \sum_{\alpha,\beta\in\mathbf{N}} i\langle\lambda, \alpha - \beta\rangle f_{\alpha,\beta}y^\alpha\eta^\beta.$$

Hence, f is resonant iff $H_{p_{0,2}}f = \mathcal{O}((x,\xi)^\infty)$.

(2) If $Q(x,\xi)$ is a quadratic form and $a \sim \sum a_j h^j \in S_{\mathrm{cl}}^0(\Omega)$, then we have

$$[Q^w(x,hD_x), a^w(x,hD_x)] = -ih\mathrm{Op}_h^w\{Q,a\} = -ih\mathrm{Op}_h^w(H_Q(a)).$$

This follows from the Weyl calculus (see Chapter 7). With $p_{0,2}$ as above we have: the symbol of $[p_{0,2}^w(x,hD_x), a^w(x,hD_x;h)]$ is equal to zero in S_t^{-1}, iff each a_j is resonant.

(3) If $\lambda_1, \lambda_2, \ldots, \lambda_n$ are **Z**-independent (i.e. $\mathcal{M}_\lambda = \{0\}$), then the resonant functions are of the form $f(\frac{1}{2}(\xi_1^2 + x_1^2), \ldots, \frac{1}{2}(\xi_n^2 + x_n^2)) + \mathcal{O}((x, \xi)^\infty)$ with $f \in C^\infty$.

In the appendix to Chapter 11, we reviewed some theory of h-fouriors, and in particular how to associate such an operator with a canonical transformation κ between two neighborhoods $(0,0)$ in $\mathbf{R}^{2n}$, with $\kappa(0,0) = (0,0)$. Let U be such an operator of order 0 with a compactly supported symbol of class S_{cl}^0. Then U^* is associated with κ^{-1}, and if we normalize the phase in U by adding a suitable constant, then $U^*U = \mathrm{Op}_h(j_1)$, $UU^* = \mathrm{Op}_h(j_2)$, with j_1, j_2 in S_{cl}^0, of compact support modulo $S^{-\infty}(\langle(x,\xi)\rangle^{-N})$ for every N. Choosing the symbol of U suitably, we can arrange so that $j_1 - 1$, $j_2 - 1$ are of class $S^{-\infty}$ near 0, and we then say that U is microlocally unitary near $(0,0)$. In most of this chapter, we only consider the symbols near $(0,0)$. If Ω is a sufficiently small neighborhood of that point and $p \sim \sum_{j=0}^\infty p_j h^j \in S_{cl}^0(\Omega)$ is real-valued, then $U^* p^w U$ is a well-defined formal h-pseudor with a real-valued symbol $\tilde{p}(x, \xi; h) \sim \sum_{j=0}^\infty \tilde{p}_j(x, \xi) h^j \in S_{cl}^0(\tilde{\Omega})$, where $\tilde{\Omega} = \kappa^{-1}(\Omega)$, $\tilde{p}_0 = p_0 \circ \kappa$.

If $d\kappa(0,0)$ is close to the identity, then κ is given by a smooth generating function φ defined near $(0,0)$, so that: $\kappa : (\frac{\partial \varphi}{\partial \eta}(x,\eta), \eta) \mapsto (x, \frac{\partial \varphi}{\partial x}(x,\eta))$. In fact, this can be seen as in the proof of Theorem 1.3, using the fact that the symplectic form $\sum d\xi_j \wedge dx_j + \sum dy_j \wedge d\eta_j$ vanishes on the graph of κ, and letting (x, η) play the role of the x-variables in Theorem 1.3. We can choose U of the form

$$Uu(x) = (2\pi h)^{-n} \iint e^{i(\varphi(x,\eta) - y\cdot\eta)/h} a(x, \eta; h) u(y) dy d\eta, \qquad (14.4)$$

with $a \in S_{cl}^0$.

Remark 14.3.

(1) Let $a(x, \xi; h) \sim \sum_{j=0}^\infty a_j(x, \xi) h^j$, $p(x, \xi; h) \sim \sum_{j=0}^\infty p_j(x, \xi) h^j \in S^0(\mathbf{R}^{2n})$ be two real valued symbols. Following the procedure of Chapters 10–11, $e^{-ita^w/h} p^w$ is a fourior, for small t, with associated canonical transformation $\phi_t = \exp tH_{a_0}$, where ϕ_t is the flow generated by the Hamiltonian field H_{a_0}.

(2) Using the fact that $e^{-i(t+s)a^w/h} = e^{-ita^w/h} e^{-isa^w/h}$ as well as the fact that the composition of two fourior is again a fourior (see the appendix of Chapter 11) we see that the above remark remains true for all $t \in \mathbf{R}$.

We have the so-called Birkhoff normal form for the principal symbol $p_0(x, \xi)$ (recalling that after linear canonical transformation p_0 satisfies (14.3) in the (y, η) variables):

Proposition 14.4. *There exists a real smooth canonical transformation κ from a neighborhood of $(0,0)$ onto a neighborhood of $(0,0) \in \mathbf{R}^{2n}$, with $\kappa(0,0) = (0,0)$, $d\kappa(0,0) = \mathrm{id}$ such that $p_0 \circ \kappa$ is resonant.*

For the proof, we shall need two lemmas.

Lemma 14.5. *Let b be a real-valued function defined in a neighborhood of $(0,0)$ such that $b(\rho) = \mathcal{O}(\rho^2)$ near that point. If q is a real-valued smooth function defined near $(0,0)$ which vanishes to the order $m \geq 2$ there, then*

$$b(\exp H_q(\rho)) - b(\rho) = H_q(b)(\rho) + \mathcal{O}(\rho^{2(m-1)}).$$

Proof. From the mean value theorem and the fact that $H_q(\rho) = \mathcal{O}(\rho^{m-1})$, we have
$$\exp t H_q(\rho) - \rho = \mathcal{O}(t\rho^{m-1}), \text{ for all } t \in [0,1]. \tag{14.5}$$

Taylor's formula and (14.5) show that

$$H_q(b)(\exp t H_q(\rho)) - H_q(b)(\rho) = \mathcal{O}(t\rho^{2(m-1)}) \text{ for all } t \in [0,1]. \tag{14.6}$$

Now the lemma follows from (14.6) and the equality

$$b(\exp H_q(\rho)) - b(\rho) = \int_0^1 \partial_t(b(\exp t H_q(\rho)))dt = \int_0^1 H_q(b)(\exp t H_q(\rho))dt.$$

$$\#$$

Lemma 14.6. *If g is a smooth real valued function defined near $(0,0) \in \mathbf{R}^{2n}$, vanishing to the order $m \geq 0$ there, then there is a smooth real valued function f vanishing to the order m at $(0,0)$ such that*

$$H_{p_0}(f) = g + r, \tag{14.7}$$

where r is a resonant function.

Proof. Let f be solution of (14.7). Then we have

$$H_{p_0}(\mathrm{Re}\, f) = g + \mathrm{Re}\, r. \tag{14.8}$$

Since a function r is resonant iff $H_{p_{0,2}}(r) \sim 0$, it follows that $\mathrm{Re}\, r$ is resonant. Hence if f is a solution of (14.7) then $\mathrm{Re}\, f$ has the properties required in the lemma. Then we can take f real valued. It remains to show the existence of a complex valued solution of (14.7). Let us introduce the (y, η) variables. We are looking for a function f defined near $(0,0)$ with Taylor-Maclaurin series

$f = \sum_{\alpha,\beta \in \mathbf{N}^n} f_{\alpha,\beta} y^\alpha \eta^\beta$. Put $g = \sum_{|\alpha|+|\beta| \geq m} g_{\alpha,\beta} y^\alpha \eta^\beta$. Using the fact that p_0 satisfies (14.3), we get

$$H_{p_0}(f) - g = \sum (i f_{\alpha,\beta} \langle \lambda, \alpha - \beta \rangle + F_{\alpha,\beta}(f_{\alpha,\beta}) - g_{\alpha,\beta}) y^\alpha \eta^\beta, \qquad (14.9)$$

where $F_{\alpha,\beta}(f_{\alpha,\beta})$ is a finite linear combination of terms $f_{\alpha',\beta'}$ with $|\alpha'|+|\beta'| < |\alpha|+|\beta|$. For $|\alpha|+|\beta| < m$, $g_{\alpha,\beta} = 0$, and we take $f_{\alpha,\beta} = 0$. For $|\alpha|+|\beta| = m$, the coefficient of $y^\alpha \eta^\beta$ becomes

$$i f_{\alpha,\beta} \langle \lambda, \alpha - \beta \rangle - g_{\alpha,\beta}.$$

If $\alpha - \beta \in \mathcal{M}_\lambda$, we can choose $f_{\alpha,\beta}$ arbitrarily and $g_{\alpha,\beta} y^\alpha \eta^\beta$ is a resonant term. If $\alpha - \beta \notin \mathcal{M}_\lambda$, we take $f_{\alpha,\beta} = -i g_{\alpha,\beta}/\langle \lambda, \alpha - \beta \rangle$. For $|\alpha| + |\beta| > m$, we are in the same situation as above with $g_{\alpha,\beta}$ replaced by a known number $g_{\alpha,\beta} - F_{\alpha,\beta}(f_{\alpha,\beta})$. Given $f_{\alpha,\beta}$ for all $\alpha, \beta \in \mathbf{N}^n$, we can construct by Borel's theorem (see Chapter 2) a C^∞ function such that $f \sim \sum_{\alpha,\beta \in \mathbf{N}^n} f_{\alpha,\beta} y^\alpha \eta^\beta$ in a neighborhood of $(0,0)$ and f has the required properties. #

Proof of Proposition 14.4. Write $p_0(\rho) = p_{0,2}(\rho) + p_3(\rho)$, so that $p_3 = \mathcal{O}(\rho^3)$. Lemma 14.6 applied to p_3 gives a C^∞ function $q_3 = \mathcal{O}(\rho^3)$ such that

$$H_{p_0}(q_3) = p_3 - r_3, \qquad (14.10)$$

where r_3 is a resonant function. From Lemma 14.5 we have

$$p_0(\exp H_{q_3}(\rho)) - p_0(\rho) = -H_{p_0}(q_3) + p_4, \qquad (14.11)$$

with $p_4 = \mathcal{O}(\rho^4)$. (14.10) combined with (14.11) gives:

$$p_0(\exp H_{q_3}(\rho)) = p_{0,2} + r_3 + p_4.$$

Consider the sequence $m_1 = 3, m_2 = 4, \ldots, m_{j+1} = 2(m_j - 1), \ldots$ Assume by induction that we have constructed smooth real functions $q_{m_1}, q_{m_2}, \ldots, q_{m_k}$, with $q_{m_j} = \mathcal{O}(\rho^{m_j})$ such that $p_0 \circ \exp H_{q_{m_1}} \circ \ldots \circ \exp H_{q_{m_k}} = p_{0,2} + r_{3,k} + p_{m_{k+1}}$ with $r_{3,k} = \mathcal{O}(\rho^3)$ resonant and $p_{m_{k+1}} = \mathcal{O}(\rho^{m_{k+1}})$. Using again the two preceding lemmas, we find $q_{m_{k+1}} = \mathcal{O}(\rho^{m_{k+1}})$ with

$$p \circ \exp H_{q_{m_1}} \circ \ldots \circ \exp H_{q_{m_{k+1}}} = p_{0,2} + r_{3,k+1} + p_{m_{k+2}}.$$

From $q_{m_j} = \mathcal{O}(\rho^{m_j})$ and (14.5) we have

$$\exp H_{q_{m_j}}(\rho) - \rho = \mathcal{O}(\rho^{m_j - 1}). \qquad (14.12)$$

If $\kappa_j = \exp H_{q_{m_1}} \circ \ldots \circ \exp H_{q_{m_{k+1}}}(\rho)$, then (14.12) implies that $\kappa_{j+k} - \kappa_j = \mathcal{O}(\rho^{m_{j+1} - 1})$ for $k \geq 0$, and if we consider generating functions for our canonical transformations, we see that there exists a smooth canonical transformation κ with $\kappa(0) = 0, d\kappa(0) = I$, and $\kappa - \kappa_j = \mathcal{O}(\rho^{m_{j+1}-1})$ for all $j \geq 1$. It follows that κ has the required properties.   #

Theorem 14.7. *Let* $p(x, \xi, h) \sim \sum_{j=0}^{\infty} p_j(x, \xi) h^j$ *be a real valued symbol defined in a neighborhood of* $(0,0)$ *and assume that* $p_0(x, \xi)$ *has the form (14.3) with* $\lambda_j > 0$. *Then we can find a real canonical transformation* κ *from a neighborhood of* $(0,0)$ *onto a neighborhood of* $(0,0)$ *with* $\kappa(0,0) = (0,0)$, $d\kappa(0,0) = I$ *and an associated unitary fourior* U *(defined microlocally near* $(0,0)$) *such that every term in the asymptotic expansion of the (formal) h-pseudor* $[\sum_{j=0}^{n} \frac{\lambda_j}{2}((hD_{x_j})^2 + x_j^2), U^* p^w U]$ *vanishes to infinite order at* $(x, \xi) = (0,0)$.

Proof. Let κ be the real valued canonical transformation given in Proposition 14.4. Let U_0 be a unitary fourior given by (14.4) associated with κ (we choose $a(x, \eta, h)$ in (14.4) such that U_0 becomes unitary). Then $U_0^* p^w U_0$ is a formal h-pseudor with a real valued symbol $\tilde{p} = \sum_{j=0}^{\infty} \tilde{p}_j(x, \xi) h^j \in S_{cl}^0(\tilde{\Omega})$ for some neighborhood $\tilde{\Omega}$ of $(0,0)$ such that $\tilde{p}_0 = p_0 \circ \kappa$, is resonant (i.e. $H_{p_{0,2}} \tilde{p}_0 = \mathcal{O}(\rho^{\infty})$). We recall that $p_{0,2}$ is the quadratic part of p_0 and $\rho = (x, \xi)$. Since $\kappa(0,0) = (0,0)$ and $d\kappa(0,0) = \mathrm{id}$, $\tilde{p}_0$ has the same quadratic part $\tilde{p}_{0,2} = p_{0,2}$. From now on, we drop the tilde in connection with this new operator, and simply denote it by p.

Let $A \sim \sum_{j=0}^{\infty} a_j h^j$ be a real valued symbol in S_{cl}^0. Then, formally, $B^w = \exp iA^w$ is a unitary elliptic h-pseudor and

$$B^{w*} p^w B^w = p^w + B^{w*}[p^w, B^w] =: \tilde{p}^w. \qquad (14.13)$$

Using the pseudor-calculus of Chapter 7 (more precisely the fact that the leading term of $B^{w*}[p^w, B^w]$ equals $hH_p(a_0)$), we see that $\tilde{p} \sim \sum_{j\geq 0} \tilde{p}_j h^j$ with $\tilde{p}_0 = p_0$, $\tilde{p}_1 = p_1 + H_{p_0}(a_0)$. Using Lemma 14.6 we can then construct a real valued symbol a_0, such that $\tilde{p}_1$ is resonant. Assume inductively that we have found $a_0, a_1, \ldots, a_{N-1}$ such that $p \sim \sum_{j=0}^{\infty} p_j h^j$, where p_j is resonant for $j \leq N$. We then look for $B^w = \exp(ih^N A_N^w)$ with $A_N \in S_{cl}^0(\Omega)$, and a_N as its principal symbol. Thanks again to (14.13), the leading term of $B^{w*}[p^w, B^w]$ is $h^{N+1} H_{p_0}(a_N)$, so $B^{w*} p^w B^w = \tilde{p}^w$, with $\tilde{p} \sim \sum_{j=0}^{\infty} \tilde{p}_j h^j$, where $\tilde{p}_j = p_j$ for $j \leq N$ and with $\tilde{p}_{N+1} = p_{N+1} + H_{p_0}(a_N)$. We choose a_N such that $\tilde{p}_{N+1}$ becomes resonant. We finally obtain a microlocally unitary fourior as an infinite (asymptotically convergent) product $U = U_0 \circ \exp(ia_0^w) \circ \exp(iha_1^w) \circ \exp(ih^2 a_2^w) \circ \ldots$ such that if the symbol of $U^* p^w U$ is $\sim \sum_{j=0}^{\infty} \tilde{p}_j h^j$, then $H_{p_{0,2}} \tilde{p}_j = \mathcal{O}(\rho^{\infty})$ for every $j \in \mathbf{N}$, where $p_{0,2} = \sum_{j=1}^{n} \frac{1}{2} \lambda_j(x_j^2 + \xi_j^2)$. From Remark 14.2 we obtain the result.   #

Using Remark 14.3 and Theorem 14.7 we get:

Corollary 14.8. *Let* $p(x,\xi;h) \sim \sum p_j(x,\xi)h^j \in S^0_{\mathrm{cl}}$ *with* $p_0(x,\xi)$ *as above and assume that* $\mathcal{M}_\lambda = \{0\}$. *Then there is a real valued smooth symbol* $f(\tau_1,\dots,\tau_n;h) \sim \sum_{j\geq 0} f_j(\tau_1,\dots,\tau_n)h^j$ *defined for* $(\tau_1,\dots,\tau_n)$ *in a neighborhood of* $(0,\dots,0)$, *with* $f_0(\tau_1,\dots,\tau_n) = \sum \lambda_j \tau_j + \mathcal{O}(|\tau|^2)$, *and a formal fourior* U, *which is unitary, and whose associated canonical transformation is defined in a neighborhood of* $(0,0)$, *and maps this point onto itself, such that* $U^*PU = \widetilde{P} = \widetilde{p}^w(x,hD_x;h)$ *microlocally near* $(0,0)$, *where* $\widetilde{p}(x,\xi;h) \sim \sum_{j\geq 0} \widetilde{p}_j(x,\xi)h^j \in S^0_{\mathrm{cl}}(\widetilde{\Omega})$, *with* $\widetilde{p}_j(x,\xi) = f_j(\frac{1}{2}(\xi_1^2 + x_1^2),\dots,\frac{1}{2}(\xi_n^2 + x_n^2)) + \mathcal{O}((x,\xi)^\infty)$. *Here* $\widetilde{\Omega}$ *is a neighborhood of* $(0,0)$.

As mentioned at the end of Chapter 4, we will study more precisely the asymptotic behaviour of the lowest eigenvalues of a Schrödinger operator when the potential has a single non-degenerate minimum. Let $p(x,\xi;h) \in S^0(\mathbf{R}^{2n})$ be real valued with asymptotic expansion $p(x,\xi;h) \sim \sum_{j\geq 0} p_j(x,\xi)h^j$. We assume that $p_0(x,\xi) \geq 0$ with equality only at $(0,0)$, and $\lim\inf_{|(x,\xi)|\to\infty} p_0(x,\xi) > 0$. After a linear canonical transformation (implemented as in the appendix of Chapter 7), we may assume that

$$p_0(x,\xi) = \sum_{j=1}^n \frac{1}{2}\lambda_j(\xi_j^2 + x_j^2) + \mathcal{O}((x,\xi)^3). \tag{14.14}$$

We know (by the semi-classical sharp Gårding inequality, Theorem 7.11) that $p^w(x,hD_x;h)$ is semi-bounded from below by $-Ch$ for some $C > 0$. Combining this with results of Chapter 9, we see that the spectrum of the operator $P = p^w(x,hD_x;h)$ is discrete and of total multiplicity $\mathcal{O}(h^{-n})$ in an interval $]-\infty,\eta]$ with $\eta < \lim\inf_{|(x,\xi)|\to\infty} p_0(x,\xi)$.

Theorem 14.9. *Assume that* $\mathcal{M}_\lambda = \{\alpha \in \mathbf{Z}^n;\, \alpha \cdot \lambda = 0\} = \{(0,\dots,0)\}$. *Then there exists a real valued smooth function*

$$F(\tau_1,\dots,\tau_n;h) \sim F_0(\tau_1,\dots,\tau_n) + F_1(\tau_1,\dots,\tau_n)h + \dots,\quad (\tau_1,\dots,\tau_n) \in \mathbf{R}^n$$

with $F_j = $ const. *for* $\tau_1 + \dots + \tau_n \geq 1$, $F_0(\tau) = \sum \lambda_j \tau_j + \mathcal{O}(|\tau|^2)$, $F_0 > 0$ *when* $\tau_j \geq 0$, $\tau \neq 0$, *such that for every fixed* $\delta > 0$, *the eigenvalues of* $P = p^w(x,hD_x;h)$ *in* $]-\infty,h^\delta]$ *are of the form*

$$F((k_1 + \frac{1}{2})h,\dots,(k_n + \frac{1}{2})h;h) + \mathcal{O}(h^\infty),\quad k \in \mathbf{N}. \tag{14.15}$$

More precisely, if we let $\alpha_1 \leq \alpha_2 \leq \dots$ *be the increasing sequence of eigenvalues of* P *and if we let* $\beta_1 \leq \beta_2 \leq \dots$ *be the increasing sequence of values of the form* $F((k_1 + \frac{1}{2})h,\dots,(k_n + \frac{1}{2})h;h)$ *with* $k \in \mathbf{N}^n$, *then* $\alpha_j - \beta_j = \mathcal{O}(h^\infty)$ *uniformly, as long as* α_j *(or* β_j*) is* $\leq h^\delta$.

As a preparation we need the following lemma:

Lemma 14.10. *Fix $\epsilon \in [0, \frac{1}{4}[$. Let $\chi \in C_0^\infty(\mathbf{R}^{2n}; [0, 1])$ be equal to 1 in a neighborhood of $(0, 0)$. We assume that the support of χ is large enough such that $p_0(x, \xi) + \chi(h^{-\epsilon}(x, \xi)) \geq (1 + \frac{1}{C})h^{2\epsilon}$, for some $C > 0$. Let $\lambda \leq h^{2\epsilon}$ and $u \in L^2(\mathbf{R}^n)$ such that $\|u\| = 1$ and $(P - \lambda)u = 0$. Then*

$$u = \chi^{w,\epsilon}u + \mathcal{O}(h^\infty), \quad in \ L^2, \tag{14.16}$$

uniformly with respect to λ and u. Here $\chi^{w,\epsilon} = \chi^w(h^{-\epsilon}x, h^{-\epsilon}hD_x)$.

Proof. Let $\widetilde{\chi} \in C_0^\infty(\mathbf{R}^{2n}; [0, 1])$ be equal to 1 near $(0, 0)$ and have its support contained in the interior of the set where $\chi = 1$. We choose the support of $\widetilde{\chi}$ large so that

$$p(x, \xi) + \widetilde{\chi}(h^{-\epsilon}x, h^{-\epsilon}\xi) \geq (1 + \frac{1}{2C})h^{2\epsilon}.$$

Introducing $\widetilde{x} = h^{-\epsilon}x$, we can apply the semi-classical sharp Gårding inequality with $h^{1-2\epsilon}$ as the new parameter 'h', and get:

$$\widetilde{P}_\epsilon := P + \widetilde{\chi}^{w,\epsilon} \geq (1 + \frac{1}{3C})h^{2\epsilon}, \text{ for } h \text{ small enough,}$$

in the sense of selfadjoint operators. Hence for $\lambda \leq h^{2\epsilon}$;

$$(\widetilde{P}_\epsilon - \lambda)^{-1} = \mathcal{O}(h^{-2\epsilon}), \text{ in } \mathcal{L}(L^2, L^2). \tag{14.17}$$

Let $\chi_0, \chi_1, \ldots, \chi_N \in C_0^\infty(\mathbf{R}^{2n}; [0, 1])$ with $\chi_j = 1$ near $\text{supp}\,\chi_{j-1}$ for $j = 1, \ldots, N$ and with $\chi_0 = \widetilde{\chi}$, $\chi_N = \chi$. Then $[\chi_j^{w,\epsilon}, \chi_k^{w,\epsilon}] = \mathcal{O}(h^\infty)$ in $\mathcal{L}(L^2, L^2)$ when $k \neq j$, and $\chi_j^{w,\epsilon}(1 - \chi_k^{w,\epsilon}) = \mathcal{O}(h^\infty)$, $[\chi_k^{w,\epsilon}, \widetilde{P}_\epsilon]\chi_j^{w,\epsilon} = \mathcal{O}(h^\infty)$ in $\mathcal{L}(L^2, L^2)$ when $k > j$. From $(P - \lambda)u = 0$ we get $u = (\widetilde{P}_\epsilon - \lambda)^{-1}\widetilde{\chi}^{w,\epsilon}u$. Then

$$(1 - \chi^{w,\epsilon})u = [(1 - \chi^{w,\epsilon}), (\widetilde{P}_\epsilon - \lambda)^{-1}]\widetilde{\chi}^{w,\epsilon}u + (\widetilde{P}_\epsilon - \lambda)^{-1}(1 - \chi^{w,\epsilon})\widetilde{\chi}^{w,\epsilon}u.$$
$$= (\widetilde{P}_\epsilon - \lambda)^{-1}[\chi^{w,\epsilon}, \widetilde{P}_\epsilon](\widetilde{P}_\epsilon - \lambda)^{-1}\widetilde{\chi}^{w,\epsilon}u + \widetilde{R}_1 u,$$

where $\|\widetilde{R}_1\| = \mathcal{O}(h^\infty)$. Using the fact that $\chi_{N-1}^{w,\epsilon} \cdots \widetilde{\chi}^{w,\epsilon} = \widetilde{\chi}^{w,\epsilon} + \mathcal{O}(h^\infty)$ as well as the fact that $[\chi_k^{w,\epsilon}, \widetilde{P}_\epsilon]\chi_{k-1}^{w,\epsilon} = \mathcal{O}(h^\infty)$ in $\mathcal{L}(L^2, L^2)$, we get

$$(1 - \chi^{w,\epsilon})u = (\widetilde{P}_\epsilon - \lambda)^{-1}[\chi^{w,\epsilon}, \widetilde{P}_\epsilon](\widetilde{P}_\epsilon - \lambda)^{-1}[\chi_{N-1}^{w,\epsilon}, \widetilde{P}_\epsilon](\widetilde{P}_\epsilon - \lambda)^{-1} \cdots$$
$$[\chi_1^{w,\epsilon}, \widetilde{P}_\epsilon](\widetilde{P}_\epsilon - \lambda)^{-1}\widetilde{\chi}^{w,\epsilon}u + \widetilde{R}_N u,$$

where $\|\widetilde{R}_N\| = \mathcal{O}(h^\infty)$. As $[\chi_k^{w,\epsilon}, \widetilde{P}_\epsilon] = [\chi_k^{w,\epsilon}, P] + \mathcal{O}(h^\infty)$, then

$$(1 - \chi^{w,\epsilon})u = (\widetilde{P}_\epsilon - \lambda)^{-1}[\chi^{w,\epsilon}, P](\widetilde{P}_\epsilon - \lambda)^{-1}[\chi_{N-1}^{w,\epsilon}, P](\widetilde{P}_\epsilon - \lambda)^{-1} \cdots$$
$$[\chi_1^{w,\epsilon}, P](\widetilde{P}_\epsilon - \lambda)^{-1}\widetilde{\chi}^{w,\epsilon}u + R_N u, \quad \|R_N\| = \mathcal{O}(h^\infty). \tag{14.18}$$

Using the fact that the principal symbol of P vanishes to the second order at $(0,0)$ we see by the symbolic calculus of Chapter 7 (more precisely Proposition 7.7) that the symbols of $[\chi, P]$, $[\chi_j, P]$ are of the form $hr(x, \xi; h)$ with $r(x, \xi; h) \in S_\epsilon^0(1)$, and Proposition 7.11 shows that $[\chi, P]$, $[\chi_j, P] = \mathcal{O}(h)$ as bounded operators on L^2. Now (14.17), (14.18) imply

$$\|(1 - \chi^{w,\epsilon})u\| = (h^{N(1-2\epsilon)-2\epsilon}).$$

Here we may choose N as large as we like and this gives (14.16).   #

Remark 14.11. Fix $\epsilon = 0$ and let δ be a positive constant. Following the proof of the lemma, when $\lambda \le \delta$ we can choose the support of χ as small as we like provided that we take δ small enough.

In the following proposition we will reduce the spectral study of P in $]-\infty, \eta[$ for η small enough to that of $\widetilde{P} = U^* P U$, defined in Theorem 14.7. Recall that $\widetilde{P} = U^* P U = \widetilde{p}^w(x, hD_x; h)$ is only defined near $(0,0)$ and that

$$\widetilde{p}(x, \xi; h) \sim \sum_{j \ge 0} \widetilde{p}_j(x, \xi) h^j \in S^0(\widetilde{\Omega}), \qquad (14.19)$$

with $\{p_{0,2}, \widetilde{p}_j\} = \mathcal{O}((x, \xi)^\infty)$. Here $p_{0,2}$ is the common quadratic part of the Taylor expansions of $p_0(x, \xi)$ and of $\widetilde{p}_0(x, \xi)$. We may assume that ;

(i) The symbol (14.19) is globally defined and of class $S^0(\mathbf{R}^{2n})$.

(ii) For every $(x, \xi) \in \mathbf{R}^{2n}$ and every $h \in]0, h_0]$ ($0 < h_0$ is a small constant) $\widetilde{p}(x, \xi; h) \in \mathbf{R}$ and $\widetilde{p}_0(x, \xi) \ge 0$ with equality only at $(0,0)$.

(iii) $\liminf_{(x,\xi) \to \infty} \widetilde{p}_0(x, \xi) > 0$.

Then $\widetilde{P} = \widetilde{p}^w(x, hD_x; h)$ is a well defined selfadjoint operator with the same general properties as P. In particular it is semi-bounded from below by $-Ch$ and has discrete spectrum in $]-\infty, \eta]$ for some $\eta > 0$, independent of h.

Proposition 14.12. *Fix η small enough. Let $\alpha_1 \le \alpha_2 \le \ldots$ and $\beta_1 \le \beta_2 \le \ldots$ be the eigenvalues of P and $\widetilde{P}$ respectively in the interval $]-\infty, \eta]$. Then $\alpha_j - \beta_j = \mathcal{O}(h^\infty)$ uniformly for all j such that $\alpha_j, \beta_j \le \eta_1$, where η_1 is any fixed number in $]0, \eta[$.*

Proof. If $\alpha \in \sigma(P) \cap]-\infty, \eta]$, $Pu = \alpha u$ and $\|u\| = 1$ then by Proposition 14.10 $u = \chi^w(x, hD_x)u + \mathcal{O}(h^\infty)$ in L^2, where $\chi \in C_0^\infty(\mathbf{R}^{2n})$ is equal to 1 in a neighborhood of $(0,0)$. If η is small enough, then by Remark 14.11 we can take χ with supp χ contained in the region where U and U^* are defined as unitary operators and U^*u is then well defined in L^2 mod $\mathcal{O}(h^\infty)$, and $\|U^*u\| = 1$. We have $\widetilde{P}U^*u = \alpha U^*u + \mathcal{O}(h^\infty)$, and hence

$$\text{dist}\,(\alpha, \sigma(\widetilde{P})) = \mathcal{O}(h^\infty). \tag{14.20}$$

As $\widetilde{P}$ has the same general properties as P, then by the same argument we show that if $\alpha \in \sigma(\widetilde{P}) \cap\,] - \infty, \eta]$, then

$$\text{dist}\,(\alpha, \sigma(P)) = \mathcal{O}(h^\infty). \tag{14.21}$$

In particular, if $[f - h^{N_1}, f + h^{N_1}] \cap \sigma(P) = \emptyset$ or $[f - h^{N_1}, f + h^{N_1}] \cap \sigma(\widetilde{P}) = \emptyset$ for some fixed N_1 and for $f \leq \eta$, then

$$[f - \frac{1}{2}h^{N_1}, f + \frac{1}{2}h^{N_1}] \cap (\sigma(P) \cup \sigma(\widetilde{P})) = \emptyset, \tag{14.22}$$

when h is sufficiently small. Let $-Ch \leq f_1 \leq f_2 \leq \eta$ be two values satisfying (14.22) and let E, F be the spectral subspaces associated with $[f_1, f_2] \cap \sigma(P)$ and $[f_1, f_2] \cap \sigma(\widetilde{P})$ respectively. Let $e_1, \ldots, e_N$ be an orthonormal basis of eigenvectors in E: $Pe_j = \alpha_j e_j$. Then with $\widetilde{e}_j = U^*e_j$, we have $(\widetilde{e}_j, \widetilde{e}_k) = \delta_{j,k} + \mathcal{O}(h^\infty)$, and

$$\widetilde{P}\widetilde{e}_j = \alpha_j \widetilde{e}_j + \widetilde{r}_j, \text{ with } \|\widetilde{r}_j\| = \mathcal{O}(h^\infty). \tag{14.23}$$

Let $\widetilde{E}$ be the N-dimensional space spanned by the $\widetilde{e}_j$. From (14.23) we have

$$(z - \widetilde{P})^{-1}\widetilde{e}_j = (z - \alpha_j)^{-1}\widetilde{e}_j + (z - \alpha_j)^{-1}(z - \widetilde{P})^{-1}\widetilde{r}_j,$$

integrating this with respect to z over the rectangular contour Γ with the corners $f_j \pm ih$, we conclude that

$$\pi_F \widetilde{e}_j = \widetilde{e}_j + \mathcal{O}(h^\infty). \tag{14.24}$$

We recall that $\pi_F = \frac{1}{2\pi i} \int_\Gamma (z - \widetilde{P})^{-1}dz$. Thanks to the non-symmetric distance, $\overrightarrow{d}$, introduced in Chapter 6, we deduce from (14.23) that $\overrightarrow{d}(\widetilde{E}, F) = \mathcal{O}(h^\infty)$. Hence from Lemma 6.10 we get $\dim F \geq \dim \widetilde{E} = \dim E$, when h is sufficiently small. Similarly, $\dim F \leq \dim E$, so $\dim E = \dim F$. Combining this with (14.20), (14.21) we get the result.

$$\#$$

Proof of Theorem 14.9. The proof is now reduced to the study of $\sigma(\widetilde{P})$. By Corollary 14.8, we have:

$$\widetilde{p}_j(x, \xi) = f_j(\frac{x_1^2 + \xi_1^2}{2}, \ldots, \frac{x_n^2 + \xi_n^2}{2}) + \mathcal{O}((x, \xi)^\infty),$$

with $f_0(\tau_1, \ldots, \tau_n) = \sum \lambda_j \tau_j + \mathcal{O}(|\tau|^2)$. Following the construction of $\tilde{p}(x, \xi; h)$, we can assume that $f_j = $ const. in the region $\tau_1 + \ldots + \tau_n \geq 1$ and $f_0 \geq 0$ with equality only when $\tau = 0$.

By the functional calculus for functions of several commuting h-pseudors discussed at the end of Chapter 8, we can construct a real valued function $F(\tau_1, \ldots, \tau_n; h)$ in $S^0(\mathbf{R}^{2n})$, with asymptotic expansion

$$F(\tau_1, \ldots, \tau_n; h) \sim F_0(\tau_1, \ldots, \tau_n) + F_1(\tau_1, \ldots, \tau_n)h + \ldots, \qquad (14.25)$$

where F_j have the same general properties as the f_j above and where $f_0 = F_0$, such that if

$$R := R^w(x, hD_x; h) = F(\frac{1}{2}((hD_{x_1})^2 + x_1^2), \ldots, \frac{1}{2}((hD_{x_n})^2 + x_n^2); h),$$
$$(14.26)$$

then $R(x, \xi; h) \sim \sum_0^\infty r_j(x, \xi)h^j \in S^0(\mathbf{R}^{2n})$, and

$$\tilde{p}_j(x, \xi) = r_j(x, \xi) + \mathcal{O}((x, \xi)^\infty), \text{ near } (0, 0). \qquad (14.27)$$

(Notice that pseudors of the form (14.26) must have resonant symbols.)

Fix $\delta \in]0, \frac{1}{2}[$. Let $\lambda \leq h^\delta$ and $u \in L^2(\mathbf{R}^n)$ such that $\|u\| = 1$ and $(\tilde{P} - \lambda)u = 0$. Then

$$(R - \lambda)u = (R - \tilde{P})u = (R - \tilde{P})\chi^{w, \frac{\delta}{2}}u + (R - \tilde{P})(1 - \chi^{w, \frac{\delta}{2}})u,$$

where $\chi \in C_0^\infty(\mathbf{R}^{2n})$ and $\chi^{w, \frac{\delta}{2}} = \chi^w(h^{-\frac{\delta}{2}}x, h^{-\frac{\delta}{2}}hD_x)$. From Lemma 14.10 we have $(R - \tilde{P})(1 - \chi^{w, \frac{\delta}{2}})u = \mathcal{O}(h^\infty)$. Using the fact that the symbol of $(R - \tilde{P})$ is $\sim \sum k_j(x, \xi)h^j$ with $k_j(x, \xi) = \mathcal{O}((x, \xi)^\infty)$, as well as the fact that $(x, \xi) = \mathcal{O}(h^{\frac{\delta}{2}})$ on $\operatorname{supp}\chi(h^{-\frac{\delta}{2}}(x, \xi))$, we get, using the symbolic calculus: $(R - \tilde{P})\chi^{w, \frac{\delta}{2}}u = \mathcal{O}(h^\infty)$, so

$$(R - \tilde{P})u = \mathcal{O}(h^\infty).$$

This result holds after permutation of $\tilde{P}$, R. Using the same arguments as in the proof of Proposition 14.12 we get:

Lemma 14.13. *Fix $\eta > 0$ small enough. Let $\alpha_1 \leq \alpha_2 \leq \ldots$ and $\beta_1 \leq \beta_2 \leq \ldots$ be the eigenvalues of $\tilde{P}$ and R respectively in the interval $] - \infty, \eta]$. Then $\alpha_j - \beta_j = \mathcal{O}(h^\infty)$ uniformly for all j such that $\alpha_j, \beta_j \leq h^\delta$.*

Now Theorem 14.9 follows from Proposition 14.13, Lemma 14.14 and the fact that in $] - \infty, \eta[$ the spectrum of R is given by the values $F(h(k_1 + \frac{1}{2}), \ldots, h(k_n + \frac{1}{2}); h)$, $k \in \mathbf{N}^n$.

#

Remark 14.15. We have seen in Theorem 4.23 that all the eigenvalues in an interval $[0, C_0 h]$ of the Schrödinger operator, $-h^2 \Delta + V$, described at the end of Chapter 4, have asymptotics of the form

$$h \left(E_0 + \mathcal{O}(h^{\frac{1}{2}}) \right), \qquad (*)$$

where $E_0 < C_0$ runs through the eigenvalues of the associated harmonic oscillator. As indicated at the end of Chapter 4, if E_0 is a simple eigenvalue then $(*)$ has an asymptotic expansion $\sim h(E_0 + E_1 h + E_2 h^2 + \ldots)$. This follows from Theorem 14.9 when $\mathcal{M}_\lambda = \{0\}$.

Notes

Proposition 14.4 and further references can be found in [GDFGS]. Birkhoff normal forms for small perturbations of harmonic oscillators were given by Bellissard–Vittot [BeVi]. The relationship between the classical Birkhoff theorem and corresponding quantum perturbation theory is discussed in the paper of Graffi–Paul [GrPa]. More about Birkhoff normal forms can be found in Gutzwiller [Gut], Francoise–Guillemin, [FrGu], Zelditch [Ze2] and Guillemin [Gu].

In the one-dimensional case Theorem 14.9 is due to [HeRo3,4], [HeSj7], [CdV4]. A similar result in the case when the symbol has a saddle point was given by [HeSj7]. In multiple dimensions the theorem is due to Sjöstrand [Sj5], which we have followed in this chapter. More recent results in the same direction can be found in Kaidi–Kerdelhue [KaKer], Iantchenko [Ia], Popov [Po] and Bambusi–Graffi–Paul [BaGrPa].

15. Spectrum of operators with periodic bicharacteristics

In this chapter we treat scalar h-pseudors with periodic Hamiltonian flow. Recall the eigenvalue asymptotics of Theorem 10.1 (under suitable assumptions, and in particular that E_1 and E_2 are not critical values):

$$N([E_1, E_2]; h) = (2\pi h)^{-n} \iint_{p_0(x,\xi) \in [E_1, E_2]} dx d\xi + \mathcal{O}(h^{-n+1}). \qquad (15.1)$$

As we have seen, the analysis of the remainder term $\mathcal{O}(h^{-n+1})$ depends on the set of periodic trajectories in $p_0^{-1}(\{E_1, E_2\})$. If the Liouville measure of this set is equal to zero then we can replace $\mathcal{O}(h^{-n+1})$ in (15.1) by $(c_1(E_1, E_2)h^{-n+1} + o(h^{-n+1}))$. (See Theorem 11.1.) There are, however, many natural and interesting situations when all the H_{p_0} solution curves are periodic. In such cases, after a functional reduction to the case when all trajectories have the same period, we prove the existence of two constants δ, α such that the spectrum of $P(h) = \mathrm{Op}_h^w(p_0)$ is concentrated near the points $(\frac{2\pi}{T_E}l + \alpha)h + \delta$, $l = 1, 2, 3, \ldots$ (a result due to Colin de Verdière [CdV4], Duistermaat–Guillemin [DuGu], Helffer–Robert [HeRo3]). We will give an asymptotic expansion in powers of h of the counting function $N(I_l; h)$ when I_l is a subinterval of length $\mathcal{O}(h^2)$ centered at τ ($\tau = (\frac{2\pi}{T_E}l + \alpha)h + \delta \in]E_1, E_2[$). In the one dimensional case we get a more precise Bohr–Sommerfeld quantization condition (Theorem 15.10).

Let $E_1 < E_2$ be two real numbers, and let $p = p(x, \xi) \in S^0(\mathbf{R}^{2n})$ be independent of h for simplicity. We assume that p is real valued and that for $\epsilon > 0$ small enough, we have

(H1) $\liminf_{|(x,\xi)| \to \infty} d(p(x, \xi), [E_1 - \epsilon, E_2 + \epsilon]) > 0$,

(H2) $dp \neq 0$, for all $(x, \xi) \in p^{-1}([E_1 - \epsilon, E_2 + \epsilon])$,

(H3) $p^{-1}(E)$ is connected for every $E \in [E_1 - \epsilon, E_2 + \epsilon]$,

(H4) there exists a smooth function $T = T(x, \xi) > 0$ defined in $p^{-1}([E_1 - \epsilon, E_2 + \epsilon])$ such that $\exp(T(x, \xi)H_p)(x, \xi) = (x, \xi), \forall (x, \xi) \in p^{-1}([E_1 - \epsilon, E_2 + \epsilon])$.

Remark 15.1. Assumption (H1) implies that if $h > 0$ is small enough, then the spectrum of $P(h) = \mathrm{Op}_h^w(p)$ is discrete in a neighborhood of $[E_1 - \epsilon, E_2 + \epsilon]$.

Let $\gamma(x, \xi)$: $[0, T(x, \xi)] \ni t \mapsto \exp t H_p(x, \xi)$, so that $\gamma(x, \xi)$ is a closed curve.

202 Spectral Asymptotics in the Semi-Classical Limit

Lemma 15.2. $T(x,\xi)$ and $J(x,\xi) := \int_{\gamma(x,\xi)} \xi dx$ only depend on $p(x,\xi)$: writing $T(p(x,\xi))$, $J(p(x,\xi))$, $(x,\xi) \in p^{-1}([E_1 - \epsilon, E_2 + \epsilon])$, we have $J'(p) = T(p)$.

Proof. Let $[0,1] \ni s \to (x(s), \xi(s)) \in p^{-1}([E_1 - \epsilon, E_2 + \epsilon])$ be a C^1 curve and put $\varphi(t,s) = \exp(tH_p)(x(s),\xi(s))$, $0 \le s \le 1$, $0 \le t \le T(x(s),\xi(s))$ so that $\varphi(T(x(s),\xi(s)),s) = \varphi(0,s)$. Put $\gamma_s = \{\phi(t,s); 0 \le t \le T(x(s),\xi(s))\}$. Let σ be the symplectic 2 form on $\mathbf{R}^{2n}$ and let $\varphi^*(\sigma)$ be the corresponding pull-back. Then $\varphi^*(\sigma) = a(t,s)dt \wedge ds$, where (with $\varphi_* = d\varphi$),

$$a(t,s) = \langle \sigma, \varphi_*(\frac{\partial}{\partial t}) \wedge \varphi_*(\frac{\partial}{\partial s})\rangle = \langle \sigma, H_p \wedge \varphi_*(\frac{\partial}{\partial s})\rangle$$

$$= -\langle dp, \varphi_*(\frac{\partial}{\partial s})\rangle = -\frac{\partial}{\partial s}(p(\varphi(t,s))).$$

Since $d(\xi dx) = \sigma$, Stokes' formula gives

$$\int_{\gamma_s} \xi dx - \int_{\gamma_0} \xi dx = \int_0^s \int_0^{T(x(s'),\xi(s'))} \frac{\partial}{\partial s'}(p(\varphi(t,s')))dtds'. \tag{15.2}$$

Here

$$p(\varphi(t,s)) =: \overline{p}(s) \tag{15.3}$$

is independent of t, so (15.2) reduces to

$$\int_{\gamma_s} \xi dx - \int_{\gamma_0} \xi dx = \int_0^s T(x(s'),\xi(s'))d\overline{p}(s'). \tag{15.4}$$

If γ_s all belong to the same energy surface $p(x,\xi) = E$, we see that $J(x(s),\xi(s))$ is independent of s. Hence (by abuse of notation) $J(x,\xi) = J(p(x,\xi))$. Then (15.4) shows that $\frac{\partial}{\partial p}J(p) = T(p)$ with $T(p(x,\xi)) = T(x,\xi)$. #

Choose $\widetilde{J} \in S^0(\mathbf{R})$ (with order function 1 when nothing else is indicated), real valued, such that $\widetilde{J}(E) = J(E)$ for $E \in I_\epsilon$, $\widetilde{J}(E) \ge \widetilde{J}(E_2 + \epsilon)$ for $E \ge E_2 + \epsilon$ and $\widetilde{J}(E) \le \widetilde{J}(E_1 - \epsilon)$ for $E \le E_1 - \epsilon$. Here we write $I_\epsilon = [E_1 - \epsilon, E_2 + \epsilon]$, $I = I_0$. Put $Q(h) = \frac{1}{2\pi}\widetilde{J}(P(h))$. From Chapter 8, $Q(h) = \mathrm{Op}_h^w(q(h))$ with $S^0 \ni q \sim \sum_{j\ge0} q_j h^j$, q_j independent of h and of class S^0, $q_0(x,\xi) = \frac{1}{2\pi}\widetilde{J}(p(x,\xi))$, $q_1(x,\xi) = 0$. Since $H_{q_0} = \frac{1}{2\pi}\widetilde{J}'(p)H_p$ we have:

Lemma 15.3. $\exp(tH_{q_0})(x,\xi)$ is 2π-periodic for (x,ξ) in a neighborhood of $q_0^{-1}([\widetilde{J}(E_1), \widetilde{J}(E_2)])$.

As $\tilde{J}$ is a C^∞ diffeomorphism from a neighborhood of I onto a neighborhood of $\tilde{J}(I)$, the properties of the discrete spectrum of $P(h)$ in I can be deduced from those of $Q(h)$ in $\tilde{J}(I)$. Without any loss of generality we can assume:

(H5) $T_E = 2\pi$ for all $E \in I_\epsilon$.

Put

$$\delta = \frac{1}{2\pi} \int_{\gamma(E)} \xi dx - E. \tag{15.5}$$

It follows from Lemma 15.2 and assumption (H5) that δ is a constant independent of E in I_ϵ. As long as we are only interested in the eigenvalues of $P(h)$ in I, we may replace $P(h)$ by $P(h)\psi(P(h))$ with $\psi \in C_0^\infty(]E_1-\epsilon, E_2+\epsilon[)$ equal to 1 in a neighborhood of $[E_1, E_2]$. Let $U_\psi(t) = e^{-itP/h}\psi(P(h))$. Here we are interested in $U_\psi(2\pi)$. From results of Chapter 10 and for small t, $U_\psi(t)$ is a fourier with associated canonical transformation $\exp(tH_{p_0})(x, \xi)$. In our situation $\exp(2\pi H_p)(x, \xi) = (x, \xi)$ for all (x, ξ) in $\mathrm{supp}\,(\psi(p))$. Then we have the well-known result (which to a fairly large extent can be deduced from the appendix in Chapter 11, see Duistermaat [Du], Asada–Fujiwara [AsFu], Chazarain [Ch]):

Theorem 15.4. *Assume (H1) to (H5). Then $U_\psi(2\pi) = \mathrm{Op}_h^w(\tilde{q}(h))$, with $S^0(\mathbf{R}^{2n}) \ni \tilde{q} \sim \sum_{j \geq 0} \tilde{q}_j h^j$, $\tilde{q}_j$ independent of h and $\mathrm{supp}\,(\tilde{q}_j) \subset p^{-1}(]E_1 - \epsilon, E_2 + \epsilon[)$. Moreover,*

$$\tilde{q}_0(x, \xi) = \psi(p(x, \xi))e^{-2\pi i\sigma(h)}, \tag{15.6}$$

where $\sigma(h) = \frac{\alpha}{4} + \frac{\delta}{h}$, δ is defined by (15.5) and $\alpha \in \mathbf{Z}$ is the Maslov index of the trajectory $\{\exp(tH_{p_0})(x, \xi), t \in [0, 2\pi]\}$ in $p^{-1}(]E_1 - \epsilon, E_2 + \epsilon[)$.

A consequence of Theorem 15.4 is that:

$$e^{(-2\pi i/h)(P(h)-h\sigma(h))}\psi(P(h)) = \psi(P(h)) + h\mathcal{R}(h), \tag{15.7}$$

where the symbol of $\mathcal{R}(h)$ is in $S_{cl}^0(\mathbf{R}^{2n})$. Let $\psi_1 \in C_0^\infty(]E_1 - \epsilon, E_2 + \epsilon[)$ be such that $\psi\psi_1 = \psi_1$. In view of (15.7) we have:

$$e^{(-2\pi i/h)(P(h)-h\sigma(h))}\psi_1(P(h)) = \psi_1(P(h))(I + h\mathcal{R}(h)). \tag{15.8}$$

Then we can use Chapter 8, to see that for $h > 0$ small enough:

$$(I + h\mathcal{R}(h))^{-1} = I + h\tilde{\mathcal{R}}(h),$$

where $\tilde{\mathcal{R}}(h)$ is an h-pseudor with the same properties as $\mathcal{R}(h)$. For h small enough we put:

$$\widetilde{W}(h) = (1/2\pi i h)\log\,(I + h\tilde{\mathcal{R}}(h)).$$

Using Chapter 8 again, we see that $\widetilde{W}(h)$ is of class S^0_{cl}. It is clear that $\widetilde{W}(h)$ commutes with $P(h)$. Summing up we have proved:

Proposition 15.5. *Under the assumptions of Theorem 15.4, there exists* $\widetilde{W}(h) = \mathrm{Op}^w_h(\widetilde{w}(h))$, *with* $S^0_{\mathrm{cl}}(\mathbf{R}^{2n}) \ni w(h) \sim \sum_{j\geq 0} w_j h^j$, *which commutes with* $P(h)$ *such that:*

$$e^{(-2\pi i/h)(P(h)-h\sigma(h)-h^2\widetilde{W}(h))}\psi_1(P(h)) = \psi_1(P(h)).$$

We now decrease $\epsilon > 0$ so that $[E_1-\epsilon, E_2+\epsilon]$ becomes contained in the region where $\psi_1 = 1$. Let $(\lambda_j(h))_{0\leq j\leq N(h)}$ be the increasing sequence of eigenvalues of $P(h)$ in $[E_1 - \epsilon, E_2 + \epsilon]$, repeated according to their multiplicities, and let $(u_{j,h})_{0\leq j\leq N(h)}$ be a corresponding orthonormal system of eigenvectors. We can choose $u_{j,h}$ to be eigenvectors of $\widetilde{P}(h) := P(h) - h\sigma(h) - h^2\widetilde{W}(h)$ also, and we let $\nu_j(h)$ be the corresponding eigenvalues. Proposition 15.5 implies that $\nu_j(h) \in h\mathbf{Z}$ for h small enough. In view of the definition of $\widetilde{P}(h)$ we can find $C_0 > 0$ (independent of h) such that

$$|\lambda_j(h) - \nu_j(h) - h\frac{\alpha}{4} - \delta| \leq C_0 h^2. \tag{15.9}$$

Using (15.9) and the fact that $\sigma(P(h)) \cap [E_1, E_2] \subset \sigma(Q(h))$ as well as the fact that $\nu_j(h) \in h\mathbf{Z}$, we get:

Theorem 15.6. *Under the assumptions above, there exist* $C_0 > 0$, $h_0 > 0$ *such that*
$$(\sigma(P(h)) \cap [E_1, E_2]) \subset \cup_{l\in\mathbf{Z}} I_l(h),$$
for all $h \in]0, h_0]$ *with* $I_l(h) = [(l + \frac{\alpha}{4})h + \delta - C_0 h^2, (l + \frac{\alpha}{4})h + \delta + C_0 h^2]$.

Remark 15.7. If h_0 is small enough, then $I_l(h) \cap I_{l'}(h) = \emptyset$ for all $h \in]0, h_0]$ and all $l \neq l'$.

Now we assume

(H6) $\exp(tH_p)(x,\xi) \neq (x,\xi)$, for all $(x,\xi) \in p^{-1}([E_1 - \epsilon, E_2 + \epsilon])$ and all $t \in]0, 2\pi[$.

Let $\psi \in C^\infty_0(]E_1 - \epsilon, E_2 + \epsilon[$. We define for $l \in \mathbf{Z}$:

$$N_{l,\psi}(h) = \sum_{\lambda\in\sigma(P(h))\cap I_l(h)} \psi(\lambda).$$

When ψ is equal to 1 in a neighborhood of $[E_1, E_2]$ we get the number $N_\lambda(h)$ of eigenvalues in $I_l(h)$.

Theorem 15.8. *Under the assumptions (H1-6), we have*

$$N_{l,\psi}(h) \sim \sum_{j \geq 1} \Gamma_j(h(l + \frac{\alpha}{4}) + \delta)h^{j-n}, \quad h \to 0,$$

where $\Gamma_j \in C_0^\infty(]E_1 - \epsilon, E_2 + \epsilon[)$, and for $j = 1$ we have

$$\Gamma_1(\lambda) = \psi(\lambda)(2\pi)^{-n} \int_{p(\omega)=\lambda} L_\lambda(d\omega),$$

where L_λ is the Liouville measure of the hypersurface $p = \lambda$ (cf. (10.15)).

Proof. Recall that

$$\sigma(\widetilde{P}(h)) \cap [E_1 - \epsilon, E_2 + \epsilon] \subset h\mathbf{Z}, \quad \text{for all } h \in]0, h_0]. \tag{15.10}$$

Then

$$\text{tr}\left(\psi(P(h))e^{-ith^{-1}\widetilde{P}(h)}\right) = \sum_{l \in \mathbf{Z}}(\sum_{\lambda \in \sigma(P(h)) \cap I_l} \psi(\lambda))e^{-itl},$$

so $N_{l,\psi}(h)$ are the Fourier coefficients of the preceding quantity,

$$N_{l,\psi}(h) = \frac{1}{2\pi} \int_0^{2\pi} e^{itl}\text{tr}(\psi(P(h))e^{-ith^{-1}\widetilde{P}(h)})dt.$$

Let $\chi \in C_0^\infty(]\frac{-3\pi}{2}, \frac{3\pi}{2}[)$ be such that $\sum_{m \in \mathbf{Z}} \chi(t - 2\pi m) = 1$. Thanks to the expression of $\sigma(h)$ we get

$$N_{l,\psi}(h) = \frac{1}{2\pi} \int_{\mathbf{R}} e^{it\tau/h}\chi(t)\text{tr}(\psi(P(h))e^{ith\widetilde{W}(h)}e^{-itP(h)/h})dt, \tag{15.11}$$

where $\tau = \delta + (l + \frac{\alpha}{4})h \in]E_1 - \epsilon, E_2 + \epsilon[$. Expressions of the form (15.11) have been studied in Chapters 10-12. Assumption (H6) implies that $t = 0$ is the unique period in supp χ of the H_p solution curves in $p^{-1}(]E_1 - \epsilon, E_2 + \epsilon[)$, and the arguments of Chapter 11 show that $\text{tr}(\psi(P(h))e^{ith\widetilde{W}(h)}e^{-itP(h)/h})$ is $\mathcal{O}(h^\infty)$ for $t \in \text{supp}\,\chi \setminus \Omega$, if Ω is any neighborhood of 0. Then the analysis of Chapter 10 can be applied and Theorem 15.8 follows from (15.11),(10.13) and (10.15).   #

Now we restrict our attention to pseudors in one dimension. We assume that p satisfies (H1-6). We further notice in this case that $p^{-1}(E)$ is a closed smooth curve, $\gamma(E)$, with Maslov index equal to 2 (see Maslov [Ma1]) for all $E \in]E_1 - \epsilon, E_2 + \epsilon[$. In particular we have

$$J(E) = \int_{\gamma(E)} \xi dx = \int_{\{(x,\xi) \in \mathbf{R}^{2n}; E_1 - \epsilon \leq p(x,\xi) \leq E\}} d\xi dx + \text{const.},$$

and

$$J'(E) = \int_{p(\omega)=E} L_E(d\omega).$$

Combining this with Theorem 15.8, we get

Corollary 15.9. $N_l(h) = 1$ *for all* $h \in]0, h_0]$ *such that* $((l + \frac{1}{2})h + \delta) \in [E_1, E_2]$.

Let $(l, h) \in \mathbf{Z} \times]0, h_0]$ be such that $(l + \frac{1}{2})h + \delta \in [E_1, E_2]$ and let $\lambda_l(h)$ be the unique eigenvalue of $P(h)$ in $I_l(h)$ (given by Corollary 15.9). We have:

Theorem 15.10. *For every integer* $N \geq 2$ *we have*

$$\lambda_l(h) + \sum_{k=2}^{N} h^k f_k(\lambda_l(h)) = ((l + \frac{1}{2})h + \delta) + \mathcal{O}(h^{N+1}),$$

uniformly for $l \in \mathbf{Z}$. *Here* $f_k \in C^\infty(]E_1 - \epsilon, E_2 + \epsilon[; \mathbf{R})$.

Proof. Let $\psi \in C_0^\infty(]E_1 - \epsilon, E_2 + \epsilon[)$ be such that $\psi(\lambda) = \lambda$ on a neighborhood of $[E_1, E_2]$. Applying Theorem 15.8 to

$$N_{l,\psi}(h) = \sum_{\lambda \in \sigma(P(h)) \cap I_l(h)} \psi(\lambda) = \lambda_l(h),$$

we get

$$\lambda_l(h) \sim (h(l + \frac{1}{2}) + \delta) + \sum_{j \geq 2} \Gamma_j(h(l + \frac{1}{2}) + \delta)h^{j-1}, \quad h \to 0.$$

Coming back to the formula (15.11), using the fact that the subprincipal symbol of $p(x, \xi; h)$ equals zero (we recall that here $p(x, \xi; h) = p(x, \xi)$ is independent of h) we get, using formula (11.32), that $\Gamma_2(\tau) = 0$. Consequently,

$$\lambda_l(h) \sim (h(l + \frac{1}{2}) + \delta) + \sum_{j \geq 3} \Gamma_j(h(l + \frac{1}{2}) + \delta)h^{j-1}, \quad h \to 0. \tag{15.12}$$

Now Theorem 15.10 follows from (15.12). ⌗

Notes

Results on clustering of eigenvalues for compact manifolds are due to Colin de Verdière [CdV2], Wenstein [We] and J.J.Duistermaat–Guillemin [DuGu]. They showed that most of the eigenvalues are concentrated near the lattice points $\frac{2\pi}{T}k+\beta$, $k = 1, 2, \ldots$, where T is the common period of the Hamiltonian flow and β is a constant. Theorem 15.6 is due to Chazarain [Ch] in the case of a Schrödinger operator, $-h^2\Delta + V$, in the general case the theorem is due to Helffer–Robert [HeRo3]. The proofs of Theorem 15.6 and Theorem 15.8 are taken from [HeRo3]. If the assumption (H4) is only satisfied at a fixed energy level E, then Theorem 15.6 remains true provided that we restrict the interval $[E_1, E_2]$ to $[E - \mathcal{O}(h), E + \mathcal{O}(h)]$. That result is due to Brummelhuis–Uribe [BruUri] in the case of Schrödinger operators, $-h^2\Delta + V$, when $1/h \in \mathbf{N}$, and to Dozias [Doz] in the general case. We refer to Petkov–Popov [PePo] for the more general case when the union of closed trajectories is of non-vanishing measure.

References

[AbMa] R. Abraham, J. Marsden, *Foundations of mechanics*. W. A. Benjamin, Inc., New York–Amsterdam (1967).

[AbRo] R. Abraham, J. Robbin, *Transversal mappings and flows*. Appendix by A. Kelley. W. A. Benjamin, Inc., New York–Amsterdam (1967).

[Ag] S. Agmon, *Lectures on exponential decay of solutions of second-order elliptic equations*. Princeton Univ. Press, Princeton, NJ (1982).

[ADH] S. Alama, P. Deift, R. Hempel, Eigenvalue branches of the Schrödinger operator $H - \lambda W$ in a gap of $\sigma(H)$. *Comm. Math. Phys.*, **121**, 291–321 (1989).

[AlGé] S. Alinhac, P. Gérard, *Opérateurs pseudo-différentiels et théorème de Nash-Moser*. Interéditions, Savoirs actuels (Paris) (1991).

[An] M. Andersson, Taylor's functional calculus for commuting operators with Gauchy-Fantappie-Leray formulas. *Internat. Math. Res. Notices*, **6**, 247–58 (1997).

[AsFu] K. Asada, D. Fujiwara, On some oscillatory integral transformation in $L^2(\mathbf{R}^n)$. *Jap. J. Math.*, **4 (2)**, 299–361 (1978).

[AvSi] J. Avron, B. Simon, Stability of gaps for periodic potentials under variation of a magnetic field. *J. Phys. Math. Gen.*, **18 (2)** 2199–205.

[BaKo] A. Balazard-Konlein, Asymptotique semi-classique du spectre pour des opérateurs a symbole operatoriel. *C.R.A.S*, **(20)**, 903–6 (1985).

[BaGrPa] D. Bambusi, S. Graffi, T. Paul, *Normal forms and quantization formulae*, preprint (1998).

[Be] R. Beals, Characterization of pseudodifferential operators and applications, *Duke Math. J.*, **44 (1)**, 45–57 (1977).

[BeFe] R. Beals, C. Fefferman, Spatially inhomogeneous pseudo-differential operators I. *Comm. Pure Appl. Math.*, **27**, 1–24 (1974).

[BeVi] J. Bellissard, M. Vittot, Heisenberg's picture and non commutative geometry of the semiclassical limit in quantum mechanics. *Ann. Inst. Poincaré (physique théor.)*, **52**, 175–236 (1990).

[Bi1] M. Birman, Discrete spectrum in the gaps of the continuous one for perturbations with large coupling limit. *Adv. Soviet Math.*, **7**, 57–73 (1991).

[Bi2] M. Birman, On the discrete spectrum in the gaps of a perturbed periodic second order operator. *Funct. Anal. Appl.*, **25**, 158–61 (1991).

[BiRa] M. Birman, G. Raikov, Discrete spectrum in the gaps for perturbations of the magnetic Schrödinger operator. *Adv. Soviet Math.*, **7**, (1991).

[Bis] J. M. Bismut, The Witten complex and the degenerate Morse inequalities. *Journal of differential Geometry*, **23**, 207–40 (1986).

[Bon] J. M. Bony, *Caractérisations des opérateurs pseudodifférentiels*, Sém. E.D.P. Ecole Polytechnique, *Exposé no* **23** (1996-97).

[BonCh] J. M. Bony, J. Y. Chemin, Espaces fonctionnels associés au calcul de Weyl- Hörmander. *Bull. Soc. Math. France*, **122**, 77–118 (1994).

[Bo] A. Boulkhmair, Remarks on a Wiener type pseudodifferential algebra and Fourier integral operators. *Math. Res. Letters*, **4** (1), 53–67 (1997).

[BGH] L. Boutet de Monvel, A. Grigis, B. Helffer, Paramétrixes d'opérateurs pseudo-différentiels à caractéristiques multiples, *Astérisque*, **34–35**, 93–121 (1976).

[BruUri] R. Brummelhuis, A. Uribe, A semi-classical trace formula for Schrödinger operators. *Comm. Math. Phys.*, **136**, 567–84 (1981).

[Bu1] V. S. Buslaev, Semiclassical approximation for equations with periodic coefficients. *Russ. Math. Surveys.* **42** (6), 97–125 (1987).

[Bu2] V. S. Buslaev, Quasiclassical approximation for equations with periodic coefficients. *Uspekhi Mat. Nauk*, **42**, 77–98 (1987).

[Bu3] V. S. Buslaev, Adiabatic perturbation of a periodic potential. *Teoret. Mat. Fiz.*, **58**, 223–53 (1984).

[BuDi] V. S. Buslaev, A. Dimitrieva, A Bloch electron in an external field. *Algebra i Analiz*, **1** (**23**), 1–29, translated in *Leningrad Math. J.*, **1**, 287–320 (1990).

[CaNoPh] B. Candelpergher, J. C. Nosmas, F. Pham, *Approche de la résurgence*. Actualités mathématiques. Herman, Paris (1993).

[Car] U. Carlsson, An infinite number of wells in the semi-classical limit. *Asympt. Anal.*, 3 (3), 189–214 (1990).

[Char1] A. M. Charbonnel, Calcul fonctionnel à plusieurs variables pour des opérateurs pseudodifférentiels dans $\mathbf{R}^n$, *Isr. J. Math.*, 45, 69–89 (1983).

[Char2] A. M. Charbonnel, Comportement semi-classique du spectre conjoint d'opérateurs pseudodifférentiels qui commutent. *Asympt. Anal.*, 227–61 (1988).

[CharPo] A. M. Charbonnel, G. Popov, A semi-classical trace formula for several commuting operators. *Comm. in P.D.E.*, 24 (1 & 2), 283–323 (1999).

[Ch] J. Chazarain, Spectre d'un hamiltonien quantique et mécanique classique, *Comm. in P.D.E*, 5, 595–644 (1980).

[CdV1] Y. Colin de Verdière, Sur la multiplicité de la première valeur propre non nulle du Laplacien, *Comment. Math. Helv.*, 61, 254–70 (1986).

[CdV2] Y. Colin de Verdière, Spectre du laplacien et longueur des géodésiques périodiques. *Compos. Math.*, 27, 83–106 (1973).

[CdV3] Y. Colin de Verdière, Spectre conjoint d'opérateurs pseudodifferentiels qui commutent I, le cas non integrable. *Duke math. J.*, 46, 169–82 (1979).

[CdV4] Y. Colin de Verdière, Sur les spectres des opérateurs elliptiques à bicaractéristiques toutes périodiques. *Comm. Math. Helv.*, 54, 508–22 (1979).

[CDS] J. M. Combes, P. Duclos, R. Seiler, Krein's formula and one-dimensional multiple well. *J. Funct. Anal.*, 52, 257–301 (1983).

[CoFe] A. Cordoba, C. Fefferman, Wave packets and Fourier integral operators. *Comm. P.D.E.*, 979–1005 (1978).

[CFKS] H. L. Cycon, R. G. Froese, W. Kirsch, B. Simon, *Schrödinger operators, with application to quantum mechanics and global geometry.* Springer-Verlag, New York (1986).

[Da1] E. B. Davies, *Spectral theory and differential operators.* Cambridge Studies in Adv. Math. 42, Cambridge Univ. Press, Cambridge (1995).

[Da2] E. B. Davies, The functional calculus. *J. London Math. Soc.*, (52) 166–76 (1995).

[Da3] E. B. Davies, L^p spectral independence and L^1 analyticity. J. London Math. Soc., 52 177–84 (1995).

[DH] P. Deift, R. Hempel, On the existence of eigenvalues of the Schrödinger operator $H - \lambda W$ in a gap of $\sigma(H)$. Comm. Math. Phys., 103, 461–90 (1986).

[Di1] M. Dimassi, Développements asymptotiques des perturbations lentes de l'opérateur de Schrödinger périodique, Comm. P.D.E., 18 (5 & 6), 771–803 (1993).

[Di2] M. Dimassi, Développements asymptotiques des perturbations fortes de l'opérateur de Schrödinger périodique. Ann. Inst. Poincaré, 61, 189–204. (1994).

[Di3] M. Dimassi, Trace asymptotics formulas and some applications. Asym. Anal., 18, 1–32 (1998).

[DiSj] M. Dimassi, J. Sjöstrand, Trace asymptotics via almost analytic extensions. Partial differential equations and Mathematical physics. Prog. Nonlin. Diff. Eq. Appl. 21, 126-142. Birkhäuser, Boston (1996).

[DoKoMa] S. Y. Dobrokotov, V. N. Kolokol'tsov, V. P. Maslov, Splitting of the low energy levels of the Schrödinger equation and the asymptotic behavior of the fundamental solution of the equation $hu_t = \frac{1}{2}h^2\Delta u - v(x)u$. Theor. and Math. Phys., 87, 561–99 (1991).

[Doz] S. Dozias, Clustering for the spectrum of h-pseudodifferential operators with periodic flow on an energy surface. J. Funct. Anal., 145, 296–311 (1997).

[Dr] B. Droste, Extension of the functional calculus mappings and duality by $\bar{\partial}$-closed forms with growth. Math. Ann., 261, 185–200 (1982).

[Du] J. J. Duistermaat, Fourier integral operators. Progress in mathematics, Birkhäuser, Boston (1996).

[DuGu] J. J. Duistermaat, V. Guillemin, The spectrum of positive elliptic operators and periodic bicharacteristics. Invent. Math., 29, 39–79 (1975).

[Dy1] E. M. Dyn'kin, An operator calculus based upon the Cauchy-Green formula. Zapiski Nauchn. semin. LOMI, 30 (1972), 33-40 and J. Soviet Math., 4 (4), 329–34 (1975).

[Dy2] E. M. Dyn'kin, The pseudoanalytic extension, J. Anal. Math., 60, 45–70 (1993).

[FaLa] W. G. Farris, R. B. Lavine, Commutators and self-adjointness of Hamiltonian operators, *Comm. Math. Phy.*, **35**, 39–48 (1974).

[FePh] C. Fefferman, D. H. Phong, The uncertainty principle and sharp Gårding inequalities. *Comm. Pure. Appl. Math.*, **34**, 285–331 (1981).

[FrGu] J. P. Francoise, V. Guillemin, On the period spectrum of a symplectic mapping. *J. Funct. Anal.*, **100**, 317–58 (1991).

[Gå] L. Gårding, Dirichlet's problem for linear elliptic partial differential equations. *Math. Scan.*, **1**, 55–72 (1953).

[G] P. Gérard, *Mesures semi-classiques et ondes de Bloch.* Sém. E.D.P. Ecole Polytechnique, exposé no XVI, (1990–1991).

[Gé] C. Gérard, Sharp propagation estimates for N-particle systems. *Duke. Math. J.*, **67 (3)**, 483–515 (1992).

[GéGr] C. Gérard, A. Grigis, Precise Estimates of tunneling and eigenvalues near a potential barrier. *J. Diff. Eq.*, **72**, 149–77 (1988).

[GMS] C. Gérard, A. Martinez, J. Sjöstrand, A Mathematical Approach to the effective Hamiltonian in perturbed periodic Problems. *Commun. Math. Phys.*, **142**, 217–44 (1991).

[GeSi] F. Gesztezy, B. Simon, On a theorem of Deift and Hempel. *Commun. Math. Phys.*, **116**, 503–5 (1988).

[GDFGS], A. Giorgilli, A. Delshams, E. Fontich, L. Galgani, C. Simo', Effective stability for a Hamiltonian system near an equilibrium point, with application to the restricted three body problem. *Diff. Eq.*, **77**, 167–98 (1989).

[GoKr] I. C. Gohberg, M. G. Krein, *Introduction to the theory of linear non-selfadjoint operators.* Amer. Math. Soc., Providence, RI (1969).

[GrPa] S. Graffi, T. Paul, The Schrödinger equation and canonical perturbation theory. *Comm. Math. Phys.*, **108**, 25–40 (1987).

[Gr] A. Grigis, Estimations asymptotiques des intervalles d'instabilité pour l'équation de Hill. *Ann. Sci. Éc. Norm. Sup. 4 ème sér.*, **20**, 641–72 (1987).

[GrSj] A. Grigis, J. Sjöstrand, *Microlocal analysis for differential operators.* London Math. Soc. Lect. Notes series 196, Cambridge Univ. Press (1994).

[Gu] V. Guillemin, Wave trace invariants, *Duke Math. J.*, **83** (2), 287–352 (1996).

[GRT] J. C. Guillot, J. Ralston, E. Trubowitz, Semi-classical methods in solid state physics. *Comm. Math. Phys.*, **116**, 401–15 (1988).

[Gut] M. Gutzwiller, *Chaos in classical and quantum mechanics*, Springer Verlag (1990).

[Ha1] E. M. Harrell, Double wells. *Comm. Math. Phys.*, **75**, 239–61 (1980).

[Ha2] E. M. Harrell, The band structure of a one dimensional periodic system in the scaling limit. *Ann. Phys.*, **119**, 351–69 (1974).

[He] B. Helffer, *Semi-classical analysis for the Schrödinger operator and applications.* Springer Lect. Notes in Math., 1336 (1988).

[HeRo1] B. Helffer, D. Robert, Calcul fontionnel par la transformation de Mellin et opérateurs admissibles, *J. Funct. Anal.*, **53** (3), 246–68 (1983).

[HeRo2] B. Helffer, D. Robert, Comportement semi-classique du spectre des hamiltoniens quantiques elliptiques, *Ann. Inst. Fourier, Grenoble*, **31** (3), 169–223 (1981).

[HeRo3] B. Helffer, D. Robert, Puits de potentiel généralisés et asymptotique semi-classique. *Ann. Inst. Poincaré*, **41**, 291–331 (1984).

[HeRo4] B. Helffer, D. Robert, Asymptotique des niveaux d'énèrgie pour des Hamiltoniens à un degré de liberté. *Duke Math. J.*, **49**, 853–68 (1982).

[HeSj1] B. Helffer, J. Sjöstrand, Puits multiples en limite semi-classique II. Interaction moléculaire-symétries-perturbation. *Ann. Inst. Poincaré*, **42**, 127-212 (1985).

[HeSj2] B. Helffer, J. Sjöstrand, Multiple wells in the semi-classical limit I. *Comm. in P.D.E*, **9** (4), 337–408 (1984).

[HeSj3] B. Helffer, J. Sjöstrand, Puits multiples en mecanique semi-classique IV-Etude du complexe de Witten. *Comm. in P.D.E*, **10**, 245–340 (1985).

[HeSj4] B. Helffer, J. Sjöstrand, On diamagnetism and de Haas-van Alphen effect, *Ann. Inst. Poincaré*, **52**, 303–75 (1990).

[HeSj5] B. Helffer, J. Sjöstrand, Analyse semi-classique pour l'équation de

Harper. *Bull. de la SMF,* **116** (**4**), mémoire no 34 (1988).

[HeSj6] B. Helffer, J. Sjöstrand, *Equation de Schrödinger avec champ magnétique et équation de Harper.* Springer Lect. Notes in Physics 345, 118–97, Springer, Berlin (1989).

[HeSj7] B. Helffer, J. Sjöstrand, Analyse semi-classique pour l'équation de Harper II. Comportement semi-classique prés d'un rationnel. *Bull. de la SMF* **118** (**1**), mémoire no 40 (1990).

[HeSj8] B. Helffer, J. Sjöstrand, Effet tunnel pour l'équation de Schrödinger avec champ magnétique. *Ann. Sc. Norm. Sup. Pisa.*, **14**, 625–57 (1987).

[Hem1] R. Hempel, Eigenvalue branches of the Schrödinger operator $H \pm \lambda W$ in a spectral gap of H. *J. Reine Angew. Math.*, **399**, 38–59 (1989).

[Hem2] R. Hempel, Eigenvalues in gaps and decoupling by Neumann boundary conditions. *J. of Math. Anal. and Appl.*, **169**, 229–59 (1992).

[HoZe] H. Hofer, R. E. Zehnder, *Symplectic invariants and Hamiltonian dynamics.* Birkhäuser Verlag, Basel (1994).

[Höl] L. Hörmander, Fourier integral operators I. *Acta Math.*, **127**, 79–183 (1971).

[Hö2] L. Hörmander, *Lecture notes at the Nordic Summer School of Mathematics* (1968).

[Hö3] L. Hörmander, The spectral function of an elliptic operator. *Acta Math.*, **124**, 173–218 (1968).

[Hö4] L. Hörmander, *The analysis of linear partial differential operators I-IV.* Grundlehren, Springer, 256 (1983), 257 (1983), 274 (1985), 275 (1985).

[Hö5] L. Hörmander, The Weyl calculus of pseudodifferential operators. *Comm. Pure Appl. Math.*, **32**, 359–443 (1979).

[Hö6] L. Hörmander, Pseudodifferential operators and non-elliptic boundary problems. *Ann. of Math.*, **83**, 129–209 (1966).

[Ia] A. Iantchenko, La forme normale de Birkhoff pour un opérateur intégral de Fourier, *Asympt. Anal.*, **17** (**1**), 71–92 (1998).

[I1] V. Ivrii, *Microlocal analysis and precise spectral asymptotics*, Springer monographs in Math. (1998).

[I2] V. Ivrii, On sharp quasi-classical spectral asymptotics for the Schrödinger operator on manifold with boundary. *Soviet Math. Dokl.*, **26** (2), 285–9 (1982).

[I3] V. Ivrii, On quasi-classical spectral asymptotics for the Schrödinger operator on manifolds with boundary. *Dokl. Akad. Nauk SSSR*, 14–8 (1982).

[I4] V. Ivrii, On the second term of the spectral asymptotics for the Laplace–Beltrami operator in manifolds with boundary. *Funk. Anal. i pril*, **14**, 25–34 (1982).

[JN] A. Jensen, S. Nakamura, Mapping properties of functions of Schrödinger operators between L^p-spaces and Besov spaces. *Adv. Stud. Pure Math.*, **23**, 187–209 (1994) .

[Kap] T. Kappeler, Multiplicities of the eigenvalues of the Schrödinger equation in any dimension. *J. Funct. Anal.*, **77** (2), 346–51 (1988).

[Ka1] T. Kato, Schrödinger operator with singular potentials, *Isr. J. Math.*, **13**, 135–48 (1972).

[Ka2] T. Kato, *Perturbation theory*, Springer-Verlag, New York (1966).

[Ke] J. B. Keller, Corrected Bohr–Sommerfeld quantum conditions for non-separable systems, *Ann. Phys.*, **4**, 180–8 (1958).

[KaKer] N. Kaidi, P. Kerdelhue, *Forme normale de Birkhoff et résonances*, preprint (1998)

[KiSi1] W. Kirsch, B. Simon, Universal lower bounds on eigenvalue splittings for one dimensional Schrödinger operators. *Comm. Math. Phys.*, **97**, 453–60 (1985).

[KiSi2] W. Kirsch, B. Simon, Comparison theorems for the gap of Schrödinger operators. *J. Funct. An.*, **75**, 396–410 (1987).

[KoNi] J. J. Kohn, L. Nirenberg, An algebra of pseudo-differential operators. *Comm. Pure Appl. Math.*, **18**, 269–305 (1965).

[Ku] Kucherenko, Maslov's canonical operator on a germ of complex almost analytic manifold. *Dokl. Akad. Nauk SSSR* **213**, 1251–4 (1973).

[La] P. D. Lax, Asymptotic solutions of oscillatory initial value problems. *Duke Math. J.*, **24**, 627–46 (1957).

[LaNi] P. D. Lax, L. Nirenberg, On stability for difference schemes; a sharp form of Gårding's inequality. *Comm. Pure Appl. Math.*, **19**, 473–92 (1966).

[Le] S. Z. Levendorskii, Lower bounds for the number of eigenvalue branches for the Schrödinger operator $H - \lambda W$ in a gap of H: The case of indefinite W. *Comm. P.D.E.*, **20**, 827–54 (1995).

[Li] L. Lithner, A theorem of the Phragmén-Lindelöf type for econd order elliptic operators, *Ark. f. Mat.*, **5** (**18**), 281–5 (1963).

[M1] A. Martinez, Estimations de l'effet tunnel pour le double puits, I. *J. Math. Pures Appl.*, **66**, 195–215 (1987).

[M2] A. Martinez, Estimations de l'effet tunnel pour le double puits, II. Etats hautement exités. *Bull. Soc. Math. Fr.*, **116**, 199–229 (1988).

[MR] A. Martinez, M. Rouleux, Effet tunnel entre puits dégénerés. *Comm. P.D.E.*, **13**, 1157–87 (1988).

[Ma1] V. P. Maslov, *Théorie des perturbations et méthodes asymptotiques.* Dunod, Paris (1972).

[Ma2] V. P. Maslov, The characteristics of pseudo-differential operators and difference schemes. *Congr. Int. Math. Nice,* **2**, 755–69 (1970).

[Mat] J. L. Mather, *On Nirenberg's proof of Malgrange's preparation theorem.* Springer Lect. Notes in Math., 192, 116–32 (1971).

[MeSj] A. Melin, J. Sjöstrand, *Fourier integral operators with complex valued phase functions,* Springer Lect. Notes in Math., 459, 120–223 (1974).

[Mi] J. Milnor, *Morse theory,* Princeton Univ. Press (1963).

[NaRi] B. Sz. Nagy, F. Riesz, *Leçons d'analyse fonctionnelle.* 4 ème édition, Gauthier-Villars (Paris), (1965).

[Na] S. Nakamura, A remark on eigenvalue splittings for one-dimensional double-well Hamiltonians. *Letters in Math. Phys.*, **11**, 337–40 (1986).

[Ne1] G. Nenciu, Bloch electrons in a magnetic field: Rigorous justification of the Peierls-Onsager effective Hamiltonian. *Lett. Math. Phys.*, **17**, 247–52

(1989).

[Ne2] G. Nenciu, Stability of energy gaps under variations of the magnetic field. *Letters in Math. Phy.*, **11**, 127–32 (1986).

[Ni] L. Nirenberg, *A proof of the Malgrange preparation theorem*. Springer Lect. Notes in Math., 192, 97–105, (1971).

[Ou] A. Outassourt, Comportement semi-classique pour l'opérateur de Schrödinger à potentiel périodique. *J. Funct. Anal.*, **72**, 65–93 (1987).

[Pe] A. Person, Bounds for the discrete part of the spectrum of a semi-bounded Schrödinger operator. *Math. Scand.*, **8**, 143–53 (1960).

[PePo] V. Petkov, G. Popov, Semi-classical trace formula and clustering of eigenvalues for Schrödinger operator. *Ann. Inst. Poincaré*, **63**, 17–83 (1998).

[PeRo] V. Petkov, D. Robert, Asymptotique semi-classique du spectre d'hamiltoniens quantiques et trajectoires classiques périodiques. *Comm. P.D.E* **10** (4), 365–90 (1985).

[Po] G. Popov, *Invariant tori effective stability and quasimodes with exponentially small error terms*, preprint, Univ. de Nantes (1997).

[Ra] J. Ralston, *Magnetic breakdown*. Methodes semi-classiques, Vol. 2. Astérisque 210, 263–82 (1992).

[ReSi] M. Reed, B. Simon, *Methods of modern mathematical physics*, II-IV. Academic Press, New York, II (1975), III (1979), IV (1978).

[Ro1] D. Robert, *Autour de l'approximation semi-classique*. Progress in Mathematics 68, Birkhäuser (1987).

[Ro2] D. Robert, Calcul fonctionnel sur les opérateurs admissibles et application. *J. Funct. Anal.*, **45**, 74–94 (1982).

[SaVa] Yu. Safarov, D. Vassiliev, *The asymptotic distribution of eigenvalues of partial differential operators*, Transl. of Math. Monographs 155, Amer. Math. Soc. (Providence, RI) (1997).

[Se] J. P. Serre, *Représentations linéaires de groupes finis*, Herman, Paris (1967)

[Si1] B. Simon, Semiclassical analysis of low lying eigenvalues, I. Nondegenerate minima: Asymptotic expansions. *Ann. Inst. Poincaré*, **38**, 296–307 (1983).

[Si2] B. Simon, Semiclassical analysis of low lying eigenvalues, II. Tunneling. *Ann. Math.*, **120**, 89–118 (1984).

[Si3] B. Simon, Semiclassical analysis of low lying eigenvalues, III. Width of the ground state band in strongly coupled solids. *Ann. Phys.*, **158**, 415–20 (1984).

[Sj1] J. Sjöstrand, *Wiener type algebras of pseudodifferential operators*. Sém. E.D.P. Ecole Polytechnique, exposé no 4 (1994–1995).

[Sj2] J. Sjöstrand, An algebra of pseudodifferential operators. *Math. Res. Lett.*, (**1**), 185–92 (1994).

[Sj3] J. Sjöstrand, *Microlocal analysis for the periodic magnetic Schrödinger equation and related questions*. Springer Lect. Notes in Math., 1495, 237–332 (1991).

[Sj4] J. Sjöstrand, *A traceformula and review of some estimates for resonances*, Microlocal Analysis and Spectral Theory, NATO ASI Series C, vol. 490, 377–437, Kluwer (1997).

[Sj5] J. Sjöstrand, Semi-excited states in non-degenerate potential wells, *Asympt. Anal.*, **6**, 29–43 (1992).

[SjZw] J. Sjöstrand, M. Zworski, Complex scaling and the distribution of scattering poles. *J. Amer. Math. Soc.*, (**4**), 729–69 (1991).

[Sk] E. Skibsted, Smoothness of N-Body scattering amplitudes. *Rev. Math. Phys.*, (**4**), 619–58 (1992).

[So] A. V. Sobolev, Weyl asymptotics for the discrete spectrum of the perturbed Hill operator. *Adv. in Soviet Math.*, (**7**), (1991).

[St] S. Sternberg, *Lectures on differential geometry*, Prentice Hall (1965).

[Str] R. S. Strichartz, A functional calculus for elliptic pseudodifferential operators, *Amer. J. Math.*, **94**, 711–22 (1972).

[Ta] M. Taylor, *Pseudodifferential operators and spectral theory*, Princeton Univ. Press (Princeton) (1981).

[Tr] F. Treves, *Introduction to pseudodifferential and Fourier integral operators*. Vol 1, 2, Plenum Press (New York) (1980).

[UrZe] A. Uribe, S. Zelditch, Spectral statistics on Zoll surfaces. *Comm. Math. Phys.*, **154**, 313–46 (1993).

[Vo] A. Voros, The return of the quartic oscillator. The complex WKB-method. *Ann. Inst. Poincaré*, **29**, 211–338 (1983).

[We] A. Weinstein, Asymptotics of eigenvalue clusters for the Laplacian plus a potential. *Duke. Math. J.*, **44**, 883–92 (1977).

[Wil] M. Wilkinson, An example of phase holonomy in WKB theory, *J. Phys. A. Math. Gen.* **17**, 3459–76 (1984).

[W] E. Witten, Super symmetry and Morse theory, *J. Diff. Geom.*, **17**, 661–92 (1982).

[Ze1] S. Zelditch, On the rate of quantum ergodicity II: lower bounds. *Comm. P.D.E.* **(9 & 10)**, 1565–79 (1994).

[Ze2] S. Zelditch, Wave invariants at elliptic closed geodesics, *GAFA, Geom. Funct. Anal.*, **7**, 145–213 (1997).

Index

Notation-index

Printed in the United States
By Bookmasters